NUMBER THEORY AND ANALYSIS

E. Landau

NUMBER THEORY AND ANALYSIS

A COLLECTION OF PAPERS IN HONOR OF EDMUND LANDAU (1877–1938)

EDITED BY PAUL TURÁN

E. Bombieri · B. M. Bredihin · J. G. van der Corput · H. Davenport · M. Deuring
P. Erdös · H. Heilbronn · E. Hlawka · A. E. Ingham† · V. Jarník · S. Knapowski†
J. P. Kubilius · E. Landau† · J. V. Linnik · J. E. Littlewood · L. J. Mordell · G. Pólya
J. Popken · H. Rademacher · A. Rényi · A. Sárközi · I. J. Schoenberg · C. L. Siegel
E. Szemerédi · N. G. Tschudakoff · P. Turán · Anna Walfisz · Arnold Walfisz†

SPRINGER-SCIENCE+BUSINESS MEDIA, LLC

Library of Congress Catalog Card No: 68–8991

Published by VEB DEUTSCHER VERLAG DER WISSENSCHAFTEN
as
Abhandlungen aus Zahlentheorie und Analysis
Zur Erinnerung an Edmund Landau (1877–1938)

ISBN 978-1-4613-7184-7 ISBN 978-1-4615-4819-5 (eBook)
DOI 10.1007/978-1-4615-4819-5

FOREWORD

February 14, 1968 marked the thirtieth year since the death of Edmund Landau. The papers of this volume are dedicated by friends, students, and admirers to the memory of this outstanding scholar and teacher. To mention but one side of his original and varied scientific work, the results and effects of which cannot be discussed here, Edmund Landau performed one of his greatest services in developing the analytic theory of prime numbers from a subject accessible only with great difficulty even to the initiated few to the general estate of mathematicians. With the exception of the work of Chebyshev, Riemann, and Mertens, before Landau the problems of this theory were attempted only in a number of papers which were filled with gaps and errors. These problems were such that even Gauss abandoned them after several attempts in his youth, and they were described by N. H. Abel in a letter of 1823 and by O. Toeplitz in a lecture in 1930 as the deepest part of mathematics. Clarification first began with the papers of Hadamard, de la Vallée Poussin, and von Mangoldt. At the end of the foreword to his work "Handbuch der Lehre von der Verteilung der Primzahlen" which appeared in 1909, Landau could thus remark with complete justification: "... The difficulty of the previously unsolved problems has frightened nearly everyone away from the theory of prime numbers. I hope that I have succeeded in paving the path to the boundaries which have now been reached in such a way that this path will now be travelled and extended by many." Subsequent developments showed that his wish was realized to full extend: while perhaps the results contained in that book have since been surpassed, the effects of his pioneering work are evident in the more recent results, as they are in most of the papers of the present volume.

Prof. I. J. Schoenberg is to be thanked for the photograph at the beginning and for the list of publications at the end of the volume. Thanks are also due to the VEB Deutscher Verlag der Wissenschaften and the VEB Leipziger Druckhaus for the careful preparation and organization of the book.

Budapest, May 1968

PAUL TURÁN

CONTENTS

E. Bombieri in Pisa and H. Davenport in Cambridge

ON THE LARGE SIEVE METHOD

1. Let

$$S(x) = \sum_{n=M+1}^{M+N} a_n e(nx) \quad (e(\Theta) = e^{2\pi i \Theta}), \tag{1}$$

where the a_n are real or complex numbers and M, N are integers with $N > 0$. Let $x_1, \ldots, x_R$ be real numbers satisfying[1])

$$\|x_r - x_s\| \geqq \delta \quad \text{for} \quad r \neq s, \tag{2}$$

where $0 < \delta \leqq \frac{1}{2}$. In a recent paper (which was inspired by papers of Roth and Bombieri) Davenport and Halberstam[2]) proved that

$$\sum_{r=1}^{R} |S(x_r)|^2 \leqq \frac{11}{5} \max(N, \delta^{-1}) \sum_{n=M+1}^{M+N} |a_n|^2. \tag{3}$$

They also showed that if $N\delta$ is sufficiently small the right hand side can be improved to

$$(1 + \varepsilon)\, \delta^{-1} \sum_{n=M+1}^{M+N} |a_n|^2$$

for any fixed $\varepsilon > 0$.

In the present paper we are concerned primarily with the opposite case: that in which $N\delta$ is sufficiently large. We shall show that in this case the right hand side can be improved to

$$(1 + \varepsilon)\, N \sum_{n=M+1}^{M+N} |a_n|^2.$$

However, it is desirable to have results which apply for all values of N and δ, and accordingly we prove the following theorem.

---

[1]) We denote by $\|\Theta\|$ the distance from Θ to the nearest integer.

[2]) The values of a trigonometrical polynomial at well spaced points, Mathematika **13** (1966), 91–96.

Theorem 1. *If $S(x)$ is defined by* (1), *and $x_1, \ldots, x_R$ satisfy* (2), *then*

$$\sum_{r=1}^{R} |S(x_r)|^2 \leqq (N^{1/2} + \delta^{-1/2})^2 \sum_{n=M+1}^{M+N} |a_n|^2, \tag{4}$$

and

$$\sum_{r=1}^{R} |S(x_r)|^2 \leqq 2 \max (N, \delta^{-1}) \sum_{n=M+1}^{M+N} |a_n|^2. \tag{5}$$

Here (4) includes both the results mentioned above for the cases when $N\delta$ is large or small, and (5) represents a slight improvement on (3) which we include because it calls for little extra work.

The case when $N\delta$ is large is of particular arithmetical interest. In applications to number theory one commonly takes the numbers x_r to consist of all rationals a/q, in their lowest terms, with $q \leqq X$, and so obtains an upper bound for

$$\sum_{q \leqq X} \sum_{\substack{a=1 \\ (a,q)=1}}^{q} |S(a/q)|^2.$$

In this application $\delta = X^{-2}$, and then $N\delta$ is large if $X = o(N^{1/2})$.

Theorem 1 implies an analogous result for character sums, restricted, however,[1] to primitive characters, by an argument due to Gallagher[2]. In fact this result, which we now state, follows at once from (4) and inequality (5) of Gallagher's paper.

Theorem 2. *For each positive integer q let $\sum^*$ denote a summation over the primitive characters χ to the modulus q. Then*

$$\sum_{q \leqq X} \frac{q}{\varphi(q)} \sum_{\chi}{}^{*} \left| \sum_{n=M+1}^{M+N} a_n \chi(n) \right|^2 \leqq (N^{1/2} + X)^2 \sum_{n=M+1}^{M+N} |a_n|^2. \tag{6}$$

The improved constant in (4), in the case when $N\delta$ is large, has some interesting consequences. It enables us to prove by the large sieve method some estimates that have previously been obtained only by upper bound sieve methods. To illustrate this, we prove the following general theorem.

Theorem 3. *Let $p_1, \ldots, p_r$ be distinct primes, and let Q denote the set of all positive integers, including* 1, *up to X that have no prime factors other than these. Let the a_n be any real or complex numbers satisfying the condition*

$$a_n = 0 \text{ if } n \text{ is divisible by any } p_j. \tag{7}$$

---

[1]) Theorem 3 of Davenport and Halberstam, relating to a sum over all characters, is unfortunately erroneous. A Corrigendum has appeared in Mathematika **14** (1967), 229—232.

[2]) P. X. Gallagher, The large sieve, Mathematika **14** (1967), 14—20.

Then

$$\sum_{q \in Q} \frac{1}{\varphi(q)} \sum_{\chi} |\tau(\chi)|^2 \left| \sum_{n=M+1}^{M+N} a_n \chi(n) \right|^2 \leqq (N^{1/2} + X)^2 \sum_{n=M+1}^{M+N} |a_n|^2, \tag{8}$$

where the sum here is over all characters χ *to the modulus* q, *and where*

$$\tau(\chi) = \sum_{m=1}^{q} \chi(m)\, e(m/q).$$

More generally, if k *is any positive integer and* $(l, k) = 1$, *then*

$$\sum_{\substack{q \in Q \\ (q,k)=1}} \frac{1}{\varphi(q)} \sum_{\chi} |\tau(\chi)|^2 \left| \sum_{\substack{n=M+1 \\ n \equiv l (\mathrm{mod}\, k)}}^{M+N} a_n \chi(n) \right|^2$$

$$\leqq \{(Nk^{-1} + 1)^{1/2} + X\}^2 \sum_{\substack{n=M+1 \\ n \equiv l (\mathrm{mod}\, k)}}^{M+N} |a_n|^2. \tag{9}$$

If in (8) we take only the principal character χ_0 for each q, and note that $\tau(\chi_0) = \mu(q)$ and that $a_n \chi_0(n) = a_n$ by (7), we obtain in particular

$$\left(\sum_{q \in Q} \frac{\mu^2(q)}{\varphi(q)} \right) \left| \sum_{n=M+1}^{M+N} a_n \right|^2 \leqq (N^{1/2} + X)^2 \sum_{n=M+1}^{M+N} |a_n|^2. \tag{10}$$

If further we specialize a_n to be 1 whenever it is not required to be 0 by (7), we obtain:

Corollary 1. *The number of integers between* M *and* $M + N$ *that are not divisible by any of* $p_1, \ldots, p_r$ *does not exceed*

$$\frac{(N^{1/2} + X)^2}{Z}, \quad \textit{where} \quad Z = \sum_{q \in Q} \frac{\mu^2(q)}{\varphi(q)}. \tag{11}$$

This is an estimate of the type previously proved by Selberg's upper bound sieve method. If we simply take $p_1, \ldots, p_r$ to consist of all the primes up to X, so that Q consists of all positive integers up to X, and if we suppose that $M \geqq X$, we obtain

$$\pi(M + N) - \pi(M) \leqq \frac{(N^{1/2} + X)^2}{Z}.$$

Here, by a well known argument,

$$Z = \sum_{q \leqq X} \frac{\mu^2(q)}{\varphi(q)} = \sum_{q \leqq X} \frac{\mu^2(q)}{q} \prod_{p | q} \left(1 + \frac{1}{p} + \frac{1}{p^2} + \cdots \right)$$

$$\geqq \sum_{m \leqq X} \frac{1}{m} > \log X.$$

Taking $X = N^{1/2}/\log N$, we deduce:

Corollary 2. *For any fixed positive δ and all sufficiently large N, we have*

$$\pi(M+N) - \pi(M) < \frac{2N}{\log N}\left(1 + \frac{(2+\delta)\log\log N}{\log N}\right). \tag{12}$$

We have omitted the condition $M \geqq X$, and this we are entitled to do since the number of primes up to $X = N^{1/2}/\log N$ is negligible compared with the estimate on the right.

The inequality (12) was given by A. Selberg.[1] The present method of proof suggests that if the number of primes between M and $M + N$ *can* approach $2N/\log N$, then some further assertion can be made about the distribution of these primes in arithmetic progressions. We shall prove:

Theorem 4. *Let*

$$\pi(x, \chi) = \sum_{p \leqq x} \chi(p). \tag{13}$$

Then if N is large, and $X = N^{1/2}/\log N$, and $M \geqq X$, we have

$$\sum_{q \leqq X} \left(\log \frac{X}{q}\right) \sum_{\chi}{}^{*} |\pi(M+N, \chi) - \pi(M, \chi)|^2$$
$$\leqq (N^{1/2} + X)^2 (\pi(M+N) - \pi(M)), \tag{14}$$

where $\sum^*$ denotes a summation over the primitive characters χ to the modulus q.

The condition $M \geqq X$ in the theorem could be removed at the cost of minor complications.

If we neglect all values of q except $q = 1$ on the left of (14), we obtain (12). If we suppose that the left hand side of (12) is almost as large as the right hand side, say that

$$\pi(M+N) - \pi(M) > (2 - \eta)\frac{N}{\log N}, \tag{15}$$

for some small positive constant η, then (14) implies, for each q with $1 < q \leqq N^{1/2-\delta}$,

$$\sum_{\chi}{}^{*} |\pi(M+N, \chi) - \pi(M, \chi)|^2 < C(\delta)\,\eta\left(\frac{N}{\log N}\right)^2. \tag{16}$$

This asserts, in effect, that the primes between M and $M + N$ are well distributed among the residue classes mod q, provided q is a prime. The latter restriction is needed because the sum is over primitive characters, and only for a prime modulus do these comprise all the non-principal characters.

[1]) 11th Scandinavian Mathematical Congress (1949), 13—22.

Finally, as a corollary to the more general form (9) of theorem 3, analogous to corollary 2 above, we shall deduce an inequality of the Brun-Titchmarsh type[1]) for the number of primes in a given interval which lie in a given arithmetic progression.

Corollary 3. *If N is sufficiently large, and if*

$$k < \frac{1}{3}N, \tag{17}$$

we have

$$\pi(M+N;k,l) - \pi(M;k,l) < \frac{2N}{\varphi(k)\log N/k}\left\{1 + O\left(\frac{\log\log N/k}{\log N/k}\right)\right\}. \tag{18}$$

2. Lemma 1. *For* $x > 0$,

$$\left|\int_x^\infty \frac{\sin t}{t}\,dt\right| < \frac{1}{x}. \tag{19}$$

Proof. This inequality (which is quite delicate when x is large) is easily established by contour integration. The integral is

$$\frac{1}{2i}\int_x^\infty \frac{e^{it} - e^{-it}}{t}\,dt = \frac{e^{ix}}{2i}\int_0^\infty \frac{e^{it}}{x+t}\,dt - \frac{e^{-ix}}{2i}\int_0^\infty \frac{e^{-it}}{x+t}\,dt.$$

In the first integral on the right we rotate the path of integration from the positive real axis to the positive imaginary axis. This is permissible because (i) the integrand is a regular function of t in the first quadrant, (ii) the integral along the quarter of a large circle tends to 0 by Jordan's lemma. The second integral is similarly moved to the negative imaginary axis. The result is

$$\frac{1}{2}e^{ix}\int_0^\infty \frac{e^{-u}}{x+iu}\,du + \frac{1}{2}e^{-ix}\int_0^\infty \frac{e^{-u}}{x-iu}\,du,$$

and this has absolute value

$$< \frac{1}{2}\int_0^\infty \frac{e^{-u}}{x}\,du + \frac{1}{2}\int_0^\infty \frac{e^{-u}}{x}\,du = \frac{1}{x}.$$

3. Proof of theorem 1. As in the paper of Davenport and Halberstam it is convenient to work with the symmetrical sum

$$S(x) = \sum_{n=-N}^{N} a_n e(nx) \tag{20}$$

---

[1]) See, for example, K. Prachar, *Primzahlverteilung*, p. 44, Satz 4.1. An improved result of the Brun-Titchmarsh type has been proved recently by J. H. van Lint and H.-E. Richert, On primes in arithmetic progressions, Acta Arithmetica **11** (1965), 209—216. This is slightly stronger than (18) in that there is no log log N/k in the error term.

instead of the sum (1). We shall prove that, with the definition (20),

$$\sum_{r=1}^{R} |S(x_r)|^2 \leqq \begin{cases} ((2N)^{1/2} + \delta^{-1/2})^2 \sum_{-N}^{N} |a_n|^2, \\ 2 \max(2N, \delta^{-1}) \sum_{-N}^{N} |a_n|^2. \end{cases} \tag{21}$$

This implies the results (4) and (5) for the sum defined in (1). For the range of summation in (1) can be translated to the range $-\frac{1}{2}(N-1) \leqq n \leqq \frac{1}{2}(N-1)$ if N is odd, and we can ensure that N is odd by adding an extra term with coefficient 0 to the sum in (1), if necessary.

Let

$$\psi(x) = \sum_{-\infty}^{\infty} d_n e(nx), \quad \text{with} \quad d_{-n} = d_n,$$

be a real function of integrable square which is 0 when $\|x\| > \frac{1}{2}\delta$. By (7) of the paper of Davenport and Halberstam we have

$$\sum_{r=1}^{R} |S(x_r)|^2 \leqq \left(\sum_{-\infty}^{\infty} d_n^2\right)\left(\sum_{-N}^{N} d_n^{-2} |a_n|^2\right). \tag{22}$$

Also

$$\sum_{-\infty}^{\infty} d_n^2 = \int_0^1 \psi^2(x)\,dx = 2\int_0^{\delta/2} \psi^2(y)\,dy,$$

and

$$d_n = \int_0^1 \psi(x)\,e(-nx)\,dx = 2\int_0^{\delta/2} \psi(y)\cos 2\pi ny\,dy.$$

It is convenient to make a slight change of notation. We put

$$y = \tfrac{1}{2}\delta t, \quad \psi(y) = \varphi(t).$$

Then $\varphi(t)$ is an arbitrary function of integrable square, defined for $0 \leqq t \leqq 1$, and

$$\sum_{-\infty}^{\infty} d_n^2 = \delta \int_0^1 \varphi^2(t)\,dt,$$

$$d_n = \delta \int_0^1 \varphi(t)\cos \pi\alpha t\,dt,$$

where $\alpha = n\delta$. We put

$$\xi = N\delta. \tag{23}$$

Then, for $|n| \leqq N$,

$$|d_n| \geqq \delta \min_{0 \leqq \alpha \leqq \xi} \left| \int_0^1 \varphi(t)\cos \pi\alpha t\,dt \right|.$$

It follows from (22) that

$$\sum_{r=1}^{R} |S(x_r)|^2 \leqq \delta^{-1} K(\xi) \sum_{-N}^{N} |a_n|^2, \tag{24}$$

where $K(\xi)$ is defined by

$$K(\xi) \min_{0 \leqq \alpha \leqq \xi} \left| \int_0^1 \varphi(t) \cos \pi\alpha t \, dt \right|^2 = \int_0^1 \varphi^2(t) \, dt. \tag{25}$$

Thus the proof of (21) is reduced to the proof that

$$K(\xi) \leqq \begin{cases} ((2\xi)^{1/2} + 1)^2, \\ 2 \max(2\xi, 1), \end{cases} \tag{26}$$

for a suitable choice of $\varphi(t)$.

Case 1. *Suppose that* $\xi \geqq \frac{3}{2} + \sqrt{2}$. Then

$$4\xi \geqq ((2\xi)^{1/2} + 1)^2,$$

and it suffices to prove the first of the inequalities in (26). We take

$$\varphi(t) = \frac{\sin \pi\lambda t}{t} \quad (0 \leqq t \leqq 1),$$

where λ will be chosen later as a function of ξ and will satisfy $\lambda > \xi$. We have

$$\int_0^1 \varphi^2(t) \, dt = \pi\lambda \int_0^{\pi\lambda} \left(\frac{\sin t}{t}\right)^2 dt < \pi\lambda \int_0^{\infty} \left(\frac{\sin t}{t}\right)^2 dt = \frac{1}{2}\pi^2\lambda.$$

Also

$$\int_0^1 \varphi(t) \cos \pi\alpha t \, dt = \frac{1}{2}\int_0^1 \frac{\sin \pi(\lambda + \alpha)\, t + \sin \pi(\lambda - \alpha)\, t}{t} \, dt$$

$$= \frac{1}{2}\int_0^{\pi(\lambda+\alpha)} \frac{\sin t}{t} \, dt + \frac{1}{2}\int_0^{\pi(\lambda-\alpha)} \frac{\sin t}{t} \, dt.$$

By lemma 1, this is

$$> \frac{\pi}{2} - \frac{1}{2}\,\frac{1}{\pi(\lambda + \alpha)} - \frac{1}{2}\,\frac{1}{\pi(\lambda - \alpha)} \geqq \frac{\pi}{2} - \frac{\lambda}{\pi(\lambda^2 - \xi^2)},$$

since $0 \leqq \alpha \leqq \xi$. We suppose λ sufficiently large to make the last expression positive. It follows from (25) that, for this choice of $\varphi(t)$,

$$K(\xi) < 2\lambda\left(1 - \frac{2\lambda}{\pi^2(\lambda^2 - \xi^2)}\right)^{-2}.$$

To prove the first inequality in (26) it suffices to show that we can choose $\lambda > \xi$ so that

$$\lambda^{-1/2} - \frac{2\lambda^{1/2}}{\pi^2(\lambda^2 - \xi^2)} > \frac{2^{1/2}}{(2\xi)^{1/2} + 1}.$$

We put $\lambda = \xi(1 + \gamma)^2$, and note that

$$\lambda^2 - \xi^2 = \xi^2(4\gamma + 6\gamma^2 + 4\gamma^3 + \gamma^4) > 4\xi^2\gamma(1 + \gamma).$$

Hence the inequality will be satisfied if

$$\frac{1}{1+\gamma} - \frac{1}{2\pi^2\xi\gamma} > \frac{(2\xi)^{1/2}}{(2\xi)^{1/2} + 1}.$$

We choose

$$\gamma = \frac{1}{\pi(2\xi)^{1/2} - 1},$$

since this maximizes the left hand side. On writing $2\xi = \eta^2$ the last inequality reduces to

$$(\pi\eta - 1)^2 (\eta + 1) > \pi^2\eta^3,$$

that is,

$$\pi(\pi - 2)\,\eta^2 - (2\pi - 1)\,\eta + 1 > 0.$$

This holds with an ample margin for $\eta \geqq 1 + \sqrt{2}$, corresponding to $\xi \geqq \frac{3}{2} + \sqrt{2}$.

Case 2. *Suppose that* $0 < \xi \leqq \frac{1}{2}$. We take $\varphi(t) = \cos \pi\xi t$. The minimum of

$$\int_0^1 \varphi(t) \cos \pi\alpha t \, dt = \int_0^1 \cos \pi\xi t \cos \pi\alpha t \, dt$$

for $0 \leqq \alpha \leqq \xi$ is attained when $\alpha = \xi$, since the derivative with respect to α is

$$-\pi \int_0^1 t \cos \pi\xi t \sin \pi\alpha t \, dt,$$

which is negative for $0 < \alpha \leqq \xi \leqq \frac{1}{2}$. Hence, by (25), for this choice of $\varphi(t)$,

$$K(\xi) = \left\{\int_0^1 \cos^2 \pi\xi t \, dt\right\}^{-1} = 2\left\{1 + \frac{\sin 2\pi\xi}{2\pi\xi}\right\}^{-1}.$$

Plainly $K(\xi) \leqq 2$, so the second inequality in (26) is satisfied. It remains to prove the first, which is now

$$\left(1 + \frac{\sin 2\pi\xi}{2\pi\xi}\right)((2\xi)^{1/2} + 1)^2 \geqq 2.$$

This is trivial for $\xi \geqq \frac{1}{4}$, since then $((2\xi)^{1/2} + 1)^2 > 2$. Now suppose that $\xi < \frac{1}{4}$. It will suffice if

$$\left(2 - \frac{1}{6}(2\pi\xi)^2\right)((2\xi)^{1/2} + 1)^2 \geqq 2,$$

that is, if

$$\left(1 - \frac{1}{3}\pi^2\xi^2\right)(1 + (2\xi)^{1/2})^2 \geqq 1.$$

Since $\pi^2\xi < 3$, it will suffice if

$$(1 - \xi)(1 + (2\xi)^{1/2})^2 \geqq 1,$$

and this is immediate from

$$(1 + (2\xi)^{1/2})^2 > 1 + 2\xi > 1 + \xi + \xi^2 + \cdots.$$

Case 3. *Suppose that* $\frac{1}{2} \leqq \xi \leqq \frac{3}{2} + \sqrt{2}$. In this range we have

$$2 \max(2\xi, 1) = 4\xi \leqq ((2\xi)^{1/2} + 1)^2,$$

so it suffices to prove that $K(\xi) \leqq 4\xi$ for a suitable choice of $\varphi(t)$. We take

$$\varphi(t) = \begin{cases} \cos \pi\xi t & \text{for } 0 \leqq t \leqq (2\xi)^{-1}, \\ 0 & \text{for } (2\xi)^{-1} \leqq t \leqq 1. \end{cases}$$

Then

$$\int_0^1 \varphi(t) \cos \pi\alpha t \, dt = \int_0^{(2\xi)^{-1}} \cos \pi\xi t \cos \pi\alpha t \, dt,$$

and by the same argument as before this is least for $0 \leqq \alpha \leqq \xi$ when $\alpha = \xi$. Hence

$$K(\xi) = \left\{\int_0^{(2\xi)^{-1}} \cos^2 \pi\xi t \, dt\right\}^{-1} = 4\xi,$$

and this completes the proof.

4. Proof of theorem 3. We shall prove the more general inequality (9); the less general (8) is obtained by taking k to be 1, and it will be apparent from the proof that there is then no need to replace N by $Nk^{-1} + 1$ on the right.

We have

$$\tau(\bar{\chi})\,\chi(n) = \sum_{\substack{m=1 \\ (m,q)=1}}^{q} \bar{\chi}(m)\, e(mn/q)$$

provided $(n, q) = 1$. If $q \in Q$ then $a_n = 0$ if $(n, q) > 1$, by (7). Hence

$$\tau(\bar{\chi}) \sum_{\substack{n=M+1 \\ n \equiv l(\mathrm{mod}\, k)}}^{M+N} a_n \chi(n) = \sum_{\substack{m=1 \\ (m,q)=1}}^{q} \bar{\chi}(m) \sum_{\substack{n=M+1 \\ n \equiv l(\mathrm{mod}\, k)}}^{M+N} a_n e(mn/q).$$

Forming the square of the absolute value of each side, and summing over all the characters χ to the modulus q, we obtain

$$\sum_{\chi} |\tau(\chi)|^2 \left| \sum_{\substack{n=M+1 \\ n \equiv l(\mathrm{mod}\, k)}}^{M+N} a_n \chi(n) \right|^2$$

$$= \varphi(q) \sum_{\substack{m=1 \\ (m,q)=1}}^{q} \left| \sum_{\substack{n=M+1 \\ n \equiv l(\mathrm{mod}\, k)}}^{M+N} a_n e(mn/q) \right|^2. \tag{27}$$

Write $n = kn' + l$ and $a_n = b_{n'}$. The range R for n' has length at most $Nk^{-1} + 1$. We have

$$\sum_{\substack{q \in Q \\ (q,k)=1}} \frac{1}{\varphi(q)} \sum_{\chi} |\tau(\chi)|^2 \left| \sum_{\substack{n=M+1 \\ n \equiv l(\mathrm{mod}\, k)}}^{M+N} a_n \chi(n) \right|^2$$

$$= \sum_{\substack{q \in Q \\ (q,k)=1}} \sum_{\substack{m=1 \\ (m,q)=1}}^{q} \left| \sum_{R} b_{n'} e\left(\frac{mkn'}{q}\right) \right|^2.$$

We now observe that, since $(q, k) = 1$, the numbers km, considered modulo q, are a permutation of the numbers m relatively prime to q. Hence the above sum is

$$\leqq \sum_{q \in Q} \sum_{\substack{m=1 \\ (m,q)=1}}^{q} \left| \sum_{R} b_{n'} e(mn'/q) \right|^2,$$

and by theorem 1 this does not exceed

$$(|R|^{1/2} + X)^2 \sum_{R} |b_{n'}|^2 \leqq \{(Nk^{-1} + 1)^{1/2} + X\}^2 \sum_{\substack{n=M+1 \\ n \equiv l(\mathrm{mod}\, k)}}^{M+N} |a_n|^2.$$

This proves (9).

5. Proof of theorem 4. We write for brevity

$$\pi(\chi) = \pi(M + N, \chi) - \pi(M, \chi),$$

$$\pi_0 = \pi(M + N) - \pi(M).$$

In (27), with the congruence condition mod k omitted, we take a_n to be 1 if n is a prime between M and $M + N$, and 0 otherwise. The condition (7) is satisfied if

$q \leqq X$, in virtue of the hypothesis that $M \geqq X$. We obtain

$$\sum_{\chi} |\tau(\chi)|^2 \, |\pi(\chi)|^2 = \varphi(q) \sum_{\substack{m=1 \\ (m,q)=1}}^{q} \Bigg| \sum_{M < p \leqq M+N} e(mp/q) \Bigg|^2 .$$

Dividing by $\varphi(q)$ and summing over $q \leqq X$, and appealing to (8), we get

$$\sum_{q \leqq X} \frac{1}{\varphi(q)} \sum_{\chi} |\tau(\chi)|^2 \, |\pi(\chi)|^2 \leqq (N^{1/2} + X)^2 \, \pi_0 . \tag{28}$$

In the sum on the left we group together those characters χ, to various moduli q, which are induced by the same primitive character. If the primitive character χ_1 to the modulus q_1 induces the character χ to the modulus q, then q_1 is a divisor of q, and if $q = rq_1$ then

$$|\tau(\chi)|^2 = \begin{cases} q_1 & \text{if } r \text{ is square-free and } (r, q_1) = 1 , \\ 0 & \text{otherwise.} \end{cases}$$

Note also that $\pi(\chi) = \pi(\chi_1)$, since none of the primes p counted in $\pi(\chi)$ can divide q. The sum on the left of (28) becomes

$$\sum_{\substack{rq_1 \leqq X \\ (r,q_1)=1}} \frac{\mu^2(r)}{\varphi(rq_1)} \, q_1 \sum_{\chi_1} |\pi(\chi_1)|^2 = \sum_{q_1 \leqq X} \frac{q_1}{\varphi(q_1)} \Bigg\{ \sum_{\substack{r \leqq X/q_1 \\ (r,q_1)=1}} \frac{\mu^2(r)}{\varphi(r)} \Bigg\} \sum_{\chi_1} |\pi(\chi_1)|^2 .$$

We are indebted to Professor Halberstam for drawing our attention to the following simple and elegant proof[1]) of a lower bound for the inner sum over r in the last expression. We have, for any k,

$$\sum_{n \leqq x} \frac{\mu^2(n)}{\varphi(n)} = \sum_{d|k} \sum_{\substack{n \leqq x \\ (n,k)=d}} \frac{\mu^2(n)}{\varphi(n)} = \sum_{d|k} \sum_{\substack{r \leqq x/d \\ (r,k/d)=1}} \frac{\mu^2(rd)}{\varphi(rd)}$$

$$= \sum_{d|k} \frac{\mu^2(d)}{\varphi(d)} \sum_{\substack{r \leqq x/d \\ (r,k)=1}} \frac{\mu^2(r)}{\varphi(r)} \leqq \Bigg(\sum_{d|k} \frac{\mu^2(d)}{\varphi(d)} \Bigg) \Bigg(\sum_{\substack{r \leqq x \\ (r,k)=1}} \frac{\mu^2(r)}{\varphi(r)} \Bigg) .$$

Hence

$$\sum_{\substack{r \leqq x \\ (r,k)=1}} \frac{\mu^2(r)}{\varphi(r)} \geqq \frac{\varphi(k)}{k} \sum_{n \leqq x} \frac{\mu^2(n)}{\varphi(n)} \geqq \frac{\varphi(k)}{k} \log x , \tag{29}$$

the last inequality being one noted earlier.

Using this with $x = X/q_1$, $k = q_1$, and replacing q_1 by q and χ_1 by χ (now primitive) we obtain the assertion (14).

---

[1]) van Lint and Richert, loc. cit., (1.3) and (1.5).

6. Proof of corollary 3. This is essentially the same as the proof of corollary 2, given in § 1, but we include it for the sake of completeness.

We take a_n to be 1 if n is a prime congruent to $l \pmod{k}$ in the range $M < n \leqq M+N$, and 0 otherwise; and we take $p_1, \ldots, p_r$ to consist of all primes up to X that do not divide k. Then the condition (7) is satisfied provided we suppose that $M \geqq X$.

We take on the left of (9) only the contribution of the principal character χ_0 for each of the moduli q in the sum. Then $\chi_0(n)\, a_n = a_n$, and

$$\sum_{n=M+1}^{M+N} a_n \chi_0(n) = \pi(M+N; k, l) - \pi(M; k, l),$$

$$\sum_{n=M+1}^{M+N} |a_n|^2 = \pi(M+N; k, l) - \pi(M; k, l).$$

It follows that

$$\pi(M+N; k, l) - \pi(M; k, l) \leqq \frac{\{(Nk^{-1}+1)^{1/2} + X\}^2}{Z_k},$$

where

$$Z_k = \sum_{\substack{q \leqq X \\ (q,k)=1}} \frac{\mu^2(q)}{\varphi(q)}.$$

By (29) we have

$$Z_k > k^{-1}\varphi(k) \log X.$$

Define U by $U = N/k$, so that $U > 3$ by hypothesis. Choose $X = U^{1/2}/\log U$. Then

$$\pi(M+N; k, l) - \pi(M; k, l) \leqq \frac{N\{(1+U^{-1})^{1/2} + (\log U)^{-1}\}^2}{\varphi(k) \log X}$$

$$\leqq \frac{2N}{\varphi(k) \log U}\left\{1 + O\left(\frac{\log\log U}{\log U}\right)\right\}.$$

This proves (18), subject to the condition $M \geqq X$. But this condition can be omitted, since the number of primes up to X is

$$O\left(\frac{X}{\log X}\right) = O\left(\frac{U}{\log^2 U}\right) = O\left(\frac{N}{k \log^2 U}\right),$$

and so can be absorbed in the error term.

Note added in proof. Mr. B. Berndt, of the University of Illinois, has observed that the second inequality in (21) is best possible of its kind. If $N = 1$, $a_1 = 1$, $a_0 = 0$, $a_{-1} = 1$, then $S(0) = 2$ and $S\left(\frac{1}{2}\right) = -2$. With $x_1 = 0$, $x_2 = \frac{1}{2}$, both sides of the inequality have the value 8.

B. M. Bredihin, J. V. Linnik und N. G. Tschudakoff in Leningrad

ÜBER BINÄRE ADDITIVE PROBLEME GEMISCHTER ART

Von großem Interesse für die moderne analytische Zahlentheorie ist die Untersuchung von binären Gleichungen folgender Art:

$$a_k + \varphi(\xi, \eta) = n, \tag{0.1}$$

wobei $\{a_k\}$ eine Folge von Zahlen mit gewissen arithmetischen Eigenschaften und $\varphi(\xi, \eta)$ eine gegebene positiv-definite binäre quadratische Form ist.

Für $a_k = \psi(\xi', \eta')$, wobei $\psi(\xi', \eta')$ eine beliebige, positiv-definite binäre quadratische Form ist, erhalten wir ein Problem über quaternäre Formen. Dieses Problem wurde untersucht in Arbeiten von Klosterman [1], Tartakowski [2], Malischev [3] und Eichler [4].

Bei $a_k = p$ ergibt sich ein Problem von Hardy-Littlewood [5], welches in den Arbeiten [6] und [7] betrachtet wurde. In diesen Arbeiten ist eine Methode für die asymptotische Abschätzung der Lösungsanzahl von Gleichung (0.1) enthalten, die auf ergodischen Eigenschaften der Lösungen von (0.1) beruht.

Betrachten wir den Grundgedanken der Methode. Die Form $\varphi(\xi, \eta)$ sei primitiv. Dann können wir den Zusammenhang zwischen quadratischen Formen und Idealen ausnutzen. Danach führt (0.1) auf die Gleichung

$$a_k + \mathfrak{N}(\mathfrak{a}) = n, \tag{0.2}$$

wobei $\mathfrak{N}(\mathfrak{a})$ die Norm eines ganzen Ideals $\mathfrak{a}$ aus einer gewissen Idealklasse ist, die definiert wird durch die Form $\varphi(\xi, \eta)$ und zum Geschlecht R gehört. Wir gehen zur Gleichung

$$a_k + \mathfrak{N}_R(\mathfrak{a}) = n \tag{0.3}$$

über, in welcher a_k dieselbe Zahlenfolge durchläuft wie in (0.2) und $\mathfrak{N}_R(\mathfrak{a})$ die Norm eines beliebigen ganzen Ideals aus dem Geschlecht R ist.

Die Zahlen $m = \mathfrak{N}_R(\mathfrak{a})$ $(1 \leqq m \leqq n)$ teilen wir in zwei Klassen E und F ein.

Zur Klasse E gehören alle m, die Normen von Idealen sind, die in ihrer kanonischen Zerlegung genügend viel Primideale ersten Grades aus jeder Idealklasse enthalten.

Zur Klasse F gehören die übrigen m.

Die Lösungen der Gleichung (0.3), die Zahlen der Klasse F entsprechen, werden durch die Siebmethode ausgeschieden.

Die Ideale $\mathfrak{a}$, welche der Gleichung (0.3) genügen und Zahlen der Klasse E entsprechen, sind schon gleichmäßig über die Idealklassen im Geschlecht R verteilt. Danach erhält man aus der asymptotischen Abschätzung für die Lösungsanzahl der Gleichung (0.3) eine asymptotische Abschätzung für die Lösungsanzahl von (0.2), also folglich auch von (0.1). Voraussetzung ist dabei natürlich die asymptotische Abschätzung der Lösungsanzahl von (0.3).

In dieser Arbeit werden wir die Gleichung (0.1) für $a_k = \psi(\xi', \eta') = p$ behandeln, d. h. unter der Bedingung, daß a_k die Primzahlen durchläuft, die von der gegebenen positiv-definiten quadratischen Form dargestellt werden.

Unser Ziel ist, ein möglichst klares Bild der Beweismethode zu geben. Deshalb beschränken wir uns auf die Betrachtung von primitiven Formen mit negativer zu einem quadratischen Körper gehöriger Diskriminante.

Die Beweise werden von uns vollständig durchgeführt unter der Annahme, daß in dem Formengeschlecht, zu dem die Form $\psi(\xi', \eta')$ gehört, nur eine Klasse vorhanden ist. Im Fall von mehrklassigen Formengeschlechtern > 1 werden also Satz A und B in unserer Arbeit nicht vollständig bewiesen.[1])

Ein Beweis dieser Sätze im allgemeinen Fall ist wahrscheinlich möglich, indem man nachträglich das Verhalten der Größe a_k untersucht.

Wir führen folgende Bezeichnungen und Definitionen ein:

Das Symbol O hat die übliche Bedeutung. Die Buchstaben $c_0, c_1, c_2, \ldots$ und eine Konstante in dem Ausdruck $O(\;)$ bezeichnen positive Konstanten, die nur von den gegebenen Formen $\varphi(\xi, \eta)$ und $\psi(\xi', \eta')$ abhängig sind.

$\varepsilon_1, \varepsilon_2, \ldots$ sind kleine positive Zahlen, C ist eine genügend große positive ganze Zahl, nicht immer ein und dieselbe.

n ist eine beliebig große positive ganze Zahl (Grundparameter der Arbeit),

p eine Primzahl,

$\varphi(m)$ die Eulersche Funktion, (m_1, m_2) bezeichnet den größten gemeinsamen Teiler der Zahlen m_1 und m_2.

$K(\sqrt{-d})$ ist ein quadratischer Körper, K Körper der rationalen Zahlen.

$-d$ ist eine negative ganze quadratfreie rationale Zahl,

$-D$ die Diskriminante des Körpers $K(\sqrt{-d})$,

$P = P_D = p_1 p_2 \cdots p_t$ das Produkt aller verschiedenen Primteiler der Diskriminante $-D$,

$\chi_D(x)$ ist der quadratische Charakter (mod D) des Körpers $K(\sqrt{-d})$,

$\mathfrak{a}, \mathfrak{b}, \mathfrak{c}$ sind ganze Ideale,

---

[1]) Ein Beweis für den allgemeinen Fall wird in einer anderen Arbeit gegeben, da unser Fall nur endlich viele Formen ψ erfaßt und sich auf Primzahlen in Progressionen reduzieren läßt.

$\mathfrak{p}$, $\mathfrak{p}_{jk}$ Primideale des Körpers $K(\sqrt{-d})$.

$\mathfrak{N}(\mathfrak{a})$ ist die Norm des Ideals $\mathfrak{a}$, $\sim$ das Zeichen für die Äquivalenz von Idealen, $\mathfrak{a}'$ ist absolut äquivalent zu $\mathfrak{a}$,

A, B, C bezeichnen Idealklassen.

G ist die Idealklassengruppe des Körpers,

$R = R(A)$ das bestimmte Idealgeschlecht der Klasse $A \in R$,

$R\sigma$ ein von σ abhängiges Geschlecht,

G_0 das Hauptgeschlecht,

h die Klassenzahl in G,

t_0 die Klassenzahl in G_0,

g die Zahl der Geschlechter.

$\left(\frac{a, b}{p}\right)$ ist das Hilbert-Symbol, wobei a und b rationale Zahlen sind.

In der Arbeit werden einige Ergebnisse der arithmetischen Theorie der quadratischen Formen benutzt, die man in den Monographien [8], [9] und [10] findet.

§ 1. Formulierung der Hauptergebnisse

Wir betrachten die Gleichung

$$\psi(\xi', \eta') + \varphi(\xi, \eta) = n, \tag{1.1}$$

wobei $\varphi(\xi, \eta)$ und $\psi(\xi', \eta')$ gegebene positiv-definite primitive quadratische Formen mit negativen Diskriminanten $-D$ und $-D'$ sind, die zu den quadratischen Körpern $K(\sqrt{-d})$ und $K(\sqrt{-d'})$ gehören.

Die Veränderlichen ξ und η durchlaufen unabhängig die ganzen Zahlen unter der Bedingung $0 < \varphi(\xi, \eta) < n$, die wegen der Definitheit der Form $\varphi(\xi, \eta)$ automatisch gilt, ξ' und η' durchlaufen unabhängig ganze Zahlen, für die $\psi(\xi', \eta')$ gleich einer Primzahl ist.

Es sei

$$Q(n) = \sum_{\substack{\psi(\xi', \eta') + \varphi(\xi, \eta) = n \\ \psi(\xi', \eta') = p}} 1 \tag{1.2}$$

die Anzahl der Lösungen von Gleichung (1.1) unter den angegebenen Bedingungen, und ε_0 sei die kleinste positive Lösung der Gleichung

$$\frac{1}{h} - 2\varepsilon \ln 2 - \varepsilon + \varepsilon \ln \varepsilon = 0. \tag{1.3}$$

Dann gilt

Satz A. *Für $n \to \infty$ ist*

$$Q(n) = A(D, D', n) \cdot \frac{n}{\ln n} + O(n(\ln n)^{-1-\gamma}), \ldots, \tag{1.4}$$

wobei $A(D, D', n)$ die singuläre Reihe des Problems ist,

$$\gamma = \min\{\varepsilon_1; 0{,}042\}, \quad 0 < \varepsilon_1 < \varepsilon_0 \ln 2.$$

Wir unterdrücken die explizite Form des Ausdrucks $A(D, D', n)$, weil sie zu umfangreich ist (in impliziter Form kommt sie in § 2 vor). In einigen Fällen kann man die singuläre Reihe direkt berechnen. Zum Beispiel ist bei $(D, D') = 1$

$$A(D, D', n) \geqq c_0 > 0,$$

wobei die Konstante c_0 nur von den Diskriminanten der gegebenen Formen $\varphi(\xi, \eta)$ und $\psi(\xi', \eta')$ abhängt.

Im allgemeinen Fall kann man feststellen, ob $A(D, D', n) \neq 0$ wird oder nicht.

Die Aufgabe, eine asymptotische Abschätzung für $Q(n)$ zu finden, formulieren wir in den Bezeichnungen der Theorie quadratischer Körper.

Der Form $\varphi(\xi, \eta)$ entspricht die Idealklasse A_φ von $K(\sqrt{-d})$, der Form $\psi(\xi', \eta')$ die Idealklasse A_ψ von $K(\sqrt{-d'})$.

Dann entspricht der Frage nach der Darstellung einer Zahl durch die Form $\varphi(\xi, \eta)$ die Frage nach der Existenz von ganzen Idealen $\mathfrak{a}$ des Körpers $K(\sqrt{-d})$ mit der Norm m aus der Klasse $A = A_\varphi^{-1}$, genauer gilt

$$\sum_{\varphi(\xi,\eta)=m} 1 = w \sum_{\mathfrak{N}(\mathfrak{a})=m,\,\mathfrak{a}\in A} 1, \tag{1.5}$$

wobei

$$w = \begin{cases} 6 & \text{für } D = 3, \\ 4 & \text{für } D = 4, \\ 2 & \text{für } D > 4 \end{cases}$$

ist.

Analog ist

$$\sum_{\psi(\xi',\eta')=p} 1 = w' \sum_{\mathfrak{N}(\mathfrak{p})=p,\,\mathfrak{p}\in A'} 1, \tag{1.6}$$

wobei $A' = A_\psi^{-1}$ und $\mathfrak{p}$ ein Primideal ersten Grades ist und

$$w' = \begin{cases} 6 & \text{für } D' = 3, \\ 4 & \text{für } D' = 4, \\ 2 & \text{für } D' > 4. \end{cases}$$

Aus (1.2), (1.5) und (1.6) folgt

$$Q(n) = ww' \sum_{\substack{\mathfrak{N}(\mathfrak{p})+\mathfrak{N}(\mathfrak{a})=n \\ \mathfrak{p}\in A', \mathfrak{a}\in A}} 1. \tag{1.7}$$

Wir betrachten das Geschlecht $R = R(A)$. Es sei

$$\tilde{Q}(n) = ww' \sum_{\substack{\mathfrak{N}(\mathfrak{p})+\mathfrak{N}(\mathfrak{a})=n \\ \mathfrak{p}\in A', \mathfrak{a}\in R}} 1, \tag{1.8}$$

wobei, im Unterschied zu (1.7), $\mathfrak{a}$ die ganzen Ideale aller Klassen des Geschlechts R durchläuft.

Satz B. *Für $n \to \infty$ ist*

$$\tilde{Q}(n) = t_0 A(D, D', n) \frac{n}{\ln n} + O(n (\ln n)^{-1,042}), \tag{1.9}$$

wobei $A(D, D', n)$ die singuläre Reihe des Problems und t_0 die Zahl der Idealklassen vom Geschlecht R ist.

Der Beweis des Satzes B kann sich auf die neuesten Ergebnisse von E. Bombieri [11] gründen, der die Methode „des großen Siebes" wesentlich verschärft und vereinfacht hat. Der Beweis kann ebenfalls mit der Dispersionsmethode erhalten werden, die in [5] und [12] ausgearbeitet wurde.

Der Übergang von der Lösung der Gleichung (1.1) für ein Geschlecht zu der Lösung dieser Gleichung für die spezielle Form $\varphi(\xi, \eta)$ wird mit Hilfe des folgenden Satzes durchgeführt.

Satz C. *Für $n \to \infty$ gilt*

$$Q(n) = \frac{1}{t_0} \tilde{Q}(n) + O(n (\ln n)^{-1-\gamma}), \tag{1.10}$$

wobei

$$\gamma = \min\{\varepsilon_1, 0,042\}, \quad 0 < \varepsilon_1 < \varepsilon_0 \ln 2$$

und ε_0 durch (1.3) *bestimmt ist.*

Der Satz A folgt unmittelbar aus (1.9) und (1.10). Deshalb ergibt sich Satz A ohne weiteres aus dem Beweis der Sätze B und C.

§ 2. Beweis des Satzes B

Wir nehmen an, daß in dem zu $\psi(\xi', \eta')$ gehörigen Geschlecht nur eine Klasse enthalten ist. Dann gilt

$$\sum_{\mathfrak{N}(\mathfrak{p})=p', \mathfrak{p}\in A'} 1 = \sum_{x|p'} \chi_{D'}(x) = 2, \tag{2.1}$$

wobei p' alle Primzahlen durchläuft, die den Bedingungen

$$(p', D') = 1, \quad \chi_{D'}(p') = 1,$$

$$\left(\frac{p'D'}{p}\right) = e_p(R') \quad \text{für} \quad p|D' \tag{2.2}$$

genügen.

Die Zahlen $e_p(R')$ in (2.2) sind die Invarianten des Geschlechts $R' = R'(A)$.

Mit Hilfe der Eigenschaften des Hilbert-Symbols kann man den Bedingungen (2.2) folgendes Aussehen geben:

$$p' \equiv l_i \,(\mathrm{mod}\; D').$$

Dabei durchlaufen die l_i insgesamt $\dfrac{\varphi(D')}{2g'}$ Zahlen mit den Eigenschaften

$$0 < l_i < D', \quad (l_i, D') = 1, \quad \chi_{D'}(l_i) = 1.$$

Aus (1.7), (1.8) und (2.1) erhält man

$$Q(n) = 2ww' \sum_{l\in\{l_i\}} Q_i(m) + O(n^{\varepsilon_2}), \tag{2.3}$$

wobei

$$Q_i(n) = \sum_{\substack{p+\mathfrak{N}(\mathfrak{a})=n, \mathfrak{a}\in A \\ p\equiv l_i (\mathrm{mod}\, D')}} 1 \tag{2.4}$$

und

$$\tilde{Q}_i(n) = \sum_{\substack{p+\mathfrak{N}(\mathfrak{a})=n, \mathfrak{a}\in R \\ p\equiv l_i (\mathrm{mod}\, D')}} 1 \tag{2.4a}$$

gesetzt wurde.

Bekanntlich läßt sich jedes ganze Ideal $\mathfrak{a} \in R(A)$ eindeutig in der Form

$$\mathfrak{a} = B_1^2 \mathfrak{B}_2 C \tag{2.5}$$

darstellen, wobei $\mathfrak{N}(B_1^2\mathfrak{B}_2)$ nur Primteiler der Diskriminante $-D$ enthält und $(\mathfrak{N}(C), D) = 1$ ist, $\mathfrak{B}_2$ ist ein quadratfreies Ideal. Seien A, B und C Klassen, die zu den Idealen $\mathfrak{a}$, $\mathfrak{B}_2$ und $\mathfrak{C}$ gehören. Wegen $B_1^2\mathfrak{B}_2 \sim \mathfrak{B}_2$ ist $A = BC$ woraus

$$R(A) = R(B)\, R(C) \tag{2.6}$$

folgt. (Hier wird die Übertragung der Multiplikation von Klassen auf Geschlechter angewandt.)

Bei fester Norm $\sigma = \mathfrak{N}(\mathfrak{B}_2)$ und gegebenem Geschlecht $R(A)$ wird das Geschlecht $R_\sigma = R(C)$ eindeutig bestimmt durch (2.6).

Nach [8] gilt

$$f(m) = \sum_{\mathfrak{N}(\mathfrak{a}) = m, \mathfrak{a} \in R} 1 = \sum_{x|m} \chi_D(x), \tag{2.7}$$

wobei x alle Teiler der natürlichen Zahl m durchläuft. Deshalb gilt für ein m, das nur Primteiler der Diskriminante $-D$ enthält, die Gleichung

$$\sum_{\mathfrak{N}(\mathfrak{b}) = m, \mathfrak{b} \in R(B)} 1 = 1. \tag{2.8}$$

Aus (2.4), (2.5), (2.6) und (2.8) folgt

$$\tilde{Q}_i(n) = \sum_{\sigma|P} \tilde{Q}_\sigma(n), \tag{2.9}$$

wobei

$$\tilde{Q}_\sigma(n) = \sum_{\substack{p + r^2\sigma\mathfrak{N}(\mathfrak{c}) = n \\ p \equiv l_i (\mathrm{mod}\, D'), \mathfrak{c} \in R_\sigma, (\mathfrak{N}(\mathfrak{c}), D) = 1}} 1 \tag{2.10}$$

ist. In (2.10) durchläuft r die natürlichen Zahlen, welche in der Primzahlzerlegung nur Potenzen von Primteilern von P enthalten, σ ist ein fixierter Teil von P. Aus (2.7) und (2.10) erhält man

$$\tilde{Q}_\sigma(m) = \sum_{\substack{p + r^2\sigma m = n \\ p \equiv l_i (\mathrm{mod}\, D')}} \sum_{x|m} \chi_D(x), \tag{2.11}$$

wobei m die natürlichen Zahlen durchläuft, die den folgenden Bedingungen genügen:

$$(m, D) = 1, \quad \chi_D(m) = 1,$$

$$\left(\frac{mD}{p}\right) = e_p(R_\sigma) \quad \text{für} \quad p|D. \tag{2.12}$$

Die Zahlen $e_p(R_\sigma)$ in (2.12) sind die Invarianten des Geschlechts R_σ.

Mittels der Eigenschaften des Hilbert-Symbols kann man den Bedingungen (2.12) ein anderes Aussehen geben:

$$m \equiv L \,(\mathrm{mod}\, D),$$

wobei L die Menge $(L)_\sigma$ von $\dfrac{L(D)}{2g}$ Zahlen mit den Eigenschaften

$$0 < L < D, \quad (L, D) = 1, \quad \chi_D(L) = 1$$

durchläuft. Aus (2.11) folgt

$$\tilde{Q}_\sigma(m) = \sum_{L\in(L)_\sigma} \tilde{Q}_L(n), \tag{2.13}$$

wobei

$$\tilde{Q}_L(n) = \sum_{\substack{p+r^2\sigma xy=n \\ p\equiv l_i(\mathrm{mod}\,D'),\, xy\equiv L(\mathrm{mod}\,D)}} \tag{2.14}$$

ist mit festen σ und L.

Hilfssatz I [12]. *Es gilt*

$$\tilde{Q}_L(n) = 2\tilde{A}_L(n) + O(n\,(\ln n)^{-1,042}), \tag{2.15}$$

wobei

$$\tilde{A}_L(n) = \sum_{(Q),\, x<\sqrt{n}/(\ln n)^c} \chi_D(x),$$

ist und (Q) *den Komplex der Bedingungen aus* (2.14) *bezeichnet.*

Hilfssatz 1 ist ein Analogon des Hilfssatzes 2.3.1 der Arbeit [12] und wird analog mit Hilfe der Untersuchungen von C. Hooley [13] und der Bewertungen von P. Erdös [14], [15] bewiesen.

Hilfssatz 2 [11]. *Es gilt das geglättete asymptotische Gesetz der Primzahlverteilung in einer arithmetischen Progression:*

$$\sum_{q\leqq\sqrt{n}/(\ln n)^{c_2}} \max_{(a,q)=1} \left| \pi(n, q, a) - \frac{1}{\varphi(q)}\int_2^n \frac{du}{\ln u} \right| < \frac{n}{(\ln n)^{c_1}}, \tag{2.16}$$

wobei $c_1 > 0$ *eine beliebige fixierte Konstante ist,* $c_2 = c_2(c_1)$.

Hilfssatz 2 ergibt sich als offensichtliche Folge der Sätze von E. Bombieri.

Es sei

$$T(n) = \sum_{x\leqq\sqrt{n}/(\ln n)^c,\,(x,m)=1} \frac{\chi_D(x)}{\varphi(sx)}, \tag{2.17}$$

mit $s = o((\ln n)^e)$, $m = o(n)$. Dann gilt

Hilfssatz 3 [12].

$$\begin{aligned} T(n) = {} & L(1,\chi_D) \prod_p \left(1 + \frac{\chi_D(p)}{p(p-1)}\right) \prod_{p|m} \frac{(p-1)(p-\chi_D(p))}{p^2 - p + \chi_D(p)} \\ & \times \frac{1}{s} \prod_{p|(s,m)} \left(1 - \frac{1}{p}\right)^{-1} \prod_{p|(s,p+m)} \frac{p^2}{p^2 - p + \chi_D(p)} \\ & + O(n^{-1/2+\varepsilon_4}), \end{aligned} \tag{2.18}$$

wobei

$$L(1, \chi_D) = \sum_{n=1}^{\infty} \frac{\chi_D(n)}{n}$$

ist.

Elementare Rechnungen zeigen, daß

$$\tilde{A}_L(n) = \sum_{\substack{\varkappa < \sqrt{n/(\ln n)^c} \\ (\varkappa, nD) = 1}} \chi_D(x) \sum_{\substack{r \\ (r,n)=1}} \sum_{\substack{p \leqq n \\ p \equiv s \left(\bmod \frac{r^2 \sigma x D D'}{d}\right)}} 1 + O(n^{\varepsilon_5}) \tag{2.19}$$

gilt, wobei $d = (D', r^2\sigma Dx)$ ist. Außerdem müssen in (2.19) die Bedingungen

$$\left.\begin{aligned} &(n^2 - r^2\sigma L, D) = 1, \\ &n - r^2\sigma L x \varkappa_D' \equiv l_i \,(\bmod d) \end{aligned}\right\} \tag{2.20}$$

berücksichtigt werden; hier ist $\varkappa_D'$ das Inverse von $x(\bmod D)$, woraus

$$\left(s, \frac{r^2\sigma x D D'}{d}\right) = 1; \quad (\sigma, n) = 1$$

folgt.

Wir wenden das auf (2.19) an mit der ergänzenden Bedingung

$$\frac{r^2\sigma D D'}{d} < (\ln n)^{c/2},$$

die in (2.19) einen Fehler der Ordnung $O(n(\ln n)^{-c/4})$ ergibt. Dann können wir die im Hauptglied angegebenen Bedingungen (mit demselben Fehler) weglassen. Man setze dabei $q = \dfrac{r^2\sigma\varkappa D D'}{d}$, $c_2 = \dfrac{c}{2}$ und berücksichtige, daß die Anzahl der Wiederholungen von q der Abschätzung

$$\sum_{r < (\ln n)^{c/4}} 1 = O((\ln \ln n)^t) \tag{2.21}$$

genügt.

Aus (2.16), (2.19) und (2.21) erhalten wir

$$\tilde{A}_L(n) = \int_2^n \frac{du}{\ln u} \sum_{\substack{r \\ (r,n)=1}} \sum_{\substack{\varkappa \leqq \sqrt{n/(\ln n)^c} \\ (\varkappa, nD) = 1}} \frac{\chi_D(x)}{\varphi\left(\dfrac{r^2\sigma D D' x}{d}\right)} + O(n(\ln n)^{-c_1/2}). \tag{2.22}$$

Den Bedingungen in (2.19) und (2.20) kann man eine andere für die weiteren Rechnungen bequemere Form geben.

Es sei $d_1 = (D', r^2\sigma D)$ und $d_2 = (D', \varkappa)$. Dann wird

$$d = d_1 d_2, \quad (d_1, d_2) = 1;$$

$$x = d_2 x_1, \quad \left(\varkappa_1, \frac{D'}{d_2}\right) = 1.$$

Die Kongruenz in (2.20) ist gleichbedeutend mit dem Kongruenzsystem

$$l_i \equiv n - r^2\sigma L \pmod{d_1},$$

$$l_i \equiv n \pmod{d_2}.$$

Aus (2.4), (2.9), (2.13)–(2.15) und (2.22) ergibt sich nun für $\tilde{Q}(n)$ die Abschätzung

$$\tilde{Q}(n) = t_0 A(D, D', n) \frac{n}{\ln n} + O(n (\ln n)^{-1{,}042}), \tag{2.23}$$

wobei

$$A(D, D', n) = \frac{4ww'}{t_0} \sum_{\substack{d_2 | D' \\ (d_2, nD) = 1}} \chi_D(d_2) \sum_{\substack{d_1 | D' \\ (d_1, d_2) = 1}} \sum_{\substack{\sigma | P \\ (\sigma, n) = 1}}$$

$$\times \sum_{L \in (L)_\sigma} \sum_{r_1 (r, n) = 1} \sum_{l_i \in \{l_i\}}$$

$$\times \sum_{x_1 < \sqrt{n}/d_2 (\ln n)^c, \ \left(x_1, nD\frac{D'}{d_2}\right) = 1} \frac{\chi_D(x_1)}{\varphi\left(\dfrac{r^2\sigma D D'}{d_1} x_1\right)} \tag{2.24}$$

ist und die Summierung über σ, L, r und l_i zu erstrecken ist mit den folgenden Einschränkungen:

$$(n - r^2\sigma L, D) = 1, \qquad (r^2\sigma D, D') = d,$$

$$l_i \equiv n \pmod{d_2}, \qquad l_i \equiv n - r^2\sigma L \pmod{d_1}.$$

Mit dem in (2.23) angegebenen Fehler kann man die Summierung über r beschränken auf solche r, die der Bedingung

$$\frac{r^2\sigma D D'}{d_1} < (\ln n)^{c/2}$$

genügen.

Setzt man in Hilfssatz 3 $m = nD\dfrac{D'}{d_2}$, $s = \dfrac{r^2\sigma DD'}{d_1}$ und berechnet mit diesem Hilfssatz die Innensumme in (2.24), so erhält man einen expliziten Ausdruck für $A(n, D, D')$.

Damit ist Satz B bewiesen.

§ 3. Beweis von Satz C

Wir betrachten $Q_i(n)$. Aus (2.4) und (2.7) folgt

$$\tilde{Q}_i(n) = \sum_{\substack{p+m=n \\ p \equiv l_i (\mathrm{mod}\, D')}} f(m), \tag{3.1}$$

wobei $f(m)$ die Anzahl der Lösungen der Gleichung

$$\mathfrak{N}(\mathfrak{a}) = m \tag{3.2}$$

ist, $\mathfrak{a}$ ein ganzes Ideal aus dem gegebenen Geschlecht R, und für m müssen die Bedingungen

$$\left(\frac{m, D}{p}\right) = l_p(R)$$

für alle p einschließlich $p = \infty$ erfüllt sein. Die Menge der Zahlen m, welche (3.2) genügen, zerlegen wir in zwei Klassen E_i und F_i.

In die Klasse E_i kommen alle m, die in ihrer Primzahlzerlegung wenigstens k_0 Primzahlen $p_{jk} (k = 1, 2, \dots, k_0)$ ersten Grades enthalten, für welche $\chi_D(p_{jk}) = 1$ ist. Für diese gilt also

$$p_{jk} = \mathfrak{p}_{jk}\mathfrak{p}'_{jk}, \quad \mathfrak{p}_{jk} \neq \mathfrak{p}'_{jk},$$

$$\mathfrak{N}(\mathfrak{p}_{jk}) = \mathfrak{N}(\mathfrak{p}'_{jk}) = p_{jk}.$$

Dabei sei $\mathfrak{p}_{jk}$ (oder $\mathfrak{p}'_{jk}) \in c_j$ für jedes $j = 1, 2, \dots, h$, $k_0 = [\varepsilon_0 \ln \ln n]$, wobei ε_0 in (1.3) definiert ist. Mit anderen Worten, die Zahlen m aus der Klasse E_i enthalten in ihrer Zerlegung im Körper $K(\sqrt{-d})$ genügend viele Primideale ersten Grades aus jeder Idealklasse c_j dieses Körpers. Zu der Klasse F_i werden die übrigen m zugerechnet.

Auf diese Weise erhält man

$$\tilde{Q}_i(n) = \sum_{\substack{p+m=n, m \in E_i \\ p \equiv l_i (\mathrm{mod}\, D')}} f(m) + \sum_{\substack{p+m=n, m \in F_i \\ p \equiv l_i (\mathrm{mod}\, D')}} f(m) = \sum\nolimits_{E_i} + \sum\nolimits_{F_i}. \tag{3.3}$$

Wenn wir in $\sum_{F_i}$ die Bedingung, daß die Zahlen p der vorgegebenen Progression angehören, weglassen, finden wir

$$\sum\nolimits_{F_i} \leqq \sum\nolimits_{F}, \tag{3.4}$$

wobei

$$\sum\nolimits_{F} = \sum_{p+m=n, m \in F_i} f(m) \tag{3.5}$$

ist. Die Summe (3.5) stimmt mit einer ähnlichen Summe überein, die in der Arbeit [7] betrachtet wurde. Für diese Summe erhält man mit der Siebmethode folgende Abschätzung:

Hilfssatz 4 ([7], (3.30)).

$$\sum_F = O\left(\frac{n(\ln\ln n)^4 (\ln n)^{\varepsilon_0 \ln 2l - \varepsilon_0 \ln \varepsilon_1}}{(\ln n)^{1+h^{-1}}}\right) \tag{3.6}$$

Die gleiche Abschätzung gilt nach (3.4) und (3.6) auch für $\sum_{F_i}$.

Wir gehen nun zur Abschätzung der Summe $\sum_{E_i}$ über. Für $m \in E_i$ gilt bekanntlich die asymptotische Gleichverteilung der ganzen Ideale $\mathfrak{a}$, welche der Gleichung (3.2) genügen, in den Idealklassen des gegebenen Geschlechts. Es gilt folgender

Hilfssatz 5. *Es sei $f_A(m)$ die Anzahl der Lösungen der Gleichung* (3.2) *unter der Bedingung, daß die ganzen Ideale $\mathfrak{a}$ die vorgegebene Klasse A aus dem Geschlecht R durchlaufen.*

Dann gilt für $m \in E_i$ die asymptotische Gleichverteilung

$$f_A(m) = \frac{f(m)}{t_0}\left(1 + O\left(\frac{1}{(\ln n)^{\varepsilon_0 \ln 2}}\right)\right). \tag{3.7}$$

Mit Hilfe von Hilfssatz 5 kann man die Summe $\sum_{E_i}$ in folgender Weise abschätzen:

$$\sum_{E_i} = t_0 \sum_{\substack{p+m=n, m\in E_i \\ p\equiv l_i (\operatorname{mod} D')}} f_A(m)\left(1 + O\left(\frac{1}{(\ln n)^{\varepsilon_0 \ln 2}}\right)\right). \tag{3.8}$$

Wir kommen nun zum Beweis von Satz C. Aus (2.4a) folgt

$$Q_i(n) = \sum_{\substack{p+m=n \\ p\equiv l_i (\operatorname{mod} D')}} f_A(m).$$

Hieraus, aus (3.3) und der offensichtlichen Ungleichheit $f_A(m) \leqq f(m)$ ergibt sich

$$Q_i(n) = \sum_{\substack{p+m=n, m\in E_i \\ p\equiv l_i (\operatorname{mod} D')}} f_A(m) + O(\textstyle\sum_F). \tag{3.9}$$

Nach (3.8) und (3.9) gilt

$$Q_i(n) = \frac{1}{t_0} \sum_{E_i}\left(1 + O\left(\frac{1}{(\ln n)^{\varepsilon_0 \ln 2}}\right)\right) f(m) + O(\textstyle\sum_F). \tag{3.10}$$

Aus (3.3), (3.4) und (3.10) folgt

$$Q_i(n) = \frac{1}{t_0}\tilde{Q}_i(n)\left(1 + O\left(\frac{1}{(\ln n)^{\varepsilon_0 \ln 2}}\right)\right) + O(\textstyle\sum_F). \tag{3.11}$$

Sei ε_1 eine feste Zahl, die der Bedingung

$$0 < \varepsilon_1 < \varepsilon_0 \ln 2$$

genügt, wobei ε_0 die kleinste positive Lösung der Gleichung (1.3) ist.

Aus (2.3), (2.4), (3.6) und (3.11) folgt die asymptotische Gleichverteilung

$$Q(n) = \frac{1}{t_0} \tilde{Q}(n) + O(n (\ln n)^{-1-\gamma}),$$

wobei $\gamma = \min \{\varepsilon_1, 0{,}042\}$ ist.

Es ist nicht schwer festzustellen, daß $\varepsilon_1 = \dfrac{\ln 2}{4n \ln (n+1)}$ der obigen Ungleichung genügt. Satz C ist damit vollständig bewiesen.

Literatur

[1] H. D. Kloosterman, On the representation of numbers in the form $ax^2 + by^2 + cz^2 + dt^2$, Acta Math. **49** (1926), 407—464.

[2] В. А. Тартаковский, Die Gesamtheit der Zahlen die durch eine positive quadratische Form $F(x_1, \ldots, x_s)$ darstellbar sind, ИАН (1929), 111—122.

[3] А. В. Малышев, О представлении целых чисел положительными квадратичными формами, Труды матем. ин-та им. В. А. Стеклова, Изд-во АН СССР, Москва-Ленинград 1962.

[4] M. Eichler, Quaternäre quadratische Formen und die Riemannsche Vermutung für die Kongruenzzetafunktion, Arch. Math. **5** (1954), 355—366.

[5] Ю. В. Линник, Дисперсионный метод в бинарных аддитивних задачах, Пзд-во ЛГУ, Ленинград 1961.

[6] Б. М. Бредихин и Ю. В. Линник, Асимптотика в общей проблеме Гарди-Литтлвуда, ДАН СССР **168**, № 5 (1966), 975-977.

[7] Б. М. Бредихин и Ю. В. Линник, Асимптотика и эргодические свойства решений обобщенного уравнения Гарди-Литтлвуда, Матем. сб. **71** (113), № 2 (1966), 145—161.

[8] S. J. Borewicz und I. R. Šafarewič, Zahlentheorie, Birkhäuser Verlag, Basel 1966 (Übersetzung aus dem Russischen).

[9] Б. А. Венков, Элементарная теория чисел, Гостехиздат, Москва-Ленинград 1937.

[10] B. Jones, The arithmetic theory of quadratic forms, Corns Math. Monographs, 1950.

[11] E. Bombieri, On the large sieve, Mathematika **12** (1965), 201—225.

[12] Б. М. Бредихин, Дисперсионный метод и бинарные аддитивные проблемы определенного типа, Успехи матем. наук XX, вып. 2 (122) (1965), 89—130.

[13] C. Hooley, On the representation of numbers as the sum of two squares and a prime, Acta Math. **97** (1957), 189—210.

[14] П. Эрдеш, Об одном асимптотическом неравенстве в теории чисел, Вестник ЛГУ **13** (1960), 41—49.

[15] Б. М. Бредихин, Улучшение остаточного члена в проблемах типа Гарди-Литтлвуда, Вестник ЛГУ **19** (1962), 133—137.

Wir danken Herrn Professor P. Turán für seine freundliche Hilfe.

J. G. VAN DER CORPUT IN AMSTERDAM

HOW TO EXTEND A CALCULUS

Fifty years ago while writing my doctor's thesis I was already strongly under the influence of Landau although I had never met him as yet. The fact was that in his correspondence he gave as well sharp criticism as warm encouragement which altered the contents considerably. Especially the period, in which I was fortunately enough to work with him at Göttingen, contributed beyond measure to my mathematical development. The present paper would perhaps never have been written without him, even if it diverges to a great extent from the method followed always by him in thought and intent. He preferred to devote his attention to special problems which were characteristic for extensive domains. Whenever as a pioneer he had found a solution for these problems, it was made easy for his followers to generalize the results. The opposite way is taken in this article in which the general question is treated whether and in which measure a given calculus can be extended. In view of the extensiveness of the subject I must restrict myself to some indication. The method developed herewith can be applied in many diversified branches, a.o. in the analytic theory of numbers and for this reason it may find a place in a volume dedicated to Edmund Landau. Had this publication appeared fourty years earlier he certainly would have read, accepted and applied it.

As starting point a certain calculus (fundamental analysis) is chosen with notions such as $=$, $\neq$, $>$ and so on and operators, relations, rules of calculation etc. The problem is to extend this analysis. The notion "extension" can be conceived in different ways. We begin with a very simple concept. A calculus is said to be an extension of the fundamental analysis if each equality $\alpha = \beta$ valid in the fundamental analysis holds also true in the new calculus. In other words we require that the notion of equality is permanent. There are extensions in which not only equality but also inequality is permanent. In that case each inequality $\alpha \neq \beta$ valid in the fundamental analysis holds also in the new calculus.

To construct an extension we introduce a class $\mathfrak{K}$ formed by functions $\varphi(\xi)$ defined for each element ξ of an abstract set Δ (domain). The functions $\varphi(\xi)$ belong to the fundamental analysis so that $\varphi(\xi)$ represents for each element ξ of the domain Δ a mathematical object belonging to the fundamental analysis. This class is divided into a finite or infinite number of disjoint subclasses. Two functions $\varphi(\xi)$ and $\psi(\xi)$ belonging to $\mathfrak{K}$ are said to be equivalent ($\varphi(\xi) \hookrightarrow \psi(\xi)$) if and only if $\varphi(\xi)$ and $\psi(\xi)$ belong to the same subclass. The class $\mathfrak{K}$ with the said partition is termed a neutrix M^*. The set Δ is termed the domain of this neutrix.

The following convention is fundamental. The subclass of $\mathfrak{K}$ which contains a given function $\varphi(\xi)$ is designated by the symbol $\varphi(M)$. Notice that ξ is replaced by the letter which indicates the neutrix, but without star. Even if $\varphi(\xi)$ can be represented by an expression (which generally involves ξ) and if in this expression ξ is replaced by M, then the new expression represents the subclass of $\mathfrak{K}$ which contains the function $\varphi(\xi)$. The symbols $\varphi(M)$ bear the name of neutralized values. This convention implies

$$\varphi(M) = \psi(M) \tag{1}$$

if and only if

$$\varphi(\xi) \hookrightarrow \psi(\xi).$$

The new calculus is the system of the relations between the neutralized values $\varphi(M)$.

In the special case that $\varphi(\xi)$ assumes for each element ξ of the domain Δ a same value α, then the subclass of $\mathfrak{K}$ which contains the function $\varphi(\xi)$ is according to the fundamental convention represented by $\varphi(M)$ and also by α, so that $\varphi(M) = \alpha$. Consequently each constant occuring in the fundamental analysis is also a generalized value occuring in the new calculus. This calculus is an extension of the fundamental analysis in the sense formulated above. Indeed, if the equality $\alpha = \beta$ holds in the fundamental analysis, then the functions $\varphi(\xi) = \alpha$ and $\psi(\xi) = \beta$ lie in the same subclass of $\mathfrak{K}$, so that the equality $\varphi(M) = \psi(M)$, therefore $\alpha = \beta$ holds in the new calculus. The concept of inequality is not always permanent. It may happen that two functions $\varphi(\xi) = \alpha$ and $\psi(\xi) = \beta$ with $\alpha \neq \beta$ lie in a same subclass of $\mathfrak{K}$. In such a case one has $\alpha \neq \beta$ in the fundamental analysis, but $\alpha = \beta$ in the new calculus.

Theorem 1. *In the extension constructed above the notion of inequality is permanent if and only if any subclass of $\mathfrak{K}$ contains at most one constant function.*

In the papers up till now devoted to the neutrix calculus the condition formulated in theorem 1 is called the neutrix condition, because in those papers only neutrices come into discussion in which the notion of inequality is permanent.

It is possible that the class $\mathfrak{K}$ contains a function $\varphi(\xi)$ which is equal to ξ for each element ξ of the domain. In this case one has the neutrix M^* and the neutralized value $\varphi(M) = M$. In this introduction the star is used to distinguish those two different notions. However the neutrices occur so often in the text that it is recommendable to omit the star in their designation, but then we must say the neutrix M and the neutralized value M whenever misunderstanding is possible.

Neutrices are represented by Latin capital letters, apart from veiled neutrices such as $-$, i and ∞. Negative numbers, rational numbers, algebraic numbers, complex numbers are neutralized values.

Example. Let the fundamental analysis be the calculus of the real numbers. The functions $\alpha \log 2\xi + \beta$ $(0 < \xi < p)$, where α, β and p denote real numbers with $p > 0$ and where $\varphi(\xi) = \alpha \log 2\xi + \beta$ and $\psi(\xi) = \gamma \log 2\xi + \delta$ are equivalent if and only if $\beta = \delta$, form a neutrix M with the property

$$\int_M^1 \frac{dx}{x} = \log 2 .$$

Indeed

$$-\log \xi = \int_\xi^1 \frac{dx}{x} = -\log 2\xi + \log 2 \rightharpoondown \log 2 .$$

The functions $\varphi(\xi) = \alpha \log \xi + \beta$ $(0 < \xi < p)$, where $\alpha \log \xi + \beta$ and $\gamma \log \xi + \delta$ are equivalent if and only if $\beta = \delta$, form a neutrix H with the property

$$\int_H^1 \frac{dx}{x} = 0 .$$

H is termed a Hadamard neutrix since it yields, if applied on integrals, the value which Hadamard calls the finite part of the integral.

Both neutrices M and H neutralize the fatal influence of the singularity of the integrand $\frac{1}{x}$ at $x = 0$, but they do that in different ways since they do not yield the same result.

The notion of inequality is according to theorem 1 permanent in the analysis based on the neutrix M and also in the analysis based on the neutrix H.

In the previous articles on neutrices the class $\mathfrak{K}$ is always an additive group which possesses a given additive subgroup $\mathfrak{N}$; furthermore two functions belonging to $\mathfrak{K}$ are said to be equivalent if and only if their difference belongs to $\mathfrak{N}$. In those articles the name "neutrix" is reserved for this additive subgroup $\mathfrak{N}$, but with the present terminology $\mathfrak{N}$ is the neutralized value 0, namely the subclass which contains the function which is identically equal to zero. The functions belonging to this subclass are termed "negligible".

If

$$\varphi(\xi) + \psi(\xi) = \sigma(\xi), \tag{2}$$

then one has according to the fundamental convention

$$\varphi(M) + \psi(M) = \sigma(M). \tag{3}$$

Do we have the right to say that the left hand side represents the sum of the neutralized values $\varphi(M)$ and $\psi(M)$? The two neutralized values can also be denoted by $\tilde{\psi}(M)$

and $\tilde{\psi}(M)$, where $\tilde{\varphi}(\xi) \rightharpoondown \varphi(\xi)$ and $\tilde{\psi}(\xi) \rightharpoondown \psi(\xi)$. From (3) it follows that $\tilde{\varphi}(\xi) + \tilde{\psi}(\xi) \rightharpoondown \sigma(\xi)$, hence $\tilde{\varphi}(M) + \psi(M) = \sigma(M)$, so that the result indicated in (3) is independent of the choice of the representative forms of the neutrices. In that case the left hand side of (3) is termed the sum of the neutralized values $\varphi(M)$ and $\psi(M)$. In the same way the difference $\varphi(M) - \psi(M) = \delta(M)$ between the two neutralized values $\varphi(M)$ and $\psi(M)$ is defined; here $\varphi(\xi) - \psi(\xi) = \delta(\xi)$. With these definitions of addition the neutralized values $\varphi(M)$ form an additive group. For this reason the neutrices of this kind are termed additive neutrices. In an analysis based on additive neutrices addition and subtraction are always possible with the usual properties.

Consider now the case that $\mathfrak{R}$ is a commutative ring. As we have seen the neutralized values form an additive group, but in general they do not form a commutative ring, since

$$\tilde{\varphi}(\xi) \rightharpoondown \varphi(\xi); \quad \tilde{\psi}(\xi) \rightharpoondown \psi(\xi) \tag{4}$$

does not always imply

$$\tilde{\varphi}(\xi)\, \tilde{\psi}(\xi) \rightharpoondown \varphi(\xi)\, \psi(\xi). \tag{5}$$

Theorem 2. *If in an additive neutrix M the product of a negligible function and a function belonging to $\mathfrak{R}$ is always negligible, then the neutralized values form a commutative ring.*

Indeed, if (4) holds, then $\tilde{\varphi}(\xi) = \varphi(\xi) + \mu(\xi)$ and $\tilde{\psi}(\xi) = \psi(\xi) + \nu(\xi)$, where $\mu(\xi)$ and $\nu(\xi)$ are negligible, so that $\tilde{\varphi}(\xi)\, \tilde{\psi}(\xi) - \varphi(\xi)\, \psi(\xi) = \varphi(\xi)\, \nu(\xi) + (\psi(\xi) + \nu(\xi))\, \mu(\xi)$ is negligible. This implies (5), so that the neutralized values $\varphi(M)$ and $\psi(M)$ possess a product $\varphi(M)\, \psi(M)$, and it is easy to see that with this definition of multiplication the neutralized values $\varphi(M)$ form a commutative ring.

On the other hand, even if the condition formulated in theorem 2 is not satisfied then according to the fundamental convention

$$\varphi(\xi)\, \psi(\xi) = \pi(\xi) \quad \text{implies} \quad \varphi(M)\, \psi(M) = \pi(M).$$

Here the left hand side of the last relation does not represent the product of the neutralized values $\varphi(M)$ and $\psi(M)$. It is to be considered as one symbol, termed a pseudo-product of $\varphi(M)$ and $\psi(M)$. A pseudo-product can change its value if instead of $\varphi(M)$ and $\psi(M)$ other representative forms of $\varphi(M)$ and $\psi(M)$ are used. This is not so bad as it seems, since often the functions $\varphi(\xi)$ and $\psi(\xi)$ are given.

What we have done here for the multiplication and for additive neutrices can be done for arbitrary operators and arbitrary neutrices. The fundamental analysis contains operators such as addition, subtraction, multiplication, division, differentiation, integration, Fourier- and Laplacetransforms and so on. Each such operator yields in the neutrix calculus a corresponding pseudo-operator which in some cases is even an operator that can be applied on neutralized values. This new analysis with notions such as $=$, $\neq$, $>$, with operators and pseudo-operators and with general rules of calculation, consists of relations between generalized values.

The previous papers on neutrices, particularly the report "Neutrici", published by the Istituto Matematico "Guido Castelnuovo" contain a great number of applications belonging to the fundamental analysis. The explanation how a theory lying outside an analysis may be useful for the proof or results belonging to the analysis itself is easy. In the proof that two mathematical objects α and β occuring in the fundamental analysis are inequal it is sufficient to show by means of results and rules valid in the neutrix calculus that the inequality $\alpha \neq \beta$ holds in the neutrix calculus. Indeed the equality $\alpha = \beta$ in the fundamental analysis would imply the same equality in the extension. To prove that two mathematical objects α and β occuring in the fundamental analysis are equal we must restrict ourselves to extensions which satisfy the condition formulated in theorem 1, namely the condition that the notion of inequality is permanent. If one finds $\alpha = \beta$ in such an extension one knows that the equality $\alpha = \beta$ holds in the fundamental analysis, since $\alpha \neq \beta$ in the fundamental analysis would imply $\alpha \neq \beta$ in the extension.

This phenomenon explains why the introduction of (the neutrix) i is so important in the investigation of the real numbers.

There is more. Functions belonging to special classes such as analytic functions are characterized by the condition that a certain functional has for each function of the said class the same value. We can replace this condition by the weaker hypothesis that the values of the functional are equivalent with respect to a given neutrix M. Then we get more general notions such as "analytic with respect to a neutrix M" which may yield in the fundamental analysis, besides the classical theories, other more general theories of similar form but with larger domain of applicability.

Nevertheless, an important part of the neutrix calculus lies outside the fundamental analysis. In the last decennia new calculi have been introduced such as the theory of distributions and the non-standard analysis. A number of these can be considered as belonging to the neutrix calculus, so that here unification is possible.

The question arises whether two or more neutrices may be used simultaneously. This is certainly allowed if on the notion of extension only the condition of the permanence of equality is imposed, but it is not always allowed if supplementary conditions are imposed, for instance if the permanence of inequality is required. Let us consider for instance the two neutrices M and H mentioned in our example. In both extensions the inequality is permanent. The function $\log \xi$ is equivalent with $-\log 2$ in M and equivalent with 0 in H. Using both neutrices we would find $-\log 2$ is 0, so that we would have $-\log 2 \neq 0$ in the fundamental analysis and $-\log 2 = 0$ in the extension.

The answer to the question whether two or more neutrices can be used at the same time depends therefore on the question whether on the notion of an extension supplementary conditions are imposed or not. If these supplementary conditions are chosen once for all we can construct a structure. A structure is a finite or infinite set of neutrices which can be used simultaneously with maintenance of the said supplementary conditions. If the structure $\mathfrak{S}$ is given, then the analysis $\mathfrak{A}(\mathfrak{S})$, based on all the neutrices belonging to this structure is uniquely determined.

The neutralized values in $\mathfrak{A}(\mathfrak{S})$ are represented by symbols of the form $\varphi(M_1, \ldots, M_m)$, where M_h $(h = 1, \ldots, m)$ denote neutrices belonging to $\mathfrak{S}$. Here it is assumed that $\varphi(\xi_1, \ldots, \xi_m)$ is defined for each choice of the element ξ_h of the domain Δ_h of M_h $(h = 1, \ldots, m)$. Then $\varphi(M_1, \ldots, M_m)$ is the subclass formed by all the functions which are equivalent with the said function $\varphi(\xi_1, \ldots, \xi_m)$. As example we mention the neutralized value

$$\int_{M_1}^{M_2} \int_{M_3}^{M_4} \sqrt{f^2(x, y) + M_4^2}\, dx\, dy, \tag{6}$$

where M_1, M_2, M_3, M_4 denote neutrices belonging to $\mathfrak{S}$. Assume that the double integral

$$\int_{\xi_1}^{\xi_2} \int_{\xi_3}^{\xi_4} \sqrt{f^2(x, y) + \xi_4^2}\, dx\, dy \tag{7}$$

exists for each element ξ_h of Δ_h $(h = 1, 2, 3, 4)$. Then (6) is the subclass formed by the functions which are equivalent with the function (7).

The power of the neutrix calculus lies partially in the generality of the notation.

The case that a structure $\mathfrak{S}$ is empty is also admitted. Then no neutrices are admitted, so that $\mathfrak{A}(\mathfrak{S})$ is the fundamental analysis.

References

Berg, Lothar, Asymptotische Entwicklungen mit Hilfe von Neutrizen, Arch. Math. **14** (1963), 162—171.

Corput, J. G. van der, Neutrices, SIAM (1959), 253—279.

—, Introduction to the neutrix calculus, J. d'Analyse Mathématique **7** (1959—1960), 281—398.

—, Neutrix calculus I, Neutrices and Distributions, Proc. Roy. Acad. Amsterdam A **63** (1960), 115—123; Indag. Math. **22** (1960), 115—123.

—, Neutrix calculus II A and B, Special Neutrix Calculus, Proc. Roy. Acad. Amsterdam A **64** (1960), 1—14; 15—37; Indag. Math. **23** (1960), 1—14; 15—37.

—, Introduction to the residue calculus, Proc. Roy. Acad. Amsterdam A **64** (1960), 143—156; Indag. Math. **23** (1960), 143—156.

—, Distributions with compatible neutrices, J. d'Analyse Mathématique **8** (1960—1961), 185—207.

Corput, J. G. van der, Inleiding tot de neutrixrekening, Wiskunde in de XXe eeuw. **8** (1961), 245--258.
—, Toepassingen van de neutrixrekening, Wiskunde in de XXe eeuw **8** (1961), 261--269.
—, Voordrachten over neutrixrekening, Math Centrum Amsterdam, 93 pages.
—, The neutralized sum formula of Euler, J. d'Analyse Mathématique **9** (1961--1962), 205--345.
—, La valeur qu'il faut attribuer à une fonction en un point singulier, J. de Math. pures et appliquées **42** (1963), 353—366.
—, Sul calculo neutralizato dei residui, Acc. Naz. dei Lincei, Rend. di Classe di Szienzi a fisiche, mat. e naturali **8**, fasc. 2 (1965), 166—170.
—, Neutrici, Università degli Studi di Roma, Istituto Mat. „Guido Castelnuovo", 1965, 313 pages.
—, Neutralized values I; Neutralized values with one neutrix II; Neutralized values with more than one neutrix, Proc. Roy. Acad. Amsterdam A **69** (1966), 387—401; 402--411; Indag. Math. **28** (1966), 387—401; 402—411.

H. DAVENPORT IN CAMBRIDGE AND E. LANDAU[1])

ON THE REPRESENTATION OF POSITIVE INTEGERS AS SUMS OF THREE CUBES OF POSITIVE RATIONAL NUMBERS

1. Richmond proved[2]) that every integer $R > 0$ is representable as $x^3 + y^3 + z^3$, where x, y, and z are rational and positive.

It is well known that for an infinity of integers $R > 0$, these x, y, z cannot be chosen as integers; for instance, every cube of an integer being congruent to 0, 1 or -1 (mod 9), no $R \equiv 4 \pmod 9$ is a sum of three cubes of integers.

We shall prove that in

$$R = x^3 + y^3 + z^3 \tag{1}$$

it is possible to take x, y, z as positive rational numbers with a common denominator $T = T(R)$ such that

$$T = O(R^2). \tag{2}$$

2. Richmond based his proof on the identity published by Ryley in 1825; in this, x, y, z are expressed as rational functions of R and one arbitrary parameter. But in a postscript to his paper, Richmond gave another identity containing two arbitrary parameters, and it is this which we shall use.

Let p and λ be arbitrary, and let Θ be defined by

$$\Theta = \frac{3R - \lambda^3(1 + 3p - 3p^3)}{3R + \lambda^3(1 + 3p^3)}. \tag{3}$$

---

[1]) This paper was written, in a rough form, in February 1935, when Landau visited Cambridge to give the lectures which he afterwards published as a Cambridge Tract under the title „Über einige neuere Fortschritte in der additiven Zahlentheorie". § 1 of the present version is copied from the first page of the manuscript, which is in Landau's handwriting; the later sections have been rewritten but without change of substance. As far as I can recollect, the reason why the paper was not published was that we had hoped to get some results for higher powers, but did not succeed in this. [H. D.]

[2]) On analogues of Waring's Problem for rational numbers, Proc. London Math. Soc. (2), **21** (1922), 404—409.

Let ϱ be defined by

$$\varrho = \frac{\lambda}{3(1-\Theta)\{1+3p(1-\Theta^2)+3p^2(1-\Theta)\}}. \tag{4}$$

Richmond's formulae are[1])

$$x+y+z = 3\varrho(1-\Theta)\{1+\Theta+3p-3p^3(1-\Theta)\}, \tag{5}$$

$$y+z = \varrho\{1-9p^2(1-\Theta)-9p^3(1-\Theta)^2(1+\Theta)\}, \tag{6}$$

$$z = \Theta(x+y+z). \tag{7}$$

It is plain from these formulae that x, y, z are positive if (say)

$$\varrho > 0, \quad 0 < \Theta < \frac{1}{10}, \quad 0 < p < \frac{1}{100}. \tag{8}$$

3. We put $p = r/\lambda$ and take r, λ to be integers. Then (3) becomes

$$\Theta = \frac{3R-\lambda^3-3\lambda^2 r+3r^3}{Q}, \tag{9}$$

where

$$Q = 3(R+r^3)+\lambda^3; \tag{10}$$

and (4) becomes

$$\varrho = \frac{Q^3}{3\lambda(2\lambda+3r)F}, \tag{11}$$

where

$$F = Q^2+3r\lambda(2\lambda+3r)(6R-3\lambda^2 r+6r^3)+3r^2(2\lambda+3r)Q. \tag{12}$$

The formulae (5) to (7) become

$$x+y+z = \frac{QX}{F}, \tag{13}$$

$$y+z = \frac{Y}{3\lambda(2\lambda+3r)F}, \tag{14}$$

$$z = \frac{XZ}{F}, \tag{15}$$

---

[1]) The formulae (3), (4), (5), (6), (7) are (i), (ii), (iii), (vi), (vii) of Richmond's postscript, except that a misprint in (vii), where p^2 should be p^3, has been corrected. The origin of the formulae is explained in a later paper: On rational solutions of $x^3+y^3+z^3=R$, Proc. Edinburgh Math. Soc. (2) 2 (1930—1931), 92—100. In this paper, t corresponds to $(1-\Theta)^{-1}$ and h to $\varrho p^3(1-\Theta)^3$.

where

$$X = 6R\lambda - 3\lambda^3 r + 6\lambda r^3 + 3rQ - 3r^3(2\lambda + 3r), \tag{16}$$

$$Y = Q^3 - 9Q^2 r^2(2\lambda + 3r) - 9\lambda r^3(2\lambda + 3r)^2 (6R - 3\lambda^2 r + 6r^3), \tag{17}$$

$$Z = 3R - \lambda^3 - 3\lambda^2 r + 3r^3. \tag{18}$$

4. For every sufficiently large positive integer R there is a prime P satisfying

$$P \equiv 2 \pmod 3, \quad \tfrac{1}{2}\delta R^{1/3} < P < \delta R^{1/3}, \tag{19}$$

where δ is any fixed small positive number. Since all numbers are cubic residues modulo P, there is an integer r satisfying

$$R + r^3 \equiv 0 \pmod P, \quad 0 \leqq r < P. \tag{20}$$

We take λ to be an integer such that

$$\lambda \equiv 0 \pmod P, \quad 3R(1 - 6\delta) < \lambda^3 < 3R(1 - 3\delta). \tag{21}$$

This is possible if

$$\{3R(1 - 3\delta)\}^{1/3} - \{3R(1 - 6\delta)\}^{1/3} > P,$$

and this holds if δ is sufficiently small, since $P < \delta R^{1/3}$.

It follows from (10) that Q lies between $6R(1 - 3\delta)$ and $(6 - 9\delta + 3\delta^3)R$. Hence, by (9), Θ is positive and less than a fixed multiple of δ. We also have

$$0 \leqq p = r\lambda^{-1} < P\lambda^{-1} < \delta.$$

Hence the conditions (8) are satisfied if δ is sufficiently small, and then x, y, z are positive.

We have $P|\lambda$ and $P|R + r^3$, whence $P|Q$ by (10) and $P|F$ by (12). Also $P^2|Y$ by (17) and $P|Z$ by (18). It follows that if

$$\lambda = P\lambda', \quad F = PF',$$

then a common denominator for x, y, z is

$$T = 3\lambda'(2\lambda + 3r)F'.$$

Now

$$\lambda' = O(1), \quad 2\lambda + 3r = O(R^{1/3})$$

by (19), (20), (21), and since $F = O(R^2)$ by (12), we have $F' = O(R^{5/3})$. This proves (2).

M. Deuring in Göttingen

ANALYTISCHE KLASSENZAHLFORMELN

1. In Band 5 der Acta Arithmetica wurde von K. G. Ramanathan eine Formel von C. L. Siegel, eine Art Fourierentwicklung der Diskriminante eines algebraischen Zahlkörpers[1]), wieder aufgegriffen[2]), um sie neu zu beweisen und zu verallgemeinern, insbesondere auf nichtkommunative Algebren über Zahlkörpern. Im folgenden soll gezeigt werden, wie der Ansatz von Siegel und Ramanathan zu Entwicklungen der Klassenzahl eines algebraischen Zahlkörpers in Reihen führt, die nach den ganzen Idealen des Körpers fortschreiten. Einige der Klassenzahlformeln, die sich solchermaßen ergeben, seien aufgeführt:

Es sei k in dieser Einleitung und auch später ein fester algebraischer Zahlkörper. Es bezeichne

h *die Klassenzahl,*

D *den Betrag der Diskriminante,*

w *die Anzahl der Einheitswurzeln,*

R *den Regulator*

von k. In den Formeln (1), (2) und (3) ist die Summation über alle ganzen Ideale $\mathfrak{x}$ von k zu erstrecken.

Für reelles quadratisches k – wobei $e^R = \eta$ die durch $\eta > 1$ normierte Grundeinheit von k ist – gilt

$$h \ln \eta \, \frac{\sqrt{D}-1}{2} = \sum_{\mathfrak{x}} \int_0^\infty \left(\frac{\sin \pi \sqrt{\frac{N\mathfrak{x}}{D}}\, v}{\pi \sqrt{\frac{N\mathfrak{x}}{D}}\, v} \right)^2 \left(\frac{\sin \pi \sqrt{\frac{N\mathfrak{x}}{D}}\, \frac{1}{v}}{\pi \sqrt{\frac{N\mathfrak{x}}{D}}\, \frac{1}{v}} \right)^2 \frac{dv}{v}. \tag{1}$$

[1]) Vgl. [4].

[2]) Vgl. [1].

Für imaginäres quadratisches k haben wir (J_ν ist die Besselsche Funktion der Ordnung ν)

$$h \cdot \frac{1}{w}\left[\frac{\sqrt{D}}{2\pi} - \frac{1}{4}\right] = \sum_{\mathfrak{x}} \left(\frac{J_1\left(2\pi\sqrt{\frac{N\mathfrak{x}}{D}}\right)}{2\pi\sqrt{\frac{N\mathfrak{x}}{D}}} \right)^2 . \tag{2}$$

Eine allgemeinere Formel für imaginär quadratisches k ist

$$h = \frac{w\sqrt{D}}{2\pi}\, \nu \sum_{\mathfrak{x}} N\mathfrak{x}^{-1} J_\nu \left(2\pi\delta\sqrt{\frac{N\mathfrak{x}}{D}}\right)^2 ; \tag{3}$$

hier ist δ *eine beliebige Zahl des Intervalles* $0 < \delta \leqq 1$ *und* ν *eine beliebige Zahl der Folge* $w+1, 2w+1, 3w+1, \ldots$. (1), (2) und (3) ergeben sich aus der allgemeinen, *für beliebige Zahlkörper gültigen Formel* (44), indem die dort auftretenden frei wählbaren Parameter, nämlich 1. der Größencharakter χ von k mit dem Führer 1, 2. die beliebigen über $0 \leqq t \leqq 1$ integrierbaren Funktionen $A(t)$ und $B(t)$ und die auf das Intervall $0 < \delta \leqq 1$ beschränkte Variable δ speziell gewählt werden.

2. Wir stützen uns, wie Ramanathan, auf den folgenden Satz von Siegel[1]):
Es seien $x_1, \ldots, x_n$ reelle Variable, die wir zu einer Spalte

$$x = (x_1, \ldots, x_n)^t$$

zusammenfassen.[2])

Im n-dimensionalen Raum der x sei X ein konvexer, um den Nullpunkt symmetrischer Bereich, der keinen von 0 verschiedenen Gitterpunkt (Punkt mit ganzzahligen Koordinaten) enthält. Für zwei über X integrierbare Funktionen $\varphi(x)$ und $\psi(x)$ gilt dann die Gleichung

$$2^n \int_X \varphi(x)\,\bar{\psi}(x)\,dx = \sum_g \int_X \varphi(x)\,e^{ig^t x}\,dx \int_X \overline{\psi(x)\,e^{ig^t x}}\,dx; \tag{4}$$

dabei wird rechts über alle Gitterpunkte g summiert.

3. Den Bezeichnungsfestsetzungen in 1. fügen wir noch hinzu:

$n = [k : \boldsymbol{Q}]$ sei der *Grad von k*;

die n *Isomorphismen*

$\sigma_\nu : \xi \to \xi^{(\nu)}$ *von k in* $\boldsymbol{C}$

---

[1]) Vgl. [3] und [1].

[2]) A^t bedeute immer die Transponierte der Matrix A. Explizit anzugebende Spalten werden der Übersichtlichkeit halber als Transponierte von Zeilen geschrieben.

numerieren wir wie üblich so, daß $\sigma_1, \dots, \sigma_{r_1}$ die *reellen*, k in $\boldsymbol{R}$ abbildenden unter ihnen sind, während die übrigen in r_2 Paare *konjugiert komplexer* zerfallen:

$$\overline{\xi^{(\nu)}} = \xi^{(\nu+r_2)}, \quad \nu = r_1 + 1, \dots, r_1 + r_2.$$

Es ist also

$$n = r_1 + 2r_2.$$

Die *Gruppe E der Einheiten von k* hat den Rang

$$r = r_1 + r_2 - 1.$$

Den Isomorphismen $\sigma_1, \dots, \sigma_n$ ordnen wir der Reihe nach Variable $z^{(1)}, \dots, z^{(n)}$ zu; davon seien die den reellen σ_ν entsprechenden $z^{(1)}, \dots, z^{(n)}$ reell, und für konjugiert komplexe $\sigma_\nu, \sigma_{\nu+r_2}$ seien auch $z^{(\nu)}$ und $z^{(\nu+r_2)}$ konjugiert komplex. Die $z^{(\nu)}$ fassen wir zu einer Spalte $z = (z^{(1)}, \dots, z^{(n)})^t$ zusammen.

Das System

$$\xi^{(1)}, \dots, \xi^{(n)}$$

der Konjugierten einer Zahl ξ von k kann also als ein Wert der Spalte z angesehen werden; dementsprechend schreiben wir auch

$$Nz = \prod_{\nu=1}^{n} z^{(\nu)}, \quad \operatorname{Sp} z = \sum_{\nu=1}^{n} z^{(\nu)}.$$

Die Differente von k sei $\mathfrak{D}$, so daß

$$N\mathfrak{D} = D$$

ist.

Im folgenden sei $\mathfrak{a}$ ein festes, sonst beliebiges Ideal $\neq 0$ von k, $(\alpha_1, \dots, \alpha_n)^t = \alpha$ eine Basis von $\mathfrak{a}$, die ebenfalls fest gewählt wird. Die Determinante der zu $\mathfrak{a}$ gehörigen *Basismatrix*

$$A = (\alpha^{(1)}, \dots, \alpha^{(n)}) \tag{5}$$

hat den Betrag

$$|\det A| = N\mathfrak{a} \cdot D^{1/2}.$$

Die Matrix

$$\tilde{A} = (A^{-1})^t = (\tilde{\alpha}^{(1)}, \dots, \tilde{\alpha}^{(n)}) \tag{6}$$

ist die Basismatrix des zu $\mathfrak{a}$ *komplementären Ideals* $\tilde{\mathfrak{a}} = \mathfrak{a}^{-1}\mathfrak{D}^{-1}$, gebildet für die zu α *komplementäre Basis* $\tilde{\alpha}$ *von* $\tilde{\mathfrak{a}}$.

4. Für die ganzzahlige Spalte

$$g = (g_1, \ldots, g_n)^t$$

ist

$$g^t \tilde{\alpha} \tilde{\mathfrak{a}}^{-1} = \mathfrak{x} \tag{7}$$

ein ganzes Ideal der Idealklasse C von k, der $\tilde{\mathfrak{a}}$ angehört. Umgekehrt, zu jedem ganzen Ideal $\mathfrak{x}$ aus C gibt es eine ganze Spalte g_0 mit

$$g_0^t \tilde{\alpha} \tilde{\mathfrak{a}}^{-1} = \mathfrak{x};$$

bedeutet ε eine Einheit von k, so wird durch

$$g_0^t \tilde{\alpha} \varepsilon = g(\mathfrak{x}, \varepsilon)^t \, \tilde{\alpha}$$

eindeutig eine ganzzahlige Spalte $g(\mathfrak{x}, \varepsilon)$ bestimmt, und durch

$$(\mathfrak{x}, \varepsilon) \to g(\mathfrak{x}, \varepsilon)$$

sind die ganzzahligen Spalten $\neq 0$ den Paaren $(\mathfrak{x}, \varepsilon)$ – dabei ist $\mathfrak{x}$ ein ganzes Ideal der Klasse C, ε eine Einheit von k – eineindeutig zugeordnet.

Wir setzen noch

$$g^t(\mathfrak{x}, 1)\, \tilde{\alpha}^{(\nu)} = \gamma^{(\nu)}(\mathfrak{x})\, (N\tilde{\mathfrak{a}} N\mathfrak{x})^{1/n} = \gamma^{(\nu)}(\mathfrak{x}) \sqrt[n]{\frac{N\mathfrak{x}}{D}}\, N\mathfrak{a}^{-(1/n)} \tag{8}$$

und

$$\gamma(\mathfrak{x}) = (\gamma^{(1)}(\mathfrak{x}), \ldots, \gamma^{(n)}(\mathfrak{x}))^t; \tag{9}$$

$\gamma(\mathfrak{x})$ ist ein Wert der variablen Spalte z mit

$$|N\gamma(\mathfrak{x})| = 1.$$

Es ist

$$g^t(\mathfrak{x}, \varepsilon)\, \alpha^{(\nu)} = \gamma^{(\nu)}(\mathfrak{x})\, \varepsilon^{(\nu)} \sqrt[n]{\frac{N\mathfrak{x}}{D}}\, N\mathfrak{a}^{-(1/n)}. \tag{10}$$

5. Durch

$$x = (x_1, \ldots, x_n)^t = \tilde{A} z \quad \text{oder} \quad x^t A = z^t \tag{11}$$

sind reelle Variable $x_1, \ldots, x_n$ definiert. ϱ sei ein Wert von z mit positiven Komponenten, also

$$\varrho_\nu > 0; \quad \varrho^{(\nu + r_2)} = \varrho^{(\nu)} \quad \text{für} \quad \nu = r_1 + 1, \ldots, r_1 + r_2. \tag{12}$$

Die Punkte x mit

$$|x^t \alpha^{(\nu)}| = |z^{(\nu)}| < \varrho^{(\nu)} N\mathfrak{a}^{1/n}, \quad \nu = 1, 2, \ldots, n,$$

also insbesondere

$$|Nz| = N(x^t\alpha) < N\varrho\, N\mathfrak{a},$$

bilden einen konvexen, um den Nullpunkt symmetrischen Bereich $X(\varrho)$. Unter der Voraussetzung

$$N\varrho \leqq 1$$

ist $X(\varrho)$ frei von ganzzahligen Punkten $x = g$ außer $x \neq 0$; denn für ganzes $g \in X(\varrho)$ ist $|N(g^t\alpha)|$ einerseits $< N\varrho\, N\mathfrak{a} \leqq N\mathfrak{a}$, andererseits wegen $\mathfrak{a}|g^t\alpha$ teilbar durch $N\mathfrak{a}$, was nur für $N(g^t\alpha) = 0$, also $g = 0$ möglich ist. Auf $X = X(\varrho)$ kann also die Siegelsche Formel (4) angewendet werden.

Wir setzen, um die Integrale in (4) bequem berechnen zu können,

$$y^{(\nu)} = z^{(\nu)}\varrho^{(\nu)-1}N\mathfrak{a}^{-(1/n)}; \quad \nu = 1, 2, \ldots, n$$

und verwenden $y^{(1)}, \ldots, y^{(r_1)}, Ry^{(r_1+1)}, Iy^{(r_1+1)}, \ldots, Ry^{(r_1+r_2)}, Iy^{(r_1+r_2)}$ als neue Integrationsvariable. Wir erhalten

$$\int_{X(\varrho)} \varphi(x)\, e^{\pi i g^t x}\, dx = \frac{2^{r_2}N\varrho}{\sqrt{D}} \int_{|y|<1} \varphi(\tilde{A}z) \exp(\pi i g^t \tilde{A}z)\, dy, \tag{13}$$

wobei

$$dy^{(1)} \cdots dy^{(r_1)}\, dRy^{(r_1+1)}\, dIy^{(r_1+1)} \cdots dRy^{(r_1+r_2)}\, dIy^{(r_1+r_2)} = dy$$

gesetzt ist und $|y| < 1$ bedeuten soll: $|y^{(\nu)}| < 1$ für $\nu = 1, 2, \ldots, n$. Wir wollen

$$\varphi(\tilde{A}zN\mathfrak{a}^{1/n}) = F(z^{(1)}, \ldots, z^{(n)}) = F(z)$$

und

$$\psi(\tilde{A}zN\mathfrak{a}^{1/n}) = G(z^{(1)}, \ldots, z^{(n)}) = G(z)$$

und nicht φ und ψ als die gegebenen Funktionen ansehen; dabei ist zu beachten, daß wir $\mathfrak{a}$ und A fest gewählt haben, jedoch soll ϱ variabel sein.

Es sei $g = g(\mathfrak{x}, \varepsilon)$ eine von 0 verschiedene ganzzahlige Spalte.

Es ist

$$g^t Az = N\mathfrak{a}^{1/n} \sum_{\nu=1}^{n} g(\mathfrak{x}, \varepsilon)\, \tilde{\alpha}^{(\nu)} y^{(\nu)} \varrho^{(\nu)} = \sqrt[n]{\frac{N\mathfrak{x}}{D}}\, \mathrm{Sp}\,(\gamma(\mathfrak{x})\, \varepsilon\varrho y)$$

und daher

$$\int_{X(\varrho)} \varphi(x)\, e^{\pi i g^t x}\, dx = \frac{2^{r_2}N\varrho}{\sqrt{D}} \int_{|y|<1} F(\varrho y) \exp\left(i\pi \sqrt[n]{\frac{N\mathfrak{x}}{D}}\, \mathrm{Sp}\,(\gamma(\varrho)\, \varepsilon\varrho y)\right) dy,$$

statt dessen aber auch

$$\int_{X(\varrho)} \varphi(x)\, e^{\pi i g^t x}\, dx = \frac{2^{r_2}N\varrho}{\sqrt{D}} \int_{|y|<1} F\left(\varrho y \frac{|\gamma(\mathfrak{x})\varepsilon|}{\gamma(\mathfrak{x})\varepsilon}\right) \exp\left(i\pi \sqrt[n]{\frac{N\mathfrak{x}}{D}}\, \mathrm{Sp}(|\gamma(\mathfrak{x})\varepsilon|\, \varrho y)\right) dy$$

denn bei der Transformation $y \to y \dfrac{|\gamma(\mathfrak{x})\,\varepsilon|}{\gamma(\mathfrak{x})\,\varepsilon}$ wird der Bereich $|y| < 1$ in sich transformiert.

6. Wir schreiben

$$\left.\begin{aligned} &N\varrho^{1/n} = \delta, \\ &\varrho^{(\nu)}\delta^{-1} = v^{(\nu)}, \quad \text{also} \quad Nv = v^{(1)} \ldots v^{(r_1)} v^{(r_1+1)2} \ldots v^{(r_1+r_2)2} = 1. \end{aligned}\right\} \tag{14}$$

Die multiplikative Gruppe der v mit $v^{(\nu)} > 0$ und $Nv = 1$ heiße V, das invariante Volumenelement auf V schreiben wir in der Form

$$d\omega = \prod_{\nu=1}^{r_1+r_2}{}' \frac{d(v^{(\nu)e_\nu})}{v^{(\nu)e_\nu}}, \tag{15}$$

wobei

$$e_1 = \cdots = e_{r_1} = 1, \quad e_{r_1+1} = \cdots = e_{r_1+r_2} = 2$$

ist und $'$ bei $\prod$ bedeute, daß ein (beliebiger) Faktor wegzulassen ist; im Fall $r = 0$ ist also $d\omega = 1$ zu setzen.

Es sei E die Gruppe der Einheiten von k, $|E|$ die Gruppe der Spalten

$$|\varepsilon| = (|\varepsilon^{(1)}|, \ldots, |\varepsilon^{(n)}|)^t, \quad \varepsilon \in E;$$

$|E|$ ist eine diskrete Untergruppe von V mit kompakter Faktorgruppe $V/|E|$. $\mathfrak{F}$ sei ein Fundamentalbereich von $|E|$ auf V. Dann haben wir

$$\int_{\mathfrak{F}} \int_{X(\varrho)} \varphi(x)\, e^{\pi i g^t x}\, dx \int_{X(\varrho)} \bar{\psi}(x)\, e^{-\pi i g^t x}\, dx\, d\omega$$

$$= \frac{2^{2r_2}\delta^{2n}}{D} \int_{\mathfrak{F}} \int_{|y|<1} F\left(\delta v y \frac{|\gamma(\mathfrak{x})\,\varepsilon|}{\gamma(\mathfrak{x})\,\varepsilon}\right) \exp\left(i\pi\delta \sqrt[n]{\frac{N\mathfrak{x}}{D}}\, \mathrm{Sp}\,(|\gamma(\mathfrak{x})\,\varepsilon|\, v y)\right) dy$$

$$\times \int_{|y|<1} \bar{G}\left(\delta v y \frac{|\gamma(\mathfrak{x})\,\varepsilon|}{\gamma(\mathfrak{x})\,\varepsilon}\right) \exp\left(-i\pi\delta \sqrt[n]{\frac{N\mathfrak{x}}{D}}\, \mathrm{Sp}\,(|\gamma(\mathfrak{x})\,\varepsilon|\, v y)\right) dy\, d\omega$$

$$= \frac{2^{2r_2}\delta^{2n}}{D} \int_{\mathfrak{F}|\gamma(\mathfrak{x})\varepsilon|} \int_{|y|<1} F(\delta v y \gamma(\mathfrak{x})^{-1}\, \varepsilon^{-1}) \exp\left(i\pi\delta \sqrt[n]{\frac{N\mathfrak{x}}{D}}\, \mathrm{Sp}\,(v y)\right) dy$$

$$\times \int_{|y|<1} \bar{G}(\delta v y \gamma(\mathfrak{x})^{-1}\, \varepsilon^{-1}) \exp\left(-i\pi\delta \sqrt[n]{\frac{N\mathfrak{x}}{D}}\, \mathrm{Sp}\,(v y)\right) dy\, d\omega. \tag{16}$$

7. Wir setzen jetzt voraus, daß F und G bei Transformationen mit Einheiten invariant seien:

$$F(z\varepsilon) = F(z), \quad G(z\varepsilon) = G(z) \quad \text{für alle } \varepsilon \text{ aus } E.$$

Dann fällt ε in der letzten Formel in den Argumenten von F und G heraus und Summation über ε ergibt

$$\sum_{\varepsilon\in E}\int_{\mathfrak{F}}\int_{X(\varrho)}\varphi(x)\,e^{\pi i g^t x}\,dx\int_{X(\varrho)}\bar{\psi}(x)\,e^{-\pi i g^t x}\,dx\,d\omega$$

$$= w\,\frac{2^{2r_2}\delta^{2n}}{D}\int_V\int_{|y|<1}F(\delta v y\gamma(\mathfrak{x})^{-1})\exp\left(i\pi\delta\sqrt[n]{\frac{N\mathfrak{x}}{D}}\,\mathrm{Sp}\,(vy)\right)dy$$

$$\times\int_{|y|<1}\bar{G}(\delta v y\gamma(\mathfrak{x})^{-1})\exp\left(-i\pi\delta\sqrt[n]{\frac{N\mathfrak{x}}{D}}\,\mathrm{Sp}\,(vy)\right)dy\,d\omega;$$

dabei ist w die Anzahl der Einheitswurzeln in k, denn gerade diese werden bei dem Homomorphismus $E\to|E|$ auf 1 abgebildet. Rechnen wir auch die übrigen Integrale in der Formel (4) auf dy um, so erhalten wir aus ihr durch Integration über $\mathfrak{F}$

$$2^n\int_{\mathfrak{F}}\int_{|y|<1}F(\delta v y)\,\bar{G}(\delta v y)\,dy\,d\omega$$

$$=\frac{2^{r_2}\delta^n}{\sqrt{D}}\int_{\mathfrak{F}}\int_{|y|<1}F(\delta v y)\,dy\int_{|y|<1}\bar{G}(\delta v y)\,dy\,d\omega$$

$$+\,w\,\frac{2^{r_2}\delta^n}{\sqrt{D}}\sum_{\mathfrak{x}\in C}\int_{\mathfrak{F}}\int_{|y|<1}F(\delta v y\gamma(\mathfrak{x})^{-1})\exp\left(i\pi\delta\sqrt[n]{\frac{N\mathfrak{x}}{D}}\,\mathrm{Sp}\,(vy)\right)dy$$

$$\times\int_{|y|<1}\bar{G}(\delta v y\gamma(\mathfrak{x})^{-1})\exp\left(-i\pi\delta\sqrt[n]{\frac{N\mathfrak{x}}{D}}\,\mathrm{Sp}\,(vy)\right)dy\,d\omega. \tag{17}$$

$F(\delta v y\gamma(\mathfrak{x})^{-1})$ und $G(\delta v y\gamma(\mathfrak{x})^{-1})$ hängen nur von $\mathfrak{x}$ und nicht mehr von der Auswahl des zugehörigen $\gamma(\mathfrak{x})$ ab.

8. Es liegt nahe, wenn Z die Gruppe aller Spalten $z \neq 0$ bedeute, für F und G Charaktere der Gruppe Z/E einzusetzen, also Heckesche Größencharaktere. Die explizite Berechnung der Größencharaktere verläuft nach Hecke[1]) folgendermaßen:

Es sei Z_1 die *Normeinsgruppe* aller z mit $|Nz| = 1$; es ist

$$Z/Z_1 \cong \boldsymbol{R}_+^\times,\quad \text{sogar}\quad Z\cong Z_1\times\boldsymbol{R}_+^\times$$

durch

$$z\to(z|Nz|^{-(1/n)},\,|Nz|). \tag{18}$$

---

[1]) Vgl. [2].

U sei die Gruppe der z mit $|z^{(\nu)}| = 1$, $\nu = 1, 2, \ldots, n$, kurz $|z| = 1$. Die Charaktere von Z_1/UE finden wir so:

$$z \to |z|\,|E|$$

ist ein Homomorphismus von Z_1 auf $V/|E|$ mit dem Kern UE,

$$Z_1/UE \cong V/|E|.$$

V bilden wir durch komponentenweises Logarithmieren auf den r-dimensionalen cartesischen Raum $\boldsymbol{R}^r$ ab. Dabei geht $|E|$ in ein r-dimensionales Gitter über. $\varepsilon_1, \ldots, \varepsilon_r$ sei eine Einheitenbasis von k,

$$B = \begin{pmatrix} \ln|\varepsilon_1^{(1)}| & \ldots & \ln|\varepsilon_1^{(r+1)}| \\ \ln|\varepsilon_2^{(1)}| & \ldots & \ln|\varepsilon_2^{(r+1)}| \\ \vdots & & \vdots \\ \ln|\varepsilon_r^{(1)}| & \ldots & \ln|\varepsilon_r^{(r+1)}| \\ 1 & \ldots & 1 \end{pmatrix} = \begin{pmatrix} & B_0 & \\ 1 & \ldots & 1 \end{pmatrix}. \tag{19}$$

Dann ist

$$B^{-1} = \begin{pmatrix} & e_1/n \\ D & \vdots \\ & e_{r+1}/n \end{pmatrix}, \tag{20}$$

also

$$B_0 D = 1, \quad (1, \ldots, 1)\, D = 0. \tag{21}$$

Für z aus Z definieren wir

$$l(z)^t = (\ln|z^{(1)}|, \ldots, \ln|z^{(r_1+r_2)}|); \tag{22}$$

dann ist für eine Einheit ε von k

$$l(\varepsilon)^t = (m_1, \ldots, m_r)\, B_0 \quad \text{mit ganzen } m_\nu. \tag{23}$$

Daher sind die Charaktere von $V/|E|$

$$\chi(v, s) = \exp\,(l(v)^t\, Ds \cdot 2\pi i), \quad s \text{ eine ganzzahlige Spalte der Länge } r; \tag{24}$$

durch

$$\chi(\varepsilon, s) \to s$$

ist die Charaktergruppe $X(V/|E|)$ mit $\boldsymbol{Z}^r$ isomorph.

Wir berechnen die Charaktere der Gruppe UE/E: $U \cap E = W$ ist die Gruppe der Einheitswurzeln in k, wir haben daher den Isomorphismus

$$uE \to uW$$

von UE/E mit U/W.

Wir erklären jetzt auf der Gruppe Z die Funktion

$$\lambda(z)^t = (\arg z^{(1)}, \ldots, \arg z^{(r_1+r_2)}), \quad 0 \leqq \arg z^{(\nu)} < 2\pi. \tag{25}$$

Die Abbildung

$$u \to \lambda(u)$$

ist eine isomorphe Abbildung von U auf $(\boldsymbol{Z}^+/2)^{r_1} \times (\boldsymbol{R}^+/2\pi\boldsymbol{R}^+)^{r_2}$, und daraus schließen wir, daß die Charaktere von U die folgende Gestalt haben:

$$\psi_0(u, a) = \exp(\lambda(u)^t\, a \cdot i), \ a \text{ eine ganzzahlige Spalte der Länge } r + 1. \quad (26)$$

Die Abbildung

$$\psi_0(*, a) \to (a_1 \bmod 2, \ldots, a_{r_1} \bmod 2, a_{r_1+1}, \ldots, a_{r_1+r_2})$$

ist ein Isomorphismus von $X(U)$ mit $(\boldsymbol{Z}^+/2)^{r_1} \times \boldsymbol{Z}^{+r_2}$. $\psi_0(*, a)$ ist genau dann ein Charakter von U/W, also von UE/E, wenn – unter $\zeta^{(\nu)} = \zeta^{c_\nu}$, $c_\nu \bmod w$, $\nu = 1, 2, \ldots, r_1 + r_2$ werden die ersten $r + 1$ Konjugierten einer primitiven w-ten Einheitswurzel ζ verstanden –

$$\sum_{\nu=1}^{r+1} c_\nu a_\nu \equiv 0 \bmod w \quad (27)$$

gilt.

Es bleibt noch übrig, $\psi_0(*, a)$ zu einem Charakter von Z_1/E fortzusetzen. Wir führen die Matrix Λ aus den Spalten $\lambda(\varepsilon_\nu)$ ein:

$$\Lambda = (\lambda(\varepsilon_1), \ldots, \lambda(\varepsilon_r)), \quad (28)$$

mit deren Hilfe wir den Charakter

$$\psi(v, a) = \exp\big((\lambda(v)^t - l(v)^t D\Lambda^t)\, a \cdot i\big) \quad (29)$$

von Z_1/E definieren, natürlich unter der Voraussetzung (27) für a. Die Einschränkung von $\psi(*, a)$ auf UE ist gerade $\psi_0(*, a)$.

Ein Charakter χ von Z/E hat die allgemeine Form

$$\chi(z) = |Nz|^{i\eta}\, \chi(|z| \cdot |Nz|^{-(1/n)}, s)\, \psi(z|Nz|^{-(1/n)}, a) \quad (30)$$

mit reellem η; diese Zerlegung von χ ist eindeutig, wenn wir noch

$$a_\nu = 0 \text{ oder } 1 \quad \text{für} \quad \nu = 1, \ldots, r_1$$

fordern.

9. In (17) setzen wir für F und G Funktionen der Gestalt

$$\left.\begin{aligned} F(z) &= A(|Nz|)\,\chi(z), \\ G(z) &= B(|Nz|)\,\psi(z) \end{aligned}\right\} \quad (31)$$

ein; dabei sollen

$$\left.\begin{aligned}\chi(z) &= \chi(|z| \cdot |Nz|^{-(1/n)}, s)\, \psi(z|Nz|^{-(1/n)}, a), \\ \psi(z) &= \chi(|z| \cdot |Nz|^{-(1/n)}, j)\, \psi(z\,|Nz|^{-(1/n)}, b)\end{aligned}\right\} \tag{32}$$

Charaktere von Z_1/E und A und B integrable, sonst beliebige Funktionen sein. Es wird

$$F(\delta\gamma(\mathfrak{x})^{-1}\, vy) = A(\delta^n\,|Ny|)\, \chi(v\gamma(\mathfrak{x})^{-1})\, \chi(y),$$

$$G(\delta\gamma(\mathfrak{x})^{-1}\, vy) = B(\delta^n\,|Ny|)\, \psi(v\gamma(\mathfrak{x})^{-1})\, \psi(y).$$

Die in (17) auftretenden Integrale nach y sind jetzt von der Gestalt

$$\int_{|y|<1} H(|Ny|)\, \chi(y) \exp(i\beta\, \mathrm{Sp}(vy))\, dy$$

mit einem Charakter

$$\chi(y) = \chi(y, s)\, \psi(y, a) \quad \text{von} \quad Z_1/E$$

und $\beta \geqq 0$. Wir setzen

$$|y^{(\nu)}| = q^{(\nu)}, \quad y^{(\nu)} = q^{(\nu)}\, e^{i\Theta^{(\nu)}}, \quad 0 \leqq \Theta^{(\nu)} < 2\pi, \quad \nu = 1, \ldots, n,$$

ferner

$$D(h \cdot 2\pi - \Lambda^t a) = L(s, a) = (L_1(s, a), \ldots, L_{r+1}(s, a))^t;$$

dann hat χ die Form

$$\chi(y) = \exp\left(i\left(l(q)^t\, L(s, a) + \sum_{\nu=1}^{r+1} \Theta^{(\nu)} a_\nu\right)\right),$$

und wir können Sp (vy) folgendermaßen umformen:

$$\mathrm{Sp}(vy) = \sum_{\nu=1}^{r_1} v^{(\nu)} y^{(\nu)} + 2 \sum_{\nu=r_1+1}^{r_1+r_2} v^{(\nu)} q^{(\nu)} \cos\Theta^{(\nu)}.$$

Dies setzen wir in unser Integral ein und erhalten

$$\int_{|y|<1} H(|Ny|)\, \chi(y) \exp(i\beta\, \mathrm{Sp}(vy))\, dy$$

$$= \int_{|y|<1} H(|Ny|) \prod_{\nu=1}^{r+1} {q^{(\nu)}}^{L_\nu(s,a)i} \prod_{\nu=1}^{r_1} \exp(i(\beta v^{(\nu)} y^{(\nu)} + \Theta^{(\nu)} a_\nu))$$

$$\times \prod_{\nu=r_1+1}^{r_1+r_2} \exp(i(2\beta v^{(\nu)} q^{(\nu)} \cos\Theta^{(\nu)} + \Theta^{(\nu)} a_\nu))\, dy.$$

Für die hierin auszuführenden Integrale über $y^{(\nu)}$, $\nu = 1, \dots, r_1$ haben wir — $*$ stehe für eine Funktion allein von $q^{(\nu)}$ —

$$\int_{-1}^{+1} * \exp\left(i(\beta v^{(\nu)} y^{(\nu)} + \Theta^{(\nu)} a_\nu)\right) dy^{(\nu)}$$

$$= \int_0^1 * \exp\left(i\beta v^{(\nu)} q^{(\nu)}\right) dq^{(\nu)} + (-1)^{a_\nu} \int_0^1 * \exp\left(-i\beta v^{(\nu)} q^{(\nu)}\right) dq^{(\nu)}$$

$$= 2i^{a_\nu} \int_0^1 * \cos\left(\beta v^{(\nu)} q^{(\nu)} - a_\nu \frac{\pi}{2}\right) dq^{(\nu)}$$

und entsprechend für die Integrale über $dRy^{(\nu)}\, dIy^{(\nu)}$, $\nu = r_1 + 1, \dots, r_1 + r_2$

$$\int_0^1 \int_0^{2\pi} * \exp\left(i(2\beta v^{(\nu)} q^{(\nu)} \cos \Theta^{(\nu)} + \Theta^{(\nu)} a_\nu)\right) d\Theta^{(\nu)}\, dq^{(\nu)}$$

$$= 2\pi i^{a_\nu} \int_0^1 * q^{(\nu)} J_{a_\nu}(2\beta v^{(\nu)} q^{(\nu)})\, dq^{(\nu)}.$$

Auf diese Weise ergibt sich

$$\int_{|y|<1} H(|Ny|)\, \chi(y) \exp\left(i\beta \operatorname{Sp}(vy)\right) dy$$

$$= 2^{r_1+r_2} \pi^{r_2} i^{\sum_{\nu=1}^{r_1+r_2} a_\nu} \int_0^1 \cdots \int_0^1 H(Nq) \prod_{\nu=1}^{r+1} q^{(\nu)^{L_\nu(s,a)i}} \prod_{\nu=1}^{r_1} \cos\left(\beta v^{(\nu)} q^{(\nu)} - a_\nu \frac{\pi}{2}\right)$$

$$\times \prod_{\nu=r_1+1}^{r_1+r_2} q^{(\nu)} J_{a_\nu}(2\beta v^{(\nu)} q^{(\nu)})\, dq^{(1)} \cdots dq^{(r_1+r_2)}. \tag{33}$$

Betrachten wir zuerst den Fall $\beta = 0$. Wenn $\chi(y) = \chi(y, s)$ ein Charakter von Z_1/UE ist, so wird nach (33)

$$\int_{|y|<1} H(|Ny|)\, \chi(y, s)\, dy$$

$$= 2^{r_1+r_2} \pi^{r_2} \int_0^1 \cdots \int_0^1 H(Nq) \prod_{\nu=1}^{r+1} q^{(\nu)^{L_\nu(s,0)i}} \prod_{\nu=r_1+1}^{r_1+r_2} q^{(\nu)}\, dq^{(1)} \cdots dq^{(r_1+r_2)}. \tag{34}$$

Dagegen ist das Integral für einen Charakter $\chi(y) = \chi(y, h)\, \psi(y, a)$, der auf UE nicht konstant gleich 1 ist, gleich 0; denn es ist

$$\int_{|y|<1} H(|Ny|)\, \chi(y)\, dy$$

$$= \int_0^1 \cdots \int_0^1 H(Nq)\, \chi(qNq^{-(1/n)}, s)\, dq^{(1)} \cdots dq^{(r_1+r_2)} \int_{U/W} \psi_0(u, a)\, du = 0,$$

weil $\psi_0(u, a)$ nicht konstant gleich 1 ist. Man erkennt das natürlich auch unmittelbar aus (33); denn entweder ist ein a_ν mit $1 \leqq \nu \leqq r_1$ ungerade, also $\cos\left(-a_\nu \frac{\pi}{2}\right) = 0$, oder es ist ein a_ν mit $r_1 + 1 \leqq \nu \leqq r_1 + r_2$ von 0 verschieden, und dann ist $J_{a_\nu}(0) = 0$.

Das erste Glied rechts in (17) wird

$$\int_{\mathfrak{F}} \int_{|y|<1} F(\delta v y)\, dy \int_{|y|<1} \bar{G}(\delta v y)\, dy\, d\omega$$
$$= \int_{|y|<1} A(\delta^n |Ny|)\, \chi(y)\, dy \int_{|y|<1} \bar{B}(\delta^n |Ny|)\, \bar{\psi}(y)\, dy \int_{\mathfrak{F}} \chi\bar{\psi}(v)\, d\omega .$$

$R = \int_{\mathfrak{F}} d\omega$ ist der Regulator des Zahlkörpers k, für $\chi \neq \psi$ ist $\int_{\mathfrak{F}} \chi\bar{\psi}(v) d\omega = 0$ und daher

$$\int_{\mathfrak{F}} \int_{|y|<1} F(\delta v y)\, dy \int_{|y|<1} \bar{G}(\delta v y)\, d\omega$$

$$= \begin{cases} R \cdot 2^{2(r_1+r_2)} \pi^{2r_2} \int_0^1 \cdots \int_0^1 A(\delta^n N q) \prod_{\nu=1}^{r+1} q^{(\nu)^{L_\nu(s,0)i}} \prod_{\nu=r_1+1}^{r_1+r_2} q^{(\nu)}\, dq^{(1)} \cdots dq^{(r+1)} \\ \times \int_0^1 \cdots \int_0^1 \bar{B}(\delta^n N q) \prod_{\nu=1}^{r+1} q^{(\nu)^{-L_\nu(s,0)i}} \prod_{\nu=r_1+1}^{r_1+r_2} q^{(\nu)}\, dq^{(1)} \cdots dq^{(r+1)}, \\ \text{wenn } \chi = \psi = \chi(*, s) \text{ ein Charakter von } Z_1/UE \text{ ist;} \\ 0, \text{ wenn } \chi \neq \psi \text{ oder wenn } \chi = \psi \text{ auf } UE \text{ nicht konstant ist.} \end{cases} \tag{35}$$

Auf der linken Seite von (17) erhalten wir

$$\int_{\mathfrak{F}} \int_{|y|<1} F(\delta v y)\, dy \int_{|y|<1} \bar{G}(\delta v y)\, dy\, d\omega$$
$$= \int_{|y|<1} A(\delta^n |Ny|)\, \bar{B}(\delta^n |Ny|)\, dy \int_{\mathfrak{F}} \chi\bar{\psi}(v)\, d\omega ,$$

also

$$\int_{\mathfrak{F}} \int_{|y|<1} F(\delta v y)\, \bar{G}(\delta v y)\, dy\, d\omega$$

$$= \begin{cases} R \cdot 2^{r_1+r_2} \pi^{r_2} \int_0^1 \cdots \int_0^1 A(\delta^n N q)\, \bar{B}(\delta^n N q)\, q^{(r_1+1)} \cdots q^{(r_1+r_2)}\, dq^{(1)} \cdots dq^{(r_1+1)}, & \text{wenn } \chi = \psi, \\ 0, & \text{wenn } \chi \neq \psi. \end{cases} \tag{36}$$

Schließlich bekommen wir für das allgemeine Glied der $\mathfrak{x}$-Summe in (17), das wir mit $T(\mathfrak{x})$ bezeichnen wollen,

$$T(\mathfrak{x}) = 2^{2(r_1+r_2)}\pi^{2r_2} i^{\sum_{\nu=1}^{r_1+r_2}(a_\nu - b_\nu)} \bar{\chi}\psi(\gamma(\mathfrak{x}))$$
$$\times \int_V \int_0^1 \cdots \int_0^1 A(\delta^n Nq) \prod_{\nu=1}^{r_1+r_2} {q^{(\nu)}}^{L_\nu(s,a)i} \prod_{\nu=1}^{r_1} \cos\left(\pi\delta \sqrt[n]{\frac{N\mathfrak{x}}{D}}\, v^{(\nu)} q^{(\nu)} - a_\nu \frac{\pi}{2}\right)$$
$$\times \prod_{\nu=r_1+1}^{r_1+r_2} q^{(\nu)} J_{a_\nu}\left(2\pi\delta \sqrt[n]{\frac{N\mathfrak{x}}{D}}\, v^{(\nu)} q^{(\nu)}\right) dq^{(1)} \cdots dq^{(r_1+r_2)}$$
$$\times \int_0^1 \cdots \int_0^1 \bar{B}(\delta^n Nq) \prod_{\nu=1}^{r_1+r_2} {q^{(\nu)}}^{-L_\nu(j,b)i} \prod_{\nu=1}^{r_1} \cos\left(\pi\delta \sqrt[n]{\frac{N\mathfrak{x}}{D}}\, v^{(\nu)} q^{(\nu)} - b_\nu \frac{\pi}{2}\right)$$
$$\times \prod_{\nu=r_1+1}^{r_1+r_2} q^{(\nu)} J_{b_\nu}\left(2\pi\delta \sqrt[n]{\frac{N\mathfrak{x}}{D}}\, v^{(\nu)} q^{(\nu)}\right) dq^{(1)} \cdots dq^{(r_1+r_2)}. \tag{37}$$

Setzen wir noch zur Abkürzung

$$\Phi(A, \delta, \chi, \mathfrak{x})$$
$$= \int_0^1 \cdots \int_0^1 A(\delta^n Nq) \prod_{\nu=1}^{r_1+r_2} {q^{(\nu)}}^{L_\nu(s,a)i} \prod_{\nu=1}^{r_1} \cos\left(\pi\delta \sqrt[n]{\frac{N\mathfrak{x}}{D}}\, v^{(\nu)} q^{(\nu)} - a_\nu \frac{\pi}{2}\right)$$
$$\times \prod_{\nu=r_1+1}^{r_1+r_2} q^{(\nu)} J_{a_\nu}\left(2\pi\delta \sqrt[n]{\frac{N\mathfrak{x}}{D}}\, v^{(\nu)} q^{(\nu)}\right) dq^{(1)} \cdots dq^{(r_1+r_2)}, \tag{38}$$

so nimmt (17) die folgende Gestalt an:

$$\sum_{\mathfrak{x}\in C} \bar{\chi}\psi(\gamma(\mathfrak{x})) \int_V \chi\bar{\psi}(v)\, \Phi(A, \delta, \chi, \mathfrak{x})\, \bar{\Phi}(B, \delta, \psi, \mathfrak{x})\, d\omega$$

$$= \begin{cases} 0, & \text{wenn } \chi \neq \psi \text{ ist,} \\[2mm] \dfrac{R}{w}\sqrt{D}\,\delta^{-n}\pi^{-r_2} \displaystyle\int_0^1 \cdots \int_0^1 A(\delta^n Nq)\, \bar{B}(\delta^n Nq)\, q^{(r_1+1)} \cdots q^{(r_1+r_2)}\, dq^{(1)} \cdots dq^{(r_1+r_2)}, & \\ & \text{wenn } \chi = \psi, \text{ aber } \chi \text{ auf } UE \text{ nicht konstant ist,} \\[2mm] \dfrac{R}{w}\Bigl[\sqrt{D}\,\delta^{-n}\pi^{-r_2} \displaystyle\int_0^1 \cdots \int_0^1 A(\delta^n Nq)\, \bar{B}(\delta^n Nq)\, q^{(r_1+1)} \cdots q^{(r_1+r_2)}\, dq^{(1)} \cdots dq^{(r_1+r_2)} & \\ \quad - \displaystyle\int_0^1 \cdots \int_0^1 A(\delta^n Nq) \prod_{\nu=1}^{r_1+r_2} {q^{(\nu)}}^{L_\nu(s,0)i} q^{(r_1+1)} \cdots q^{(r_1+r_2)}\, dq^{(1)} \cdots dq^{(r_1+r_2)} & \\ \quad \times \displaystyle\int_0^1 \cdots \int_0^1 \bar{B}(\delta^n Nq) \prod_{\nu=1}^{r_1+r_2} {q^{(\nu)}}^{-L_\nu(s,0)i} q^{(r_1+1)} \cdots q^{(r_1+r_2)}\, dq^{(1)} \cdots dq^{(r_1+r_2)}\Bigr], & \\ \text{wenn } \chi = \psi \text{ und } \chi \text{ auf } UE \text{ konstant } = 1 - \text{und daher } s = j,\ a = b = 0 - \text{ist.} & \end{cases} \tag{39}$$

Im Fall $\chi = \psi = \chi(*, s)\,\psi(*, a)$ wollen wir die rechte Seite von (39) mit $S(A, B, \delta, s, a)$ bezeichnen.

10. Wir gehen jetzt von einem Heckeschen Größencharakter vom Führer 1 des Körpers k aus, also von einem Charakter χ der Gruppe der Ideale von k, der, auf die Gruppe der Hauptideale (ξ) in der nach (18) durch $(\xi) \to (\xi \cdot |N\xi|^{-(1/n)}, |N\xi|)\,E$ gegebenen Einbettung in Z/E eingeschränkt, eine Funktion auf Z/E ergibt, die zugleich auch die Einschränkung eines Charakters χ_0 von Z/E ist. Kurz, es soll für $\xi \in k^\times$

$$\chi((\xi)) = \chi_0(\xi) \tag{40}$$

gelten mit einem Charakter – vgl. (30) –

$$\chi_0(z) = |Nz|^{i\eta}\,\chi(|z| \cdot |Nz|^{-(1/n)}, s)\,\psi(z\,|Nz|^{-(1/n)}, a) \tag{41}$$

von Z/E. Wir berechnen $\chi_0(\gamma(\mathfrak{x}))$:

$$g^t(\mathfrak{x}, 1)\,\alpha = \gamma(\mathfrak{x}) \sqrt[n]{\frac{N\mathfrak{x}}{D}}\, N\mathfrak{a}^{-(1/n)} = \gamma(\mathfrak{x}) \sqrt[n]{N\mathfrak{x}\, N\tilde{\mathfrak{a}}}$$

erzeugt das Hauptideal $\mathfrak{x}\tilde{\mathfrak{a}}$, also ist

$$\chi(\mathfrak{x})\,\chi(\tilde{\mathfrak{a}}) = \chi_0(\gamma(\mathfrak{x}))\,\chi_0\,(\sqrt[n]{N\mathfrak{x}\,N\tilde{\mathfrak{a}}}) = \chi_0(\gamma(\mathfrak{x}))\,|N\mathfrak{x}\,N\tilde{\mathfrak{a}}|^{i\eta},$$

$$\chi_0\,(\gamma(\mathfrak{x})) = \chi(\tilde{\mathfrak{a}})\,N\tilde{\mathfrak{a}}^{-i\eta}\chi(\mathfrak{x})\,N\mathfrak{x}^{-i\eta}.$$

ψ sei ein zweiter Größencharakter des Führers 1 von k und

$$\psi((\xi)) = \psi_0(\xi),$$

$$\psi_0(z) = |Nz|^{i\Theta}\,\chi(|z| \cdot |Nz|^{-(1/n)}, j)\,\psi(z\,|Nz|^{-(1/n)}, b),$$

so daß

$$\bar{\chi}_0\psi_0(\gamma(\mathfrak{x})) = \bar{\chi}\psi(\tilde{\mathfrak{a}})\,N\tilde{\mathfrak{a}}^{i(\eta-\Theta)}\bar{\chi}\psi(\mathfrak{x})\,N\mathfrak{x}^{i(\eta-\Theta)}. \tag{42}$$

Betrachten wir die Summe

$$\sum_{\mathfrak{x}} \bar{\chi}\psi(\mathfrak{x})\,N\mathfrak{x}^{i(\eta-\Theta)} \int_V \chi_0\bar{\psi}_0(v)\,\Phi(A, \delta, \chi_0, \mathfrak{x})\,\bar{\Phi}(B, \delta, \psi_0, \mathfrak{x})\,d\omega, \tag{43}$$

erstreckt über alle ganzen Ideale $\mathfrak{x}$ von k. Sie ist gleich 0, *wenn χ und ψ verschieden sind.* Denn ist etwa $\chi_0 \neq \psi_0$, so gilt schon

$$\sum_{\mathfrak{x}\in C} \bar{\chi}\psi(\mathfrak{x})\,N\mathfrak{x}^{i(\eta-\Theta)} \int_V \chi_0\bar{\psi}_0(v)\,\Phi(A, \delta, \chi_0, \mathfrak{x})\,\bar{\Phi}(B, \delta, \psi_0, \mathfrak{x})\,d\omega = 0,$$

wobei über alle $\mathfrak{x}$ einer Idealklasse C summiert wird, nach dem ersten Fall der Formel (39).

In jedem Fall ist die Summe (43), wenn $\tilde{\mathfrak{a}}$ ein Repräsentantensystem der Idealklassen von k durchläuft, gleich einem von $\tilde{\mathfrak{a}}$ unabhängigen Faktor mal

$$\sum_{\tilde{\mathfrak{a}}} \chi\bar{\psi}(\tilde{\mathfrak{a}})\, N\tilde{\mathfrak{a}}^{i(\Theta-\eta)},$$

was aus (39) folgt. Nun ist aber nach (42) die Funktion $\chi\bar{\psi}(\mathfrak{x})\, N\mathfrak{x}^{i(\Theta-\eta)}$ von $\mathfrak{x}$ im Fall $\chi_0 = \psi_0$ eine Funktion nur der Idealklasse von $\mathfrak{x}$, und daher ist im Falle $\chi_0 = \psi_0$, aber $\chi \neq \psi$,

$$\sum_{\tilde{\mathfrak{a}}} \chi\bar{\psi}(\tilde{\mathfrak{a}})\, N\tilde{\mathfrak{a}}^{i(\Theta-\eta)} = 0.$$

Es bleibt der Fall $\chi = \psi$; er gibt

$$\sum_{\mathfrak{x}} \int_V \Phi(A, \delta, \chi_0, \mathfrak{x})\, \Phi(\bar{B}, \delta, \chi_0, \mathfrak{x}) = h \cdot S(A, B, \delta, s, a). \tag{44}$$

Dies ist eine Formel für die Klassenzahl h von k (jedenfalls, wenn $S(A, B, \delta, s, a) \neq 0$ ist), in die als Parameter die beiden Funktionen A und B, die Variable δ und der Charakter χ_0 von V/E, also s und a eingehen.

11. Wir betrachten einige Spezialfälle dieser Formel.

Zuerst sei $A = B = 1$, $\delta = 1$ und $\chi_0 = 1$. Dann ist

$$\int_0^1 \cdots \int_0^1 A(\delta^n Nq)\, \bar{B}(\delta^n Nq)\, q^{(r_1+1)} \cdots q^{(r_1+r_2)}\, dq^{(1)} \cdots dq^{(r_1+r_2)} = 2^{-r_2},$$

$$\int_0^1 \cdots \int_0^1 A(\delta^n Nq) \prod_{\nu=1}^{r_1+r_2} q^{(\nu)^{L_\nu(s,0)i}}\, q^{(r_1+1)} \cdots q^{(r_1+r_2)}\, dq^{(1)} \cdots dq^{(r_1+r_2)} = 2^{-r_2}$$

und

$\Phi\; A, \delta, \chi_0, \mathfrak{x})$

$$= \prod_{\nu=1}^{r_1} \int_0^1 \cos\left(\pi \sqrt[n]{\frac{N\mathfrak{x}}{D}}\, q^{(\nu)} v^{(\nu)}\right) dq^{(\nu)} \prod_{\nu=r_1+1}^{r_1+r_2} \int_0^1 q^{(\nu)} J_0\left(2\pi \sqrt[n]{\frac{N\mathfrak{x}}{D}}\, q^{(\nu)} v^{(\nu)}\right) dq^{(\nu)}$$

$$= \prod_{\nu=1}^{r_1} \frac{\sin \pi \sqrt[n]{\frac{N\mathfrak{x}}{D}}\, v^{(\nu)}}{\pi \sqrt[n]{\frac{N\mathfrak{x}}{D}}\, v^{(\nu)}} \prod_{\nu=r_1+1}^{r_1+r_2} \frac{J_1\left(2\pi \sqrt[n]{\frac{N\mathfrak{x}}{D}}\, v^{(\nu)}\right)}{2\pi \sqrt[n]{\frac{N\mathfrak{x}}{D}}\, v^{(\nu)}},$$

daher

$$\frac{hR}{w}\left[\frac{\sqrt{D}}{(2\pi)^{r_2}} - \frac{1}{2^{2r_2}}\right] = \sum_{\mathfrak{x}} \int_V \left[\prod_{\nu=1}^{r_1} \frac{\sin \pi \sqrt[n]{\frac{N\mathfrak{x}}{D}}\, v^{(\nu)}}{\pi \sqrt[n]{\frac{N\mathfrak{x}}{D}}\, v^{(\nu)}} \prod_{\nu=r_1+1}^{r_1+r_2} \frac{J_1\left(2\pi \sqrt[n]{\frac{N\mathfrak{x}}{D}}\, v^{(\nu)}\right)}{2\pi \sqrt[n]{\frac{N\mathfrak{x}}{D}}\, v^{(\nu)}}\right]^2 d\omega. \tag{45}$$

Ist insbesondere k imaginär quadratisch, so ist $r_1 = 0$, $r_2 = 1$, $R = 1$; und V reduziert sich auf einen Punkt:

$$\frac{h}{w}\left[\frac{\sqrt{D}}{2\pi} - \frac{1}{4}\right] = \sum_{\mathfrak{x}} \left(\frac{J_1\left(2\pi\sqrt{\frac{N\mathfrak{x}}{D}}\right)}{2\pi\sqrt{\frac{N\mathfrak{x}}{D}}} \right)^2, \tag{46}$$

wenn k imaginär quadratisch ist.

Wenn k reell quadratisch ist, so ist $r_1 = 2$, $r_2 = 0$, $w = 2$, und R ist der Logarithmus der durch $\mu > 1$ normierten Grundeinheit η von k, folglich ist

$$\frac{h \ln \eta}{2}[\sqrt{D} - 1] = \sum_{\mathfrak{x}} \int_0^\infty \left[\frac{\sin \pi\sqrt{\frac{N\mathfrak{x}}{D}}\, v}{\pi\sqrt{\frac{N\mathfrak{x}}{D}}\, v} \cdot \frac{\sin \pi\sqrt{\frac{N\mathfrak{x}}{D}\cdot\frac{1}{v}}}{\pi\sqrt{\frac{N\mathfrak{x}}{D}\cdot\frac{1}{v}}} \right]^2 \frac{dv}{v} \tag{47}$$

für reelles quadratisches k mit der Grundeinheit η.

Die hier auftretende Funktion

$$\lambda(z) = \int_0^\infty \left[\frac{\sin vz}{vz} \cdot \frac{\sin \frac{z}{v}}{\frac{z}{v}} \right]^2 \frac{dv}{v} \tag{48}$$

kann wie das Summenglied in der Reihe (46) für die Klassenzahl eines imaginären quadratischen Körpers durch Zylinderfunktionen ausgedrückt werden.

Für reelles z ist

$$\begin{aligned} z^4\lambda(z) &= \int_0^\infty \sin^2 zv \cdot \sin^2 \frac{z}{v}\,\frac{dv}{v} = 2\int_0^1 \sin^2 zv \cdot \sin^2 \frac{z}{v}\,\frac{dv}{v} \\ &= \frac{1}{4}\int_0^1 \cos 2z\left(v + \frac{1}{v}\right)\frac{dv}{v} + \frac{1}{4}\int_0^1 \cos 2z\left(v - \frac{1}{v}\right)\frac{dv}{v} \\ &\quad + \frac{1}{2}\left[\int_0^1 (1 - \cos 2zv)\,\frac{dv}{v} - \int_0^1 \cos\frac{2z}{v}\,\frac{dv}{v}\right]. \end{aligned}$$

Setzen wir also

$$\int_0^1 (1 - \cos v)\,\frac{dv}{v} - \int_1^\infty \cos v\,\frac{dv}{v} = c, \tag{49}$$

so daß

$$\int_0^1 (1 - \cos 2zv)\,\frac{dv}{v} - \int_0^1 \cos\frac{2z}{v}\,\frac{dv}{v} = c + \ln 2z \tag{50}$$

wird, so folgt

$$z^4\lambda(z) = \frac{1}{4}\int_0^1 \cos 2z\left(v + \frac{1}{v}\right)\frac{dv}{v} + \frac{1}{4}\int_0^1 \cos 2z\left(v - \frac{1}{v}\right)\frac{dv}{v} + \frac{c}{2} + \frac{1}{2}\ln 2z.$$

Die beiden Integrale sind Zylinderfunktionen, denn es ist $\left(v + \frac{1}{v} = 2u\right)$

$$-\frac{2}{\pi}\int_0^1 \cos 2z\left(v + \frac{1}{v}\right)\frac{dv}{v} = -\frac{2}{\pi}\int_0^1 \frac{\cos 4uz}{\sqrt{u^2 - 1}}\,du = N_0(4z)$$

die Neumannsche Funktion nullter Ordnung für das Argument $4z$ und $\left(v - \frac{1}{v} = 2u\right)$

$$\frac{2}{\pi i}\int_0^1 \cos 2z\left(v - \frac{1}{v}\right)\frac{dv}{v} = \frac{2}{\pi i}\int_0^\infty \frac{\cos 4uz}{\sqrt{u^2 + 1}}\,du = H_0^{(1)}(4iz)$$

$$= J_0(4iz) + iN_0(4iz)$$

die Hankelsche Funktion erster Art nullter Ordnung für das Argument $4iz$. Daher ist

$$z^4\lambda(z) = -\frac{\pi}{8}[N_0(4z) + N_0(4iz) - iJ_0(4iz)] + \frac{1}{2}\ln z + \frac{1}{2}(c + \ln 2). \tag{51}$$

Die Konstante c in dieser Formel ist die Euler-Mascheronische Konstante. Um das einzusehen, setzen wir in (51) für N_0 die Entwicklung

$$N_0(z) = J_0(z)(\ln z + C) - \sum_{l=1}^{\infty} \frac{(-1)^l}{(l!)^2}\left(\sum_{m=1}^{l}\frac{1}{m}\right) z^{2l}$$

ein, in der C jetzt die Euler-Mascheroni-Konstante bezeichne:

$$z^4\lambda(z) = -\frac{1}{4}(\ln(2z) + C)(J_0(4z) + J_0(4iz))$$

$$+ \frac{1}{2}\sum_{l=1}^{\infty}\frac{1}{((2l)!)^2}\left(\sum_{m=1}^{2l}\frac{1}{m}\right)(2z)^{4l} + \frac{1}{2}\ln(2z) + \frac{c}{2}.$$

Und hieraus folgt, da für reelles z

$$0 \leqq z^4\lambda(z) \leqq 2\int_0^1 (zv)^2 \cdot 1 \cdot \frac{dv}{v} = z^2,$$

also $\lim\limits_{z\to 0} z^4\lambda(z) = 0$ gilt, in der Tat $c = C$.

Setzen wir in (51) für $J_0(4z)$ und $J_0(4iz)$ die Entwicklungen nach Potenzen von z ein, so erhalten wir

$$\lambda(z) = \frac{1}{2} z^{-4} \sum_{l=1}^{\infty} \frac{(2z)^{4l}}{((2l)!)^2} \left[\sum_{m=1}^{2l} \frac{1}{m} - d - \ln z \right] \tag{52}$$

mit

$$d = \frac{1}{2}(c + \ln 2).$$

Um ein Beispiel mit $\chi_0 \neq 1$ zu haben, nehmen wir wieder k imaginär quadratisch. Wegen $r_1 = 0$, $r_2 = 1$ ist das allgemeinste χ_0 in diesem Falle

$$\chi_0(z) = \psi(z|Nz|^{-1/2}, a) \quad \text{mit} \quad a \equiv 0 \bmod w,$$

also

$$\chi_0(z) = \exp(iw \arg z) = \left(\frac{z}{|z|}\right)^a.$$

Es werde ferner

$$A(z) = B(z) = z^{a/2}$$

gesetzt, wobei $a \geqq 0$ vorauszusetzen ist. Wir haben dann

$$\begin{aligned} \Phi(A, \delta, \chi_0, \mathfrak{x}) &= \int_0^1 A(\delta^2 q^2)\, q J_a\left(2\pi\delta \sqrt{\frac{N\mathfrak{x}}{D}}\, q\right) dq \\ &= \delta^a \int_0^1 q^{a+1} J_a\left(2\pi\delta \sqrt{\frac{N\mathfrak{x}}{D}}\, q\right) dq \\ &= \delta^a \frac{J_{a+1}\left(2\pi\delta \sqrt{\frac{N\mathfrak{x}}{D}}\right)}{2\pi\delta \sqrt{\frac{N\mathfrak{x}}{D}}} \end{aligned}$$

und

$$\int_0^1 A(\delta^2 Nq)\, \bar{B}(\delta^2 Nq)\, q\, dq = \delta^{2a} \int_0^1 q^{2a+1}\, dq = \frac{\delta^{2a}}{2(a+1)},$$

so daß (44) im Fall $a > 0$

$$\frac{h\sqrt{D}\,\delta^{-2}\pi^{-1}}{w} \frac{\delta^{2a}}{2(a+1)} = \sum_{\mathfrak{x}} \delta^{2a} \left(\frac{J_{a+1}\left(2\pi\delta \sqrt{\frac{N\mathfrak{x}}{D}}\right)}{2\pi\delta \sqrt{\frac{N\mathfrak{x}}{D}}} \right)^2$$

ergibt, oder, wenn wir noch $a + 1 = \nu$ setzen: *Für einen imaginär quadratischen Körper k gilt*

$$h = \frac{w\sqrt{D}}{2\pi}\,\nu \sum_{\mathfrak{x}} J_\nu\left(2\pi\delta\sqrt{\frac{N\mathfrak{x}}{D}}\right)^2 N\mathfrak{x}^{-1}, \tag{53}$$

wobei $0 < \delta \leqq 1$, $\nu \equiv 1 \bmod w$ *und* $\nu > 1$ *sein muß.*

Literatur

[1] K. G. Ramanathan, The zetafunction and discriminant of a division algebra, Acta Arithm. **5** (1959), 277—288.

[2] E. Hecke, Über eine neue Art von Zetafunktionen und ihre Beziehungen zur Verteilung der Primzahlen, 1. Mitteilung: Math. Z. **1** (1918), 357—376; 2. Mitteilung: Math. Z. **6** (1920), 11—51; auch: Math. Werke, Göttingen 1959, S. 215—234 und 249—289.

[3] C. L. Siegel, Über die Diskriminanten total reeller Körper, Nachr. d. Kgl. Ges. d. Wiss. Göttingen, Math.-Phys. Kl. 1922, S. 17—24; auch Ges. Abhandl., Berlin-Heidelberg-New York 1966, Band 1, S. 157—164.

[4] C. L. Siegel, Über Gitterpunkte in convexen Körpern und ein damit zusammenhängendes Extremalproblem, Acta Math. **65** (1935), 307—323; auch Ges. Abhandl., Band 1, S. 311—325.

P. Erdös, A. Sárközi und E. Szemerédi in Budapest

ÜBER FOLGEN GANZER ZAHLEN

Eine Folge ganzer Zahlen $a_1 < \cdots$ heiße primitiv, wenn $a_i \nmid a_j$ für alle $a_i < a_j$ gilt. Behrend und Erdös [1] zeigten vor mehr als dreißig Jahren, daß für primitive Zahlenfolgen ($C, c, c', C_1, \ldots$ sind positive absolute Konstanten, die wenn sie in verschiedenen Stellen vorkommen, nicht unbedingt denselben Wert haben)

$$\sum_{a_i < x} \frac{1}{a_i} < c \log x/(\log\log x)^{1/2} \tag{1}$$

und

$$\sum_k \frac{1}{a_k \log a_k} < C \tag{2}$$

gilt. Kürzlich zeigten wir [2], daß für unendliche primitive Zahlenfolgen

$$\lim_{x=\infty} \sum_{a_i < x} \frac{1}{a_i} \left(\frac{\log x}{(\log\log x)^{1/2}} \right)^{-1} = 0 \tag{3}$$

gilt. Bekanntlich lassen sich (1) und (3) nicht verschärfen.

In einer anderen Arbeit [3] zeigten wir, daß (1) auch dann noch richtig bleibt, wenn wir nur voraussetzen, daß die Gleichungen

$$[a_i, a_j] = a_r, \quad a_i < a_j < a_r \tag{4}$$

oder

$$(a_i, a_j) = a_r, \quad a_r < a_i < a_j \tag{5}$$

unlösbar sind. Weiter zeigten wir in [3], daß eine unendliche Folge $a_1 < \cdots$ existiert, für welche (4) nicht lösbar ist, aber für jedes x

$$\sum_{a_i < x} 1 > c' x/(\log\log x)^{1/2} \tag{6}$$

gilt. Aus (6) folgt sofort, daß wenn (4) unlösbar ist, die Reihe (2) nicht konvergieren muß. Es war für uns eine große Überraschung, als wir zeigen konnten, daß wenn (5) unlösbar ist, (2) gilt; daher benimmt sich die Gleichung (5) ganz anders als (4) [4].

In dieser Arbeit zeigen wir folgenden

Satz. *Ist* (5) *unlösbar, so gilt* (3).

Bevor wir den recht komplizierten Beweis angeben, möchten wir noch einige Bemerkungen machen. Es sei $a_1 < \cdots$ eine Folge, für welche

$$a_i q = a_j, \quad p(q) > P(a_i) \tag{7}$$

unlösbar ist. ($P(n)$ ist der größte, $p(n)$ der kleinste Primfaktor von n.) Dann gilt (2). (Dies ist ein Satz von Alexander, siehe auch [4].) Wir werden jetzt zeigen, daß wenn (7) unlösbar ist, (1) nicht gelten muß. Wir werden sogar zeigen, daß wenn $\varphi(x)$ beliebig langsam gegen Unendlich strebt, immer eine Folge existiert, für welche (7) unlösbar ist und doch für unendlich viele x

$$\sum_{a_i<x} \frac{1}{a_i} > \frac{\log x}{\varphi(x)} \tag{8}$$

gilt. Dies ist offenbar die bestmögliche Abschätzung; denn aus (2) folgt

$$\sum_{a_i<x} \frac{1}{a_i} = o(\log x).$$

(Diese Tatsache erschwert den Beweis unseres Satzes.) ε_i sei eine beliebige Folge, die gegen 0 strebt, und x_i strebe genügend rasch gegen Unendlich. Unsere Folge besteht aus allen Zahlen

$$x_i < a < x_i^{1+\varepsilon_i}, \quad P(a) > x^{\varepsilon_i}, \quad i = 1, 2, \ldots,$$

die für jedes $j < i$ keinen Teiler in den Intervallen $(x_j, x_j^{1+\varepsilon})$ haben. Es ist klar, daß für diese Folge (7) unlösbar ist, und aus [5] folgt leicht, daß wenn ε_i genügend langsam gegen 0 und x_i genügend rasch gegen ∞ strebt, (8) für unendlich viele x befriedigt ist.

Nun beweisen wir unseren Satz. Der Beweis wird Methoden anwenden, die wir in [3] und [4] angewendet haben. Es sei ($\exp z = e^z$)

$$f(n) = \exp\left((\log n)^{1/10}\right)$$

Lemma 1. *Es gilt*

$$\sum{}' \frac{1}{a_k} = o\left(\frac{\log x}{\log\log x}\right),$$

wobei in $\sum'$ *die Summation über die* $a_k \leqq x$ *erstreckt ist, für welche* (p *Primzahl*)

$$\sum_{\substack{p \mid a_k \\ p < f(a_k)}} 1 > \frac{1}{2} \log\log a_k \tag{9}$$

ist.

Offenbar gilt

$$\sum' \frac{1}{a_k} = \sum_1' \frac{1}{a_k} + \sum_2' \frac{1}{a_k}, \tag{10}$$

wobei $a_k < \exp\left(\frac{\log x}{(\log\log x)^2}\right)$ in $\sum_1'$ und $\exp\left(\frac{\log x}{(\log\log x)^2}\right) \leqq a_k \leqq x$ in $\sum_2'$ ist.

Für $\sum_1$ haben wir die triviale Abschätzung

$$\sum_1' \frac{1}{a_k} < \sum_{n < \exp\left(\frac{\log x}{(\log\log x)^2}\right)} \frac{1}{n} = o\left(\frac{\log x}{\log\log x}\right). \tag{11}$$

Für die a_k in $\sum_2'$ gilt wegen (9)

$$\sum_{\substack{p \mid a_k \\ p < f(x)}} 1 > \frac{1}{2} \log\log\left(\exp\left(\frac{\log x}{(\log\log x)^2}\right)\right) > \frac{1}{3} \log\log x .$$

Daher erhalten wir durch eine leichte Rechnung (aus dem bekannten Satz von Mertens)

$$\sum_2' \frac{1}{a_k} < \prod_{f(x) < p \leqq x} \left(1 + \frac{1}{p} + \frac{1}{p^2} + \cdots\right) \sum_{t > \frac{1}{3}\log\log x} \sum_{p \leqq f(x)} \left(\frac{1}{p} + \frac{1}{p^2} + \cdots\right)^t \Big/ t!$$

$$< c(\log x)^{9/10} \sum_{t > \frac{1}{3}\log\log x} \left(\frac{\log\log x}{10} + c_1\right)^t \Big/ t! = o\left(\frac{\log x}{\log\log x}\right). \tag{12}$$

Lemma 1 folgt aus (10), (11) und (12).

Lemma 2. *Es gilt*

$$\sum'' \frac{1}{a_k} = o\left(\frac{\log x}{\log\log x}\right),$$

wobei in $\sum''$ die Summation über die a_k erstreckt ist, für welche

$$\sum_{\substack{p \mid a_k \\ p > f(a_k)}} 1 < \frac{1}{2} \log\log a_k \tag{13}$$

ist.

Der Beweis von Lemma 2 ist dem von Lemma 1 sehr ähnlich und kann dem Leser überlassen werden.

Es sei nun $b_1 < \cdots$ die Teilfolge von $a_1 < \cdots$, deren Glieder weder (12) noch (13) befriedigen. Um unseren Satz zu beweisen, genügt es wegen Lemma 1 und 2,

$$\sum_{b_i \leqq x} \frac{1}{b_i} = o\left(\frac{\log x}{\log\log x}\right) \tag{14}$$

zu zeigen.

Der Beweis von (14) wird nicht ganz leicht sein. $B(b_k)$ ist eine Teilfolge von $b_1 < \cdots$, die wie folgt definiert ist: $b_j \in B(b_k)$, wenn b_k das größte b ist, für welches

$$b_j = b_k q, \quad p(q) > f(b_k) \tag{15}$$

ist. B' sei die Teilfolge der Folge $b_1 < \cdots$, deren Glieder sich für kein k in der Form (15) darstellen lassen.

Lemma 3.

$$\sum_{\substack{b_k \in B' \\ b_k \leqq x}} \frac{1}{b_k} = o\left(\frac{\log x}{(\log\log x)^{1/2}}\right).$$

Der Beweis von Lemma 3 ist genau dem Beweis von (3) nachgebildet (siehe [2]), kann also unterdrückt werden. (Wegen Lemma 1 und 2 kann Lemma 2 von [2] angewendet werden.)

Um (14) zu beweisen, müssen wir nun

$$\sum_{\substack{b_i \in B(b_k) \\ b_i \leqq x}} \frac{1}{b_i}$$

abschätzen. (14) wird (wegen Lemma 3) bewiesen sein, wenn wir zeigen können, daß für jedes $\varepsilon > 0$ und $x > x_0(\varepsilon)$, $b_1 = b_1(\varepsilon)$

$$\sum_k \sum_{\substack{b_i \in B(b_k) \\ b_k \leqq x}} \frac{1}{b_i} < \varepsilon \log x/(\log\log x)^{1/2}, \tag{16}$$

gilt, wenn $b_1 = b_1(\varepsilon)$ genügend groß ist. (Offenbar dürfen wir $b_1 = b_1(\varepsilon)$ beliebig groß annehmen, da wir nötigenfalls die kleinen b einfach weglassen können; leider dürfen wir aber nicht $b_1 > f(x), f(x) \to \infty$ zusammen mit $x \to \infty$ voraussetzen, und die Tatsache, daß wir dies nicht dürfen, macht unseren Beweis viel komplizierter.)

Es seien $b_k q_j = b_i$ die Zahlen von $B(b_k)$. Wie in [4] gilt $(q_{j_1}, q_{j_2}) \neq 1$. Wenn nämlich $b_k q_{j_1} = b_{i_1}$, $b_k q_{j_2} = b_{i_2}$, $(q_{j_1}, q_{j_2}) = 1$ wäre, so wäre $(b_{i_1}, b_{i_2}) = b_k$, was unserer Voraussetzung widerspricht. Daher gilt für die Zahlen von $B(b_k)$

$$b_k q_j, \quad j = 1, \ldots, (q_{j_1}, q_{j_2}) \neq 1, \quad p(q_j) > f(b_k). \tag{17}$$

Die k teilen wir nun (wie in [4]) in zwei Klassen. In der ersten Klasse sind die k, für welche es ein q_j gibt, so daß

$$\sum_{p \mid q_j} \frac{1}{p} < \frac{1}{f(b_k)^{1/2}} \tag{18}$$

ist.

Es seien $p_1, \ldots, p_r$ die Primfaktoren von q_j. Wegen (17) sind alle Zahlen der Form $b_k q_j$ von der Form $b_k p_i t$, wobei i eine der Zahlen $1, \ldots, r$ ist. Es sei jetzt i fest, und wir betrachten die Zahlen

$$b_k p_i t_j^{(i)}, \quad j = 1, \ldots, \quad b_k p_i t_j^{(i)} \leqq x.$$

Offenbar gilt $(t_{j_1}^{(i)}, t_{j_2}^{(i)}) \neq t_{j_3}$ für jedes $j_3 < j_2 < j_1$, da sonst $(a_1, a_2) = a_3$ wäre mit $a_r = b_k p_i t_r^{(i)}$, $r = 1, 2, 3$, was unserer Voraussetzung widerspricht. Daher folgt nach [3] ($t_j^{(i)} \leqq x/b_k p_i$)

$$\sum_j \frac{1}{t_j^{(i)}} < c \log \frac{x}{b_k p_i} \Bigg/ \left(\operatorname{loglog} \frac{x}{b_k p_i}\right)^{1/2} < \frac{c \log x}{(\operatorname{loglog} x)^{1/2}}. \tag{19}$$

Wegen (18) und (19) folgt daher, wenn k zur ersten Klasse gehört,

$$\sum_{b_k q_j \leqq x} \frac{1}{b_k q_j} = \sum_{i=1}^{r} \sum \frac{1}{b_k q_i t_j^{(i)}} < \frac{c \log x}{b_k (f(b_k))^{1/2} (\operatorname{loglog} x)^{1/2}}. \tag{20}$$

Schließlich folgt aus (20) (in $\sum_1$ geht die Summation über alle k der ersten Klasse)

$$\sum_k{}_1 \sum_j \frac{1}{b_k q_j} < \frac{c \log x}{(\operatorname{loglog} x)^{1/2}} \sum_k{}_1 \frac{1}{b_k f(b_k)^{1/2}}$$
$$< \frac{c \log x}{(\operatorname{loglog} x)^{1/2}} \sum_{n=b_1}^{\infty} \frac{1}{n(f(n))^{1/2}} < \frac{\varepsilon}{4} \frac{\log x}{(\operatorname{loglog} x)^{1/2}} \tag{21}$$

wenn $b_1 = b_1(\varepsilon)$ genügend groß ist.

Um unseren Beweis zu beenden, müssen wir nun zeigen, daß für $b_1 = b_1(\varepsilon)$

$$\sum_k{}_2 \sum_j \frac{1}{b_k q_j} < \frac{\varepsilon}{4} \frac{\log x}{(\operatorname{loglog} x)^{1/2}} \tag{22}$$

ist. In $\sum_2$ läuft die Summation über alle k der zweiten Klasse, also über die k, für welche für alle Zahlen (17) von $B(b_k)$

$$\sum_{p \mid q_j} \frac{1}{p} \geqq \frac{1}{(f(b_k))^{1/2}} \tag{23}$$

gilt.

Wir nehmen nun an, daß (22) falsch ist, und betrachten die Zahlen der Form

$$b_k q_j t, \quad 1 \leqq t \leqq \frac{x}{b_k q_j}, \tag{24}$$

wobei $b_k q_j$ von der Form (17) ist und k alle Zahlen der zweiten Klasse durchläuft.

Es seien $u_1 < \cdots < u_s \leqq x$ die Zahlen (24), der Größe nach geordnet. Für jedes u_i gibt es wegen (23) und (24) ein $Z = Z(u_i) \geqq f(b_1)$, so daß

$$\sum_{\substack{p \mid u_i \\ p > Z}} \frac{1}{p} \geqq \frac{1}{f(Z)^{1/2}} \tag{25}$$

ist.

Es folgt leicht aus (25), daß für jedes u_i ein ganzes t existiert mit $2^t > \frac{1}{2} f(b_1)$, so daß u_i mindestens zwei Primfaktoren in $(2^t, 2^{t+1})$ hat. Daher folgt durch eine leichte Rechnung (in $\sum'$ läuft t über die Zahlen $2^t > \frac{1}{2} f(b_1)$)

$$\sum_{i=1}^{s} \frac{1}{u_i} < \sum_t{}' \left(\sum_{2^t < p < 2^{t+1}} \frac{1}{p} \right)^2 \sum_{n=1}^{x} \frac{1}{n} < 2 \log x \sum_t{}' \left(\sum_{2^t < p < 2^{t+1}} \frac{1}{p} \right)^2$$
$$< c \log x \sum{}' \frac{1}{t^2} < \eta \log x \tag{26}$$

für jedes feste η, wenn $b_1 = b_1(\varepsilon, \eta)$ genügend groß ist.

Um (22) zu zeigen, werden wir nun beweisen, daß unter der Voraussetzung, daß (22) falsch ist, für genügend kleines $\eta = \eta(\varepsilon)$ (26) ebenfalls falsch sein würde. Dieser Widerspruch wird dann (22) beweisen. Zuerst beweisen wir

Lemma 4. *Es sei S eine Menge von n Elementen und $A_1, \ldots, A_k$, $k > c_1 2^n/\sqrt{n}$ Teilmengen von S, so daß*

$$A_i \cap A_j = A_r, \quad i \neq r, \quad j \neq r \tag{27}$$

unmöglich ist. Es seien $B_1, \ldots, B_l$ diejenigen Teilmengen von S, die mindestens ein A_i, $i = 1, \ldots, k$ enthalten. Dann gilt $l > c_2 2^n$ mit $c_2 = c_2(c_1)$.

In [2] wird ein ähnliches Lemma bewiesen (Lemma 3), aber unser Lemma ist stärker, denn in Lemma 3 von [2] wird an Stelle von (27) $A_i \not\subset A_j$ verlangt, und die Bedingung (27) ist offenbar schwächer.

Unser Beweis von Lemma 4 ist dem schönen Beweis von Kleitman [6] nachgebildet. Es sei $S = S_1 \cup S_2$, $|S_1| = \left[\frac{n}{2}\right]$, $|S_2| = \left[\frac{n+1}{2}\right]$ ($|A|$ bedeute die Anzahl der Elemente von A). Es sei $A_i \cap S_1 = B_i^{(1)}$, $A_i \cap S_2 = B_i^{(2)}$. Für jedes $X \subset S_1$ betrachten wir die Familie $F_{X,1}$ der Mengen A_i mit $B_i^{(1)} = X$, für $Y = S_2$ sei $F_{Y,2}$ die Familie der Mengen

A_i mit $A_i \cap S_2 = B_i^{(2)} = Y$. Die Familie $F^*_{X,1} \supset F_{X,1}$ ist nun wie folgt definiert: $A_i \in F^*_{X,1}$ für $A_i \cap S_1 = X$, und $A_i \cap S_2$ ist in keinem $A_j \cap S_2$ mit $A_j \cap S_1 = X$ enthalten. Die Familie $F^*_{Y,2} \subset F_{Y,2}$ ist wie folgt definiert: $A_i \in F^*_{Y,2}$ für $A_i \cap S_2 = Y$, und $A_i \cap S_1$ ist in keinem $A_j \cap S_1$ mit $A_j \cap S_2 = Y$ enthalten. Die Familien $F^*_{X,1}$ und $F^*_{Y,2}$ haben offenbar die Spernersche Eigenschaft [7]:

$$\text{Ist } A_i \in F^*_{X,1}, A_j \in F^*_{X,1}, \text{ dann gilt } A_i \not\subset A_j, \text{ und dasselbe gilt für } F^*_{Y,2}. \tag{28}$$

Es ist leicht zu sehen, daß jedes A_i in

$$\bigcup_X F^*_{X,1} \bigcup_Y F^*_{Y,2} \tag{29}$$

enthalten ist. Denn wäre z. B. A_i nicht in (29) enthalten, so würden wir $A_i \cap S_1 = X$, $A_i \cap S_2 = Y$ betrachten. Da aber A_i nicht in $F^*_{X,1} \cup F^*_{Y,2}$ enthalten ist, existieren Mengen $A_{j_1} \in F^*_{X,1}$, $A_{j_2} \in F^*_{Y,2}$ mit $A_{j_1} \cap S_2 \supset A_i \cap S_2$ und $A_{j_2} \cap S_1 \supset A_i \cap S_1$. Dann gilt aber offenbar (wegen $A_{j_1} \cap S_1 = X$ und $A_{j_2} \cap S_2 = Y$) $A_{j_1} \cap A_{j_2} = A_i$, was der Voraussetzung widerspricht.

Wegen (29) können wir ohne Beschränkung der Allgemeinheit voraussetzen, daß

$$\left|\bigcup_X F^*_{X,1}\right| \geqq \frac{k}{2} > \frac{c_1}{2} 2^n/\sqrt{n} \tag{30}$$

ist. Wegen (28) folgt aus dem Satze von Sperner [7] für jedes $X \subset S_1$

$$|F^*_{X,1}| < \binom{\left[\frac{n+1}{2}\right]}{\left[\frac{n+1}{4}\right]} < c_3\, 2^{n/2}/\sqrt{n}. \tag{31}$$

Also folgt aus (30) und (31) durch eine leichte Rechnung, daß für mindestens $c_4 2^{n/2}$ Mengen $X \subset S_1$ (da $2^{[n/2]}$ die Anzahl der $X \subset S_1$ ist)

$$|F^*_{X,1}| > c_5\, 2^{n/2}/\sqrt{n} \tag{32}$$

ist. Es seien

$$A_{i_1}, \ldots, A_{i_r}, \quad r > c_5\, 2^{n/2}/\sqrt{n}$$

die Mengen von $F^*_{X,1}$. Keine zwei der Mengen $A_{i_j} \cap S_2 = B_{i_j}, j = 1, \ldots, r$ enthalten einander. Daher ist die Anzahl der Teilmengen von S_2, die mindestens ein B_{i_j} enthalten, nach Lemma 3 von [2] größer als $c_6 2^{n/2}$, und da dies wegen (32) für mindestens $c_4 2^{n/2}$ Mengen $X \subset S_1$ geschieht, erhalten wir durch eine leichte Rechnung, daß die Anzahl der Teilmengen von S, die mindestens ein A_i, $i = 1, \ldots, k$ enthalten, größer als $c_4 c_6 2^n = c_2 2^n$ ist; dies beweist Lemma 4.

Mit Hilfe von Lemma 4, Lemma 3 von [2] und der Methode von Kleitman [6] können wir folgenden Satz beweisen: Es sei $|S| = n$, $A_i \subset S$, $i = 1, \ldots, k$, $k > c_7 2^n/\sqrt{n}$, ein System von Mengen, für welche (27) unmöglich ist. Es sei $A_{i_1}, \ldots, A_{i_r}$ die Menge derjenigen A, die kein anderes A_i, $i = 1, \ldots, k$ als Teilmenge enthalten. Dann gilt $r > c_8 k$, $c_8 = c_8(c_7)$.

Auf den Beweis dieses Satzes wollen wir hier nicht eingehen.

Lemma 5. *Es sei $l_1 < \cdots < l_k \leqq x$ eine Folge ganzer Zahlen, für welche $(l_i, l_j) = l_r$ in verschiedenen Zahlen l unlösbar ist und für welche*

$$\sum_{i=1}^{k} \frac{1}{l_i} > c \log x/(\log\log x)^{1/2}$$

gilt. Es seien $u_1 < \cdots < u_s \leqq x$ die Zahlen von der Form

$$l_i t, \quad t < \frac{x}{l_i}, \quad i = 1, \ldots, k.$$

Dann gilt

$$\sum_{i=1}^{s} \frac{1}{u_i} > c' \log x,$$

wobei c' nur von c abhängt.

Lemma 5 folgt aus Lemma 4 (wenn wir von unwesentlichen Abänderungen absehen), wie Lemma 1 von [4] aus Lemma 3 von [4] folgte.

Die Ungleichung (22) folgt nun sofort aus Lemma 5. Die l seien die Zahlen der Form $\{b_k q_j\}$ von (22). c sei gleich $\varepsilon/4$, und wenn η genügend klein ist, widerspricht (26) dem Lemma 5. Dieser Widerspruch beweist (22). Die Ungleichung (16) folgt nun aus (21) und (22), und damit ist unser Satz bewiesen.

Literatur

[1] F. Behrend, On sequences of numbers not divisible one by another, J. London Math. Soc. **10** (1935), 42—44; P. Erdös, Note on sequences of integers no one of which is divisible by any other, ibid., 126—128.

[2] P. Erdös A. Sárközi and E. Szemerédi, On a theorem of Behrend, J. Australian Math. Soc. **7** (1967), 9—16.

[3] P. Erdös, A. Sárközi and E. Szemerédi, On the solvability of the equations $[a_i, a_j] = a_r$ and $(a_i', a_j') = a_r'$ in sequences of positive density, J. Math. Analysis and Applications **15** (1966), 60—64.

[4] P. Erdös, A. Sárközi and E. Szemerédi, On the solvability of certain equations in dense sequences of integers, Dokl. Akad. Nauk SSSR **176** (1967), 541—544.

[5] P. Erdös, A generalisation of a theorem of Besicovitch, J. London Math. Soc. **11** (1936), 92—98.

[6] D. Kleitman, On a combinatorial problem of Erdös, Proc. Amer. Math. Soc. **17** (1966), 139—141.

[7] E. Sperner, Ein Satz über Untermengen einer endlichen Menge, Math. Z. **27** (1928), 544—548.

H. Heilbronn in Toronto

ON THE AVERAGE LENGTH OF A CLASS OF FINITE CONTINUED FRACTIONS

1. Introduction

Many years ago Dr. J. Gillis asked me the following question: Let N and a be coprime natural integers, $1 \leqq a < N$, so that a/N can be represented by a finite continued fraction

$$a/N = 1/c_1 + 1/c_2 + \cdots + 1/c_{n(a)},$$

where the c_i are natural integers depending on N and a, and where $c_{n(a)} > 1$ (to make the representation unique). What can be said about the sum

$$L(N) = \sum_{\substack{a=1 \\ (a,N)=1}}^{N} n(a)?$$

At the time I was able to make only the trivial statement that

$$L(N) = O(N \log N).$$

Recently I discovered a connection between $L(N)$ and the number $r(N)$ of representation of N by the bilinear form $N = xx' + yy'$, where the natural integers x, x', y, y' are subject to the restrictions $x > y$, $x' > y'$, $(x, y) = 1$, $(x', y') = 1$.

Theorem 1.

$$L(N) = \frac{3}{2}\varphi(N) + 2r(N) \quad \textit{for} \quad N > 2.$$

This raises the question, what can be said about the behaviour of $r(N)$. The answer is given by

Theorem 2.

$$r(N) = 6\pi^{-2} \log 2 \; \varphi(N) \log N + O(N\sigma_{-1}^3(N)),$$

where $\sigma_{-1}(N)$ denotes the sum of the reciprocals of the positive divisors of N.

It is clear that the main term dominates the error term by a factor at least of the order $\log N(\log\log N)^{-4}$. It appears very difficult to obtain a substantially better error term, though numerical evidence suggests that the error term is much too large.

Combining the two theorems leads to

Theorem 3.

$$L(N) = 12\pi^{-2}\log 2\varphi(N)\log N + O(N\sigma_{-1}^3(N)).$$

A slight extension of the method of the proof of theorem 2 leads to the following result. Let $L_c(N)$ be the number of times that the denominator c occurs in the continued fraction of the rationals a/N, $1 < a < N$, $(a, N) = 1$. Then we have

Theorem 4.

$$L_c(N) = 12\pi^{-2}\log(1-(c+1)^{-2})^{-1}\ \varphi(N)\log N + O(N\sigma_{-1}^3(N)).$$

This theorem suggests that in some sense the frequency of the denominator c equals

$$\log(1-(c+1)^{-2})^{-1}/\log 2,$$

if we consider the continued fraction of all real numbers. This is indeed the case, as Khintchine has shown [1].

2. Preliminaries

Small roman letters with or without indices are restricted to positive rational integers. The symbols $\varphi(n)$, $\mu(n)$, $\sigma_\tau(n)$, $d(n)$ have the meaning usual in elementary number theory, i.e. $\varphi(n)$ denotes the Euler function, $\mu(n)$ the Moebius function, $\sigma_\tau(n)$ the sum of the τth powers of the positive divisors of n, $d(n) = \sigma_0(n)$. The symbol O holds uniformly in all variables except possibly in $\varepsilon > 0$.

We shall make frequent use of the Moebius inversion formula, and such well known results as

$$d(n) = O(n^\varepsilon),$$

$$\sum_{n=1}^{z}\varphi(n)\,n^{-1} = 6\pi^{-2}z + O(\log z),$$

$$\sigma_{-1}(n) = O(\operatorname{loglog} N),$$

$$6\pi^{-2} < \varphi(n)\,\sigma_{-1}(n)\,n^{-1} \leqq 1.$$

We have already defined $r(N)$ as the number of solutions of $N = xx' + yy'$ subject to

$$x > y, \quad x' > y', \quad (x, y) = (x', y') = 1.$$

We further define $R(N)$ as the number of solutions subject to

$$x > y, \quad x' > y';$$

and for each $d \geqq 1$ we define $\varrho(N, d)$ as the number of solutions subject to

$$x > y, \quad x' > y', \quad (x, y) = 1, \quad x' > dx.$$

Then

$$R(N) = \sum_{bb'/N} r(N(bb')^{-1}),$$

and by a repeated application of the Moebius formula

$$r(N) = \sum_{bb'/N} \mu(b)\, \mu(b')\, R(N(bb')^{-1}).$$

Utilizing the symmetry in the definition of $R(N)$ between the primed and the 'unprimed' variables, we obtain

$$R(N) = 2 \sum_{d/N} \varrho(N d^{-1}, d) + O \sum_{x^2 < N} d(N - x^2),$$

$$r(N) = 2 \sum_{dbb'/N} \mu(b)\, \mu(b')\, \varrho(N(dbb')^{-1}, d) + O(N^{1/2+\varepsilon}). \tag{2}$$

3. Continued fractions

We define the function $[c_1, \ldots, c_n]$ in the usual way by

$$[\,] = 1, \quad [c_1] = c_1, \quad [c_1, c_2] = c_2 c_1 + 1,$$

$$[c_1, \ldots, c_n] = c_n[c_1, \ldots, c_{n-1}] + [c_1, \ldots, c_{n-2}] \quad \text{for} \quad n \geqq 3.$$

Our first object is to prove the formula

$$[c_1, \ldots, c_n] = [c_1, \ldots, c_m]\,[c_{m+1}, \ldots, c_n] + [c_1, \ldots, c_{m-1}]\,[c_{m+2}, \ldots, c_n]$$

for $1 \leqq m < n$. (I don't believe this formula is new, but I have not been able to find it in the literature.) The formula is obviously true for $1 \leqq m = n - 1$ and also, as a little calculation shows, for $1 \leqq m = n - 2$. Hence by induction with respect to m, it is true for $m \geqq 1$ and all $n > m$. Our formula makes it evident that

$$[c_1, \ldots, c_n] = [c_n, \ldots, c_1].$$

Now we introduce a pair N, a with $a < \frac{1}{2}N$, $(N, a) = 1$ and develop a/N as a continued fraction

$$a/N = 1/c_1 + 1/c_2 + \cdots + 1/c_n, \quad c_n \geqq 2, \tag{3}$$

and we have automatically

$$c_1 \geqq 2, \quad N = [c_1, \ldots, c_n].$$

If $a \neq 1$, then $n = n(a) > 1$ and we can choose m in the interval $1 \leqq m \leqq n - 1$ in $n - 1$ different ways. Put

$$x = [c_1, \ldots, c_m], \quad y = [c_1, \ldots, c_{m-1}], \tag{4}$$

$$x' = [c_n, \ldots, c_{m+1}], \quad y' = [c_n, \ldots, c_{m+2}]. \tag{5}$$

These integers, by virtue of our identity, satisfy the relation

$$N = xx' + yy',$$

and they also fulfil the conditions

$$x > y, \quad x' > y', \quad (x, y) = (x', y') = 1.$$

(If $m = 1$ or $m = n - 1$, remember $c_1 \geqq 2$ or $c_n \geqq 2$.)

Moreover we have for $m > 2$

$$\frac{y}{x} = \frac{[c_1, \ldots, c_{m-1}]}{[c_1, \ldots, c_m]} = \frac{[c_1, \ldots, c_{m-1}]}{c_m[c_1, \ldots, c_{m-1}] + [c_1, \ldots, c_{m-2}]}$$

$$= \left(c_m + \frac{[c_1, \ldots, c_{m-2}]}{[c_1, \ldots, c_{m-1}]}\right)^{-1}$$

and hence by induction

$$y/x = 1/c_m + \cdots + 1/c_1; \tag{6}$$

similarly

$$y'/x' = 1/c_{m+1} + \cdots + 1/c_n. \tag{7}$$

Conversely, given a representation of N by our bilinear form with our restrictions, we can find a unique sequence $c_1, \ldots, c_n$ not starting or finishing with 1 such that (6), (7), (4), (5) are satisfied and

$$N = [c_1, \ldots, c_n].$$

Putting $a = [c_2, \ldots, c_n]$, it is clear that (3) holds.

To sum up, we have a $(1-1)$ relation between all suitably restricted representations of N by the bilinear form, and all pairs of sequences

$$c_1, \ldots, c_m;\quad c_{m+1}, \ldots, c_n \quad \text{with} \quad c_1 \geqq 2, \quad c_n \geqq 2, \quad 1 \leqq m < n.$$

Thus, for $N > 2$

$$r(N) = \sum_{\substack{1<a<N/2 \\ (a,N)=1}} (n(a)-1) = \sum_{\substack{a<N/2 \\ (a,N)=1}} n(a) - \frac{1}{2}\varphi(N).$$

As for $0 < \alpha < \frac{1}{2}$

$$1-\alpha = \cfrac{1}{1+\cfrac{1}{\cfrac{1}{\alpha}-1}},$$

it follows that, for $a < \frac{1}{2}N$, $(a, N) = 1$, (3) implies

$$(N-a)\,N = 1/1 + 1/(c_1-1) + 1/c_2 + \cdots + 1/c_n.$$

Hence

$$\sum_{\substack{N/2<a<N \\ (a,N)=1}} n(a) = \frac{1}{2}\varphi(N) + \sum_{\substack{a<N/2 \\ (a,N)=1}} n(a),$$

and theorem 1 is proved.

4. Proof of theorem 2

We require the following

Lemma. $\varrho(N,d) = 3\pi^{-2}\log 2\; N\log(Nd^{-1}) + O(N)$ *for* $d \leqq N$.

Proof. Fix a pair x, y with

$$y < x < (Nd^{-1})^{1/2}, \quad (x,y) = 1. \tag{8}$$

We have to count the number of positive integers x' which satisfy

$$xx' \equiv N \,(\mathrm{mod}\, y), \quad x' > x, \quad x' > y' = y^{-1}(N - xx') > 0.$$

The last two inequalities can be written as

$$N(x+y)^{-1} < x' < Nx^{-1}.$$

This means we have to find the number, say $P(N, d, x, y)$ of solutions of the congruence in the interval

$$\operatorname{Max}(xd, N(x+y)^{-1}) < x' < Nx^{-1}.$$

Clearly

$$|P(N, d, x, y) - y^{-1}(Nx^{-1} - \operatorname{Max}(xd, N(x+y)^{-1}))| < 1$$

and

$$\varrho(N, d) = \sum_{x,y} P(N, d, x, y)$$

where the sum is extended over all x, y satisfying (8). We distinguish two cases.

Case 1. $xd \geqq N(x+y)^{-1}$.

Then

$$P(N, d, x, y) < 1 + y^{-1}(Nx^{-1} - xd)$$
$$\leqq 1 + y^{-1}(xd(x+y)x^{-1} - xd) = 1 + d,$$

Hence, in this case, summing over x, y satisfying (8)

$$\sum_{x,y} P(N, d, x, y) \leqq (1+d) \sum_{x<(Nd^{-1})^{1/2}} x = O(N).$$

Case 2. $xd < N(x+y)^{-1}$.

This implies $6d < N$. If $6d \geqq N$, the lemma is trivial. Keeping y fixed, x is now restricted by

$$(x, y) = 1, \quad x > y, \quad x(x+y)\,d < N. \tag{9}$$

As

$$P(N, d, x, y) = O(1) + y^{-1}(Nx^{-1} - N(x+y)^{-1}),$$

we obtain, summing over all relevant x,

$$\sum_x P(N, d, x, y) = O(N^{1/2}d^{-(1/2)}) + y^{-1}N \sum_{\substack{y<x<2y \\ (x,y)=1}} x^{-1} - y^{-1}N \sum_{\substack{x_0 \leqq x < x_0+y \\ (x,y)=1}} x^{-1}, \tag{10}$$

where x_0 is the smallest integer for which $x_0(x_0+y)\,d \geqq N$. As $x_0 > \left(\frac{1}{2} Nd^{-1}\right)^{1/2}$, the last sum including the factor $y^{-1}N$ is at most $O(N^{1/2}\,d^{1/2})$. Further

$$\sum_{\substack{y<x<2y \\ (x,y)=1}} x^{-1} = \sum_{b/y} \mu(b)\,b^{-1} \sum_{r=yb^{-1}+1}^{2yb^{-1}-1} r^{-1} = \varphi(y)\,y^{-1} \log 2 + O(d(y)\,y^{-1}).$$

Hence

$$\sum_x P(N, d, x, y) = N\varphi(y)\,y^{-2} \log 2 + O(N^{1/2}d^{1/2}) + O(Nd(y)\,y^{-2}).$$

This expression has to be summed over all $y < \left(\frac{1}{2} Nd^{-1}\right)^{1/2}$. Thus

$$\sum_{x,y} P(N, d, x, y) = O(N) + N \log 2 \sum_{y < (Nd^{-1}/2)^{1/2}} \varphi(y)\, y^{-2},$$

and a simple summation by parts gives the lemma.

It is now easy to obtain theorem 2. As

$$\begin{aligned} r(N) &= 2 \sum_{dbb'/N} \mu(b)\, \mu(b')\, \varrho(N(dbb')^{-1}\, d) + O(N^{1/2+\varepsilon}) \\ &= 6\pi^{-2} \log 2\, N \sum_{dbb'/N} \mu(b)\, \mu(b')\, (dbb')^{-1} (\log N - \log(d^2bb')) + O(1). \end{aligned}$$

From this theorem 2 follows as the error term has the required form, and as

$$\begin{aligned} \sum_{dbb'/N} \mu(b)\, \mu(b')\, (dbb')^{-1} &= \sum_{b/N} \mu(b)\, b^{-1} \sum_{n/Nb^{-1}} n^{-1} \sum_{b'/n} \mu(b') \\ &= \sum_{b/N} \mu(b)\, b^{-1} = \varphi(N)\, N^{-1}, \end{aligned}$$

whilst

$$\begin{aligned} &\sum_{dbb'/N} (dbb')^{-1}\, \mu(b)\, \mu(b') \log(d^2bb') \\ &= 2 \sum_{dbb'/N} (dbb')^{-1}\, \mu(b)\, \mu(b') \log(db') \\ &= 2 \sum_{b/N} \mu(b)\, b^{-1} \sum_{n/Nb^{-1}} n^{-1} \log n \sum_{b/N} \varphi(b') = 0. \end{aligned}$$

5. Proof of theorem 4

Assume that a fixed number c is given. We denote by $r_c(N)$, $R_c(N)$, $\varrho_c(N, d)$ the number of representations of N by the bilinear form subject to the restriction mentioned before and the additional restriction

$$cy \leqq x < (c + 1)\, y. \tag{11}$$

Reverting to the pattern of the proof of theorem 1, we see that $r_c(N)$ equals the number of times that c is a denominator in the continued fraction of the rationals a/N, $1 < a < \frac{1}{2}N$, $(N, a) = 1$, not counting any possible $c_{n(a)}$, as $m < n$. These exceptional values are at most $\varphi(N)$. If we include the rationals a/N, $\frac{1}{2}N < a < N$, $(N, a) = 1$, every c_m will occur as often as before, except for changes in the first or second place. Hence we have

$$L_c(N) = 2r_c(N) + O(\varphi(N)).$$

Formula (1) works as before if we replace R and r by R_c and r_c respectively.

Further our argument continues to be true if we replace the new restriction (11) by

$$cy' \leqq x' < (c+1)\,y',$$

because with the sequence $c_1, \ldots, c_n$ the sequence $c_n, \ldots, c_1$ also occurs. Hence $R_c(N)$ equals twice the number of solutions subject to the restrictions

$$cy \leqq x < (c+1)\,y, \; x' > y', \quad x' > x,$$

with an error $O(N^{1/2+\varepsilon})$. This implies that (2) remains true if we replace r and ϱ by r_c and ϱ_c.

The proof of the lemma goes as before but (9) has to be replaced by (11) and $(x, y) = 1$, $x(x+y)\,d < N$. Then (10) has to be replaced by

$$\sum_x P(N, d, x, y) = O(N^{1/2}d^{-(1/2)}) + y^{-1}N \sum_{\substack{cy<x<(c+1)y \\ (x,y)+1}} (x^{-1} - (x+y)^{-1})$$

$$= O(N^{1/2}d^{-(1/2)}) + y^{-1}N \sum_{b/y} \mu(b)\,b^{-1}$$

$$\times \sum_{r=cyb^{-1}+1}^{(c+1)yb^{-1}-1} (r^{-1} - (r+yb^{-1})^{-1})$$

$$= O(N^{1/2}d^{-(1/2)}) + \varphi(y)\,y^{-2}N\,(\log(1+c^{-1}))$$

$$- \log(1+(c+1)^{-1})) + O(Nd(y)\,y^{-2})$$

$$= O(N^{1/2}d^{-(1/2)}) + O(Nd(y)y^{-2}) + N\varphi(y)y^{-2}\log(1-(c+1)^{-2})^{-1}.$$

The rest of the calculation goes as before, with $\log(1-(c+1)^{-2})^{-1}$ in place of $\log 2$.

Reference

[1] A. Khintchine, Metrische Kettenbruchprobleme, Compositio Mathematica **1** (1935), 361—382.

E. Hlawka in Wien

INTERPOLATION ANALYTISCHER FUNKTIONEN AUF DEM EINHEITSKREIS

Es sei S die abgeschlossene Kreisscheibe vom Radius 1 in der komplexen z-Ebene. Es sei weiter f analytisch auf S, d. h., sie ist analytisch auf einer größeren Kreisscheibe $|z| \leqq R\,(R > 1)$. Auf dem Einheitskreis $|z| = 1$ sei nun eine unendliche Folge $\zeta_1, \zeta_2, \ldots$ von Punkten gegeben. Nun betrachte man die ersten N Glieder dieser Folge, und es sei $L_N = L_N(f, z)$ das Polynom in z vom Grad $N - 1$, welches durch Interpolation in den Werten von f in den Punkten $\zeta_1, \ldots, \zeta_N$ gefunden wird. Sind $\zeta_1, \ldots, \zeta_N$ alle voneinander verschieden, so ist bekanntlich

$$L_N(f, z) = \sum_{h=1}^{N} f(\zeta_h) \frac{(z - \zeta_1) \cdots (z - \zeta_{h-1})(z - \zeta_{h+1}) \cdots (z - \zeta_N)}{(\zeta_h - \zeta_1) \cdots (\zeta_h - \zeta_{h-1})(\zeta_h - \zeta_{h+1}) \cdots (\zeta_h - \zeta_N)} \tag{1}$$

(mit der üblichen Konvention, daß für $h = 1$ das Glied ζ_{h-1} und für $h = N$ jenes mit ζ_{h+1} wegzulassen ist). Setzen wir $\varrho_N(f, (\zeta_i)) = \sup_{|z|=1} |f(z) - L_N(f)|$, so erhebt sich die Frage, wann $\lim_{N\to\infty} \varrho_N = 0$ gilt.

Setzen wir $\zeta_j = e^{2\pi i \varphi_j}$, so liegen die φ_j im Einheitsintervall $I: 0 \leqq \varphi < 1$.

Dann gilt folgender Satz, der von L. Fejer [1] 1918 (ergänzt durch L. Kalmár 1926) herrührt (aufbauend auf C. Runge 1908, Theorie und Praxis der Reihen):

Es ist $\lim \varrho_N = 0$ *dann und nur dann für alle* f *(analytisch in* S*), wenn die Folge der* φ_j *gleichverteilt* mod 1 *ist. Ist* f *analytisch für* $|z| \leqq R\,(R > 1)$, *dann gilt in diesem Falle*

$$\overline{\lim_{N\to\infty}}\, \varrho_N^{1/N} \leqq \frac{1}{R}. \tag{2}$$

Dabei heißt eine Folge φ_j von reellen Zahlen φ nach H. Weyl (1916) gleichverteilt, wenn folgendes gilt:

Es sei $N'(t)$ die Anzahl der $\varphi_j (j = 1, \ldots, N)$ im Intervall $I_t : \langle 0, t \rangle$ mit $0 \leqq t \leqq 1$. Gilt dann für jedes t

$$\lim_{N\to\infty} \frac{N'(t)}{N} = t,$$

dann ist die Folge gleichverteilt.

Wir können dies auch gleich so formulieren: Ist $\chi_t(\varphi)$ die charakteristische Funktion von I_t, so ist $N'(t) = \sum_{h=1}^{N} \chi_t(\varphi_h)$, und es sei $\Delta_N(t) = \frac{1}{N} \sum_{h=1}^{N} \chi_t(\varphi_h) - t$; dann besagt eben die Definition der Gleichverteilung, daß $\lim_{N \to \infty} |\Delta_N(t)| = 0$ für alle t gilt.

Wir wollen nun die Abschätzung (2) vertiefen und die feineren Eigenschaften der Folge (φ_j) ins Spiel bringen. Wir führen dazu den Begriff der Diskrepanz D_N nach van der Corput ein: Es ist $D_N = \sup |\Delta_N(t)|$. Schon von H. Weyl wurde gezeigt, daß die Bedingung $\lim_{N \to \infty} D_N = 0$ notwendig und hinreichend dafür ist, daß eine Folge gleichverteilt ist. Man kennt für viele Folgen Abschätzungen für D_N und auch die Größenordnung, z. B. für die Folge $(h-1)\alpha$ ($h = 1, 2, \ldots$; α irrational). Wir werden z. B. zeigen:[1]) Es ist für beliebige $\varphi_1, \ldots, \varphi_N$

$$\varrho_N^{1/N}(f) \leqq \frac{1}{R} C_1^{1/N} \cdot C_2^{2D_N}(R); \tag{3}$$

dabei sind C_1, C_2 Größen, die von f und R abhängen. Für die genauere Formulierung sei auf (22) verwiesen. Andererseits gibt es eine analytische Funktion f in $|z| \leqq R$, so daß

$$\varrho_N^{1/N}(f) \geqq \frac{1}{R} C_1^{1/N} C_2^{D_N^2}(R) \tag{4}$$

ist [siehe dazu (23)].

Wir werden uns bei der Ableitung dieser Abschätzungen auf die bekannte Gleichung

$$f(z) - L_N(f, z) = R_N(f, z),$$

$$R_N(f, z) = \frac{1}{2\pi i} \int_{|\varrho| = R} \frac{f(\zeta)}{\zeta - z} \frac{\omega_N(z)}{\omega_N(\zeta)} \, d\zeta \tag{5}$$

stützen, wobei $\omega_N(z) = (z - \zeta_1) \cdots (z - \zeta_N)$ ist.

Wir wollen an den einfachen Beweis erinnern:

Es ist $\frac{1}{\zeta - z} = \frac{1}{\zeta - \zeta_j} + \frac{z - \zeta_j}{\zeta - \zeta_j} \frac{1}{\zeta - z}$, und beginnend bei $j = 1$ folgt

$$\frac{1}{\zeta - z} = \sum_{j=0}^{N-1} \frac{\omega_i(z)}{\omega_j(\zeta)} \frac{1}{\zeta - \zeta_{j+1}} + \frac{\omega_N(z)}{\omega_N(\zeta)} \frac{1}{\zeta - z}.$$

---

[1]) Dieser Satz wurde ohne Beweis vom Verfasser angegeben in Nuffic International Summer Sessions in Science 1962 Breukelen, S. 104.

Multiplikation mit $f(\zeta)$ und Integration längs $|\zeta| = R$ liefert die Behauptung. Sie liefert gleichzeitig die Darstellung von L_N, wenn nicht alle $\zeta_1, \ldots, \zeta_N$ verschieden sind.

Wie das Restglied zeigt, läuft alles auf die Abschätzung von $|\omega_N(z)|$ nach oben und unten hinaus.

Wir wollen noch ein weiteres Interpolationsproblem besprechen: Es sei jetzt $f(z)$ analytisch in einem Kreisring $r \leqq |z| \leqq R$ mit $1 > r > 0$, $R > 1$, und es liegen wieder die Stellen $\zeta_1, \zeta_2, \ldots$ vor und es mögen die Interpolationspolynome L_N jetzt für ungerades $N = 2M + 1$ für f betrachtet werden. Nehmen wir an, es liege in der Form (1) vor, dann ist für

$$z = e^{2\pi i\vartheta},\quad z - \zeta_h = e^{2\pi i\vartheta} - e^{2\pi i\varphi_h} = 2ie^{i\pi(\vartheta+\varphi_h)} \sin\pi(\vartheta - \varphi_h),$$

also ist

$$(z-\zeta_1)\cdots(z-\zeta_{h-1})(z-\zeta_{h+1})\cdots(z-\zeta_{2M+1}) = z^M \prod_{j\neq h} 2i\sin\pi(\vartheta-\varphi_j)\, e^{i\pi\varphi_j},$$

also

$$L_{2M+1}(f) = \sum_{h=1}^{2M+1} f(\zeta_h)\frac{z^M}{\zeta_h^M}\prod_{j\neq h}\frac{\sin\pi(\vartheta-\varphi_j)}{\sin\pi(\varphi_h-\varphi_j)}.$$

Ersetzen wir f durch $z^M f(z)$, dann ist

$$\overline{L_M}(f) = \frac{1}{z^M} L_{M+1}(z^M f) = \sum_{h=1}^{2M+1} f(\zeta_h)\prod_{j\neq h}\frac{\sin\pi(\vartheta-\varphi_j)}{\sin\pi(\varphi_h-\varphi_j)}.$$

Ist die Funktion $f(z)$ auf dem Einheitskreis reell, so steht links die bekannte trigonometrische Interpolationsformel. In den praktischen Rechnungen wählt man $\varphi_j = \frac{1}{n}$ $(j = 1, \ldots, N)$, und dieser Fall ist auch theoretisch ausführlich untersucht worden. Es haben aber schon Euler und insbesondere U. J. Leverrier 1843 vorgeschlagen, auch hier an den Stellen $\varphi_j = (j-1)\,\alpha$ (α irrational) zu approximieren.

Dieses Verfahren ist dann in numerischer Hinsicht von G. J. Hoüel 1865 und J. F. Encke 1860 weiter ausgestaltet worden. (Vgl. dazu H. Burkhardt, Jahresberichte X 2, 1 (1908) S. 237ff.). Es ist allerdings bemerkenswert, daß das verwendete α tatsächlich nicht irrational ist. Es nimmt z. B. Hoüel $4\alpha = \dfrac{8461}{180 \cdot 60}$. Aber er sagt, da $8461 = 180 \cdot 47 + 1$ eine Primzahl ist, verhält sich α so, als wäre es irrational. Tatsächlich ist die Kettenbruchentwicklung von $4\alpha = [1, 3, 1, 1, 1, 1, 1, 1, 2, 2, 1, 1, 1, 1, 4]$ also sehr lang. Von A. Wintner 1932 wurde die Frage nach dem Fehler $\bar{\varrho}_M(f) = \sup_{|z|=1} |\overline{L_M}(f) - f|$ ($\overline{L_M}$ ist kein Polynom in z) für den Fall $\varphi_j = (j-1)\alpha$ (α irrational) aufgeworfen und von G. Pólya 1934 beantwortet. Allgemein gilt wieder: Ist die Folge (φ_j) gleichverteilt, dann gilt

$$\lim_{M\to\infty} \bar{\varrho}_M^{1/N}(f) \leqq \operatorname{Max}\left(\frac{1}{R}, r\right), \tag{6}$$

also tatsächlich $\lim \bar{\varrho}_M = 0$, und diese Bedingung ist auch notwendig. Auch hier werden wir Restabschätzungen für jedes N geben, und es gilt auch in diesem Fall

$$\bar{\varrho}_M \leqq C_1\left(R^{-(M+1)}C_2^{2D_N}(R) + r^M C_3^{2D_N}(r)\right). \tag{7}$$

Die Abschätzung stützt sich auf die Gleichung

$$f(z) - \overline{L_M}(f) = \overline{R_N},$$

$$\overline{R_N} = \frac{1}{2\pi i}\int_{|\zeta|=R} \frac{z^M f(\zeta)}{\zeta^M(\zeta - z)}\frac{\omega_N(z)}{\omega_N(\zeta)}\,d\zeta - \frac{1}{2\pi i}\int_{|\zeta|=r} \frac{z^M f(\zeta)}{z^M(\zeta - z)}\frac{\omega_N(z)}{\omega_N(\zeta)}\,d\zeta. \tag{6'}$$

Der Gedankengang der vorliegenden Arbeit beruht darauf, zunächst $\omega_N(z)$ für $|z| \neq 1$ nach oben und unten abzuschätzen nach einer Methode, welche in der Theorie der Gleichverteilung oft verwendet wird und die auf H. Behnke und J. F. Koksma zurück geht. Dann benutzen wir diese Abschätzungen, um $\omega_N(z)$ für $|z| = 1$ abzuschätzen.

Dabei ergibt sich Gelegenheit, einen Satz von Erdös-Turán auf einfache Art herzuleiten. Es sei bemerkt, daß es eine wichtige Aufgabe ist (auf die Erdös auch hingewiesen hat), möglichst scharfe Abschätzungen für $\omega_N(z)$ zu geben. Hier liegen noch viele Probleme offen. Dann leiten wir die zitierten Sätze her und besprechen einige Beispiele. Der weitere Teil der Arbeit beschäftigt sich mit folgender Fragestellung: Wenn eine Folge (φ_i) gleichverteilt ist, dann ist sie auch überall dicht im Einheitsintervall bzw. die Interpolationsstellen ζ_j sind dicht auf dem Einheitskreis. Die Umkehrung gilt jedoch nicht. Wir zeigen aber, daß die Folgen dann so umgeordnet werden können, daß für große N die Diskrepanz $D_N = O\left(\frac{\log N}{N}\right)$ ist. Andererseits läßt sie sich auch so umordnen, daß $D_N \geqq \frac{1}{2}$ wird. Die Güte der Interpolation hängt also von der Reihenfolge der Interpolationsschritte ab, was bei endlich vielen Stellen natürlich nicht der Fall ist. Dies dürfte auch bei numerischen Anwendungen von Bedeutung sein. In diesem Zusammenhang geben wir noch eine Methode an, um D_N bei gegebenen Stellen $\varphi_1, \ldots, \varphi_N$ in endlich vielen Schritten beliebig genau berechnen zu können.

§ 1

Wir schicken zunächst einen Hilfssatz vorweg:

Hilfssatz 1. *Es sei $f(\varphi)$ periodisch mit der Periode* 1 *und stetig differenzierbar; dann ist für jede endliche Folge von reellen Zahlen $\varphi_1, \ldots, \varphi_N$*

$$\left|\frac{1}{N}\sum_{h=1}^{N} f(\varphi_h) - \int_0^1 f(\varphi)\,d\varphi\right| \leqq D_N \int_0^1 |f'(\varphi)|\,d\varphi \tag{8}$$

mit

$$D_N = \sup_t \left|\frac{1}{N}\sum_{h=1}^{N} \chi_t(\varphi_h) - t\right|.$$

Beweis. Es ist

$$f(\varphi) = f(1) - \int_\varphi^1 f'(t)\,dt.$$

Nun ist

$$\int_\varphi^1 f'(t)\,dt = \int_0^1 f'(t)\,\chi_t(\varphi)\,dt,$$

also

$$f(\varphi) = f(1) - \int_0^1 f'(t)\,\chi_t(\varphi)\,dt,$$

d. h.

$$\frac{1}{N}\sum_{h=1}^{N} f(\varphi_h) - \int_0^1 f(\varphi)\,d\varphi = -\int_0^1 f'(t)\left(\frac{1}{N}\sum_{h=1}^{N}\chi_t(\varphi_h) - t\right)dt,$$

und daraus folgt bereits die Behauptung.

Wir setzen nun im weiteren stets $\zeta_h = e^{2\pi i\varphi}$ $(h = 1, \ldots, N)$ und für jedes komplexe z $\omega_N(z) = \prod_{h=1}^{N} (z - {}_h)$; dann gilt

Hilfssatz 2. *Für jedes $|z| > 1$ ist*

$$|z|\left(\frac{|z|-1}{|z|+1}\right)^{2D_N} \leqq |\omega_N(z)|^{1/N} \leqq |z|\left(\frac{|z|+1}{|z|-1}\right)^{2D_N}. \tag{9}$$

Beweis. Wir setzen $\zeta = \dfrac{1}{z} = r\,e^{2\pi i\vartheta}$ $(r < 1)$ – es ist dann $|\zeta| < 1$ – und nehmen in Hilfssatz 1

$$f(\varphi) = \ln|1 - \zeta\,e^{2\pi i\varphi}| = \frac{1}{2}\ln\left(1 + r^2 - 2r\cos 2\pi(\varphi + \vartheta)\right).$$

Dann ist

$$f'(\varphi) = \frac{2r\pi\sin 2\pi(\varphi + \vartheta)}{1 + r^2 - 2r\cos 2\pi(\varphi + \vartheta)}.$$

Nun ist $f(\varphi)$ eine periodische Funktion, wir können daher bei der Abschätzung von $\int_0^1 f'(\varphi)\,d\varphi$ annehmen, daß $\vartheta = 0$ ist. Dann ist $f'(\varphi) \geqq 0$ im φ-Intervall $\left[0, \frac{1}{2}\right]$, $f(\varphi) \leqq 0$ im Intervall $\left[\frac{1}{2}, 1\right]$. Es ist also

$$\int_0^1 f'(\varphi)\,d\varphi = 2\int_0^{1/2} f'(\varphi)\,d\varphi = 2\left(f\left(\frac{1}{2}\right) - f(0)\right) = 2\ln\frac{1+r}{1-r}$$

und daher wegen $\int_0^1 f(\varphi)\,d\varphi = 0$

$$\left|\frac{1}{N}\sum_{h=1}^{N} \ln|1 - \zeta\zeta_h|\right| \leqq 2D\ln\frac{1+r}{1-r}$$

oder

$$-2D\ln\frac{1+r}{1-r} \leqq \frac{1}{N}\sum_{h=1}^{N} \ln|(1-\zeta\zeta_h)| \leqq 2D\ln\frac{1+r}{1-r},$$

also

$$\left(\frac{1-r}{1+r}\right)^{2D} \leqq \left|\prod_h (1-\zeta\zeta_h)\right|^{1/N} \leqq \left(\frac{1+r}{1-r}\right)^{2D}, \tag{10}$$

und daraus folgt die Behauptung.

Bemerkung. Wegen $|\zeta_h| = 1$ ist trivialerweise

$$|z| - 1 \leqq |\omega_N(z)|^{1/N} \leqq 1 + |z|.$$

Die Abschätzungen von Hilfssatz 2 sind also nur für kleine D_N interessant.

Hilfssatz 3. *Für $|z| < 1$ ist*

$$\left(\frac{1-|z|}{1+|z|}\right)^{D} \leqq |\omega_N(z)|^{1/N} \leqq \left(\frac{1+|z|}{1-|z|}\right)^{2D}. \tag{11}$$

Beweis. Die Folge $\varphi_1' = 1 - \varphi_1, \ldots, \varphi_N' = 1 - \varphi_N$ besitzt ebenfalls die Diskrepanz D_N. Setzen wir $\zeta_h' = e^{2\pi i(1-\varphi_h)} = \zeta_h^{-1}$ und wenden (10) mit $\zeta = z$ und den ζ_h' an, so folgt die Behauptung.

Für $|z| = 1$ sind die Abschätzungen von $|\omega_N(z)|$ wertlos. Nun ist $\omega_N(z)$ eine analytische Funktion, also ist $1 \leqq \operatorname{Max}_{|z|=1} |\omega_N(z)| \leqq \operatorname{Max}_{|z|=R} |\omega_N(z)|$, wenn $R > 1$ ist. Wählen wir $R_1 = 2D + \sqrt{4D^2 + 1}$ und wenden (9) an, so erhalten wir

$$|\omega_N(z)|^{1/N} \leqq R_1\left(\frac{R_1+1}{R_1-1}\right)^{2D} = f(D) = R_1\left(\frac{1+\sqrt{4D^2+1}}{2D}\right)^{2D} = \beta(D). \tag{12}$$

Wir können aber eine Verschärfung angeben:

Hilfssatz 4. *Für* $|z| = 1$ *ist*

$$|\omega_N(z)|^{1/N} \leqq \gamma(D), \tag{13}$$

$$\gamma(D) = \sqrt{R_2}\left(\frac{1 + \sqrt{16D^2 + 1}}{4D}\right)^{2D}, \quad R_2 = 4D + \sqrt{16D^2 + 1}. \tag{14}$$

Bemerkung. Es ist tatsächlich $\gamma < \beta$, denn es ist

$$\frac{1}{4} + \sqrt{D^2 + \frac{1}{16}} < \frac{1}{2} + \sqrt{D^2 + \frac{1}{4}}$$

und

$$R_2 = 4D + \sqrt{16D^2 + 1} \leqq 4D\sqrt{4D^2 + 1} + 8D^2 + 1 = R_1^2.$$

Um Hilfssatz 4 zu beweisen, genügt es zu zeigen: Für $|z| = 1$ ist

$$|\omega_N(z)|^{1/N} \leqq \sqrt{R}\left(\frac{R + 1}{R - 1}\right)^{2D}, \quad R > 1. \tag{15}$$

Wir geben für (15) zwei Beweise. Zunächst einen algebraischen Beweis, dessen Idee von I. Schur stammt: Ist $\zeta = Re^{i\vartheta}$ $(R > 1)$, dann ist für jedes $\zeta_h = e^{2\pi i\varphi_h}$

$$|\zeta - \zeta_h|^2 = R^2 + 1 - 2R\cos 2\pi(\vartheta - \varphi_h) = R\left(R + \frac{1}{R} - 2\cos 2\pi(\vartheta - \varphi_h)\right)$$

$$\geqq R(2 - 2\cos 2\pi(\vartheta - \varphi_h)) = R|e^{2\pi i\vartheta} - e^{2\pi i\varphi_h}|^2,$$

also

$$\prod_h |\zeta - \zeta_h| \geqq R^{N/2} \prod_{h=1}^{N} |e^{2\pi i\vartheta} - \zeta_h|, \tag{16}$$

und da ϑ beliebig ist, gilt für $z = e^{2\pi i\vartheta}$

$$|\omega_N(z)|^{1/N} \leqq R^{-(1/2)}|\omega_N(Rz)|^{1/N}.$$

Wendet man (9) auf $\omega_N(Rz)$ an, so folgt (15).

Der zweite Beweis benutzt den Hadamardschen Dreikreisesatz, angewendet auf die Kreise $|z| = \varrho$ mit $\varrho = \frac{1}{R}$, 1, R. Setzen wir $\max\limits_{|z|=1} |\omega_N(z)|^{1/N} = M$, dann erhalten wir

$$M^{2\log R} \leqq \left(\frac{R + 1}{R - 1}\right)^{2D\log R}\left(R\left(\frac{R + 1}{R - 1}\right)^{2D}\right)^{\log R},$$

also

$$M^2 \leqq R\left(\frac{R + 1}{R - 1}\right)^{4D}.$$

Wir wollen nun eine einfache Abschätzung für γ geben: Es ist

$$1 + \sqrt{1 + 16D^2} \leqq 2(1 + 2D) \leqq 2e^{2D},$$

also

$$\frac{\gamma}{\sqrt{R_2}} < \left(\frac{1 + 2D}{2D}\right)^{2D} < \frac{e^{4D^2}}{D^D} < e^{4D(\ln(1/D)+1)},$$

da $D \leqq 1$ ist, und somit

$$\gamma < (1 + 4D)\, e^{4D(\ln(1/D)+1)};$$

und wegen $1 + 4D < e^{4D}$ ist dann

$$|\omega_N(z)|^{1/N} < e^{8D(\ln(1/D)+1)} \quad \text{für alle } |z| = 1. \tag{17}$$

Andererseits gilt stets $\operatorname{Max} |\omega_N(z)| \geqq 1$, d. h., es existieren z, für die der Ausdruck links in (17) sicher $\geqq 1$ ist.

Es gilt aber nach einem Satz von Erdös–Turán schärfer für $M = \operatorname*{Max}_{|z|=1} |\omega_N(z)|^{1/N}$: Es existiert eine absolute Konstante $\alpha > 0$, so daß

$$M \geqq e^{\alpha D^2} \tag{18}$$

ist.

Wir wollen für diesen Satz einen einfachen Beweis geben und setzen $\prod_{h=1}^{N} (1 - z\zeta_h) = P(z)$. Dann ist, da $P(z)$ analytisch ist,

$$\operatorname*{Max}_{|z|=r} |P(z)| \leqq M^N \quad \text{für} \quad r < 1.$$

$\left(\text{Es ist } \operatorname*{Max}_{|z|=1} |P(z)| = \operatorname*{Max}_{|z|=1} |\omega_N(z)|\right)$, also ist für alle z mit $|z| = r < 1$

$$\frac{1}{N} \log |P(z)| = \frac{1}{N} \sum_{h=1}^{N} \log |1 - z\zeta_h| \leqq \log M.$$

Setzen wir $z = re^{i\vartheta}$ und entwickeln den Logarithmus in seine Potenzreihe, so ist also für alle ϑ

$$\sum_{h=1}^{\infty} \frac{r^h}{h} (R_h \cos 2\pi h\vartheta + I_h \sin 2\pi h\vartheta) \leqq \log M \tag{19}$$

mit

$$R_h = \operatorname{Re} S_h,\ I_h = \operatorname{Im} S_h,\ S_h = \frac{1}{N} \sum_{h=1}^{N} e^{2\pi i \varphi_h}.$$

Dies ist eine absolut und gleichmäßig konvergente Fourierreihe in ϑ. Multiplizieren wir (19) für ein natürliches l mit der nichtnegativen Funktion $1 + \cos 2\pi l\vartheta$ und integrieren nach ϑ von 0 bis 1, so bleibt in (19) nur das Glied mit $h = l$ stehen, und wir erhalten

$$\frac{1}{2}\frac{r^l}{l}R_l \leqq \log M.$$

(Wir könnten uns auch auf die Parsevalsche Gleichung berufen.) Nehmen wir $1 - \cos 2\pi l\vartheta$, so erhalten wir analog

$$-\frac{1}{2}\frac{r^l}{l}R_l \leqq \log M, \quad \text{d. h.} \quad r^l\frac{|R_l|}{l} \leqq 2\log M.$$

Diese Ungleichung gilt für jedes $r < 1$, also ist

$$|R_l| \leqq 2l\log M.$$

Nehmen wir $1 + \sin 2\pi l\vartheta$ bzw. $1 - \sin 2\pi l\vartheta$, so erhalten wir analog

$$|I_l| \leqq 2l\log M,$$

also

$$|S_l| \leqq 2\sqrt{2}\,l\log M.$$

Nun gilt nach Erdös und Turán (vgl. [4])

$$D_N \leqq 1000\left(\frac{1}{K} + \sum_{0<|l|\leqq K}\frac{|S_l|}{l}\right) \quad \text{für jedes } K > 1;$$

wir erhalten somit

$$D_N \leqq 10^3\left(\frac{1}{K} + 4\sqrt{2}\,K\log M\right).$$

Nun ist trivialerweise $M \leqq 2$. Nehmen wir also

$$K = \left[\sqrt{\frac{4}{\log M}}\right] > 2, \quad \text{so ist} \quad K > \sqrt{\frac{1}{\log M}}$$

und

$$D_N \leqq 10^3\left(\sqrt{\log M} + 8\sqrt{2}\sqrt{\log M}\right) = 10^3\left(1 + 4\sqrt{2}\right)\sqrt{\log M}$$
$$< 2\cdot 10^4\sqrt{\log M}.$$

Daraus folgt (18) mit einem $\alpha = \frac{1}{4}10^{-8}$.

Es folgt sofort

Hilfssatz 5. *Für jedes $R > 1$ existiert stets ein z mit $|z| = R$, so daß*

$$|\omega_N(z)|^{1/N} \geqq \sqrt{R}\,e^{\alpha D^2} \tag{20}$$

ist.

Beweis. Dies folgt aus (16) und (18).

Hilfssatz 6. *Für jedes $R > 1$ existiert stets ein z mit $|z| = R$, so daß*

$$|\omega_N(z)|^{1/N} < R \tag{21}$$

gilt.

Beweis. Es ist

$$\frac{1}{N}\log|\omega_N(z)| = \log R + \frac{1}{N}\sum_{h=1}^{N}\log\left|1 - \frac{1}{z}\zeta_h\right|.$$

Nun ist

$$\int_0^{2\pi}\log\left|1 - \frac{e^{-i\vartheta}}{R}\zeta_h\right| d\zeta = 0,$$

also ist für $z = Re^{i\vartheta}$

$$\int_0^{2\pi}\frac{1}{N}\lg|\omega_N(R\,e^{i\vartheta})|\,d\vartheta = \log R,$$

und daraus folgt die Behauptung.

Jetzt können wir leicht den folgenden Satz beweisen.

Satz 1. *Es sei $f(z)$ analytisch für $|z| \leqq R\,(R < 1)$, und $L_N(z,f)$ sei das zugehörige Interpolationspolynom in den Stellen $\zeta_h = e^{2\pi i\varphi_h}$ $(h = 1, \ldots, N)$; dann ist*

$$\sup_{|z|=1}|f(z) - L_N(z,f)| \leqq \frac{K}{R-1}\left(\gamma(D)\frac{1}{R}\left(\frac{R+1}{R-1}\right)^{2D}\right)^N \tag{22}$$

mit

$$\gamma_N < e^{8D(\ln(1/D)+1)}, \quad K = \frac{1}{2\pi}\int_0^{2\pi}|f(R\,e^{i\varphi})|\,d\varphi.$$

Für $D \to 0$ geht der Ausdruck in der Klammer (22) gegen $\frac{1}{R}$.

Beweis. Nach (5) ist

$$f(z) - L_N(z) = \frac{1}{2\pi i}\int_{|\zeta|=R}\frac{f(\zeta)}{\zeta - z}\,\frac{\omega_N(z)}{\omega_N(\zeta)}\,d\zeta.$$

Wendet man für $|z| = 1$ und $|\zeta| = R$ (13), (17) und die linke Abschätzung in (9) an, so folgt die Behauptung.

Wir zeigen nun

Satz 2. *Ist $R > 1$, dann existiert ein ζ mit $|\zeta| = R$, so daß für $f_0(z) = \frac{1}{z-\zeta}$*

$$\sup_{|z|=1}|f_0(z) - L_N(z,f_0)| \geqq \frac{1}{R+1}\left(\frac{e^{\alpha D^2}}{R}\right)^N \tag{23}$$

gilt.

Beweis. Es ist $f_0(z) - L_N(z, f_0) = \frac{1}{z-\zeta}\frac{\omega_N(z)}{\omega_N(\zeta)}$.

Nun gibt es nach (21) ein ζ mit $|\omega_N(\zeta)| < R^N$. Weiter ist $|z-\zeta| \leqq R+1$, also folgt aus (18) die Behauptung.

Zusatz. Für $|z| = \sigma \geqq 1$ gilt

$$\sup_{|z|=\sigma} |f_0(z) - L_N(z, f_0)| \geqq \frac{1}{R+\sigma}\left(\frac{\sqrt{\sigma}}{R}e^{\alpha D^2}\right)^N. \tag{24}$$

Dies folgt sofort aus (20), also für $\sigma = R^2$ sicher Divergenz.

Analog folgt

Satz 3. *Ist $f(z)$ analytisch in $R_1 \leqq |z| \leqq R_2$ $(R_2 > 1, 0 < R_1 < 1)$, ist $N = 2M+1$, $\overline{L_M}(f) = \frac{1}{z^M} L_n(z^M f)$, wobei L_N das Interpolationspolynom von $z^M f$ in den ζ_h $(h = 1, \ldots, N)$ ist, dann ist*

$$\overline{\varrho_M} = \sup_{|z|=1} |f(z) - \overline{L_M}(f)| \leqq S$$

mit

$$S = \gamma^N(D)\left[\frac{K_2}{R_2-1}\frac{1}{R_2^{M+1}}\left(\frac{R_2+1}{R_2-1}\right)^{2ND} + \frac{K_1}{1-R_1}R_1^M\left(\frac{1+R_1}{1-R_1}\right)^{2ND}\right]$$

und

$$K_r = \frac{1}{2\pi}\int_0^{2\pi} |f(R_i e^{i\varphi})|\, d\varphi \quad (r = 1, 2).$$

Dies folgt aus (6'), (9), (10), (13).

§ 2

Wir wollen nun annehmen, daß α eine reelle Zahl ist und $\varphi_h = (h-1)\alpha$ $(h = 1, \ldots, N)$, also $\zeta_h = e^{2\pi i(h-1)\alpha}$. Es werde α in einen Kettenbruch entwickelt, der endlich oder unendlich sein kann: $\alpha = [a_0, a_1, a_2, \ldots]$, und es seien $\frac{p_0}{q_0}, \frac{p_1}{q_1}, \ldots$ die Näherungsbrüche. Es sei $i(N)$ durch $q_i \leqq N < q_{i+1}$ definiert, dann gilt (vgl. [4])

$$D_N < 4\left(\frac{1}{q_i} + \frac{i + a_1 + \cdots + a_i}{N}\right). \tag{25}$$

Ist insbesondere $N = q_i$, dann ist $D_{q_i} \leqq \frac{2}{N}$. In diesem Fall ist also

$$\gamma_N < e^{16(\log N + 2)/N} \leqq e^{c(\log N/N)}$$

und

$$\varrho_N(f) < \frac{K}{R-1} \frac{e^{c_1 \lg N}}{R^N} \left(\frac{R+1}{R-1}\right)^4. \tag{22'}$$

In diesem Fall könnte man noch schärfer abschätzen.

Es werde nun $A = \operatorname{Max}_{h=1,\ldots,i} a_h$ gesetzt. Ist $A \neq 1$, dann gilt bekanntlich $\sum_{h=1}^{i} a_h < \frac{2A}{\log A} \log q_i + 4A$; ist $A = 1$, dann ist diese Summe $\leqq \frac{2}{\log 2} \log q_i$. Weiter gibt es zu jedem i ein N, so daß $q_i < N < q_{i+1}$ und

$$D_N > K \frac{(a_1 + \cdots + a_{i+1})}{N} \tag{26}$$

ist. Es sei nun α eine positive Irrationalzahl, dann kann man aus (26) folgern, daß es unendlich viele N gibt, so daß

$$D_N > \frac{1}{720} \frac{\log N}{N}$$

gilt, also wegen (23)

$$\varrho_N(f_0) \geqq \frac{1}{R+1} \frac{e^{c_1(\log^2 N/N)}}{R^N}. \tag{23'}$$

Sind insbesondere die a_i beschränkt, also $a_i \leqq K$ für alle i, dann ist dies richtig für die N_i, definiert durch

$$N_i = q_{4i+1} + N_{i-1}, \quad N_1 = 95 \quad (i = 2, \ldots),$$

und dann ist

$$D_{N_i} \geqq \frac{\log N_i}{8(A+1)^6}.$$

Andererseits ist für jedes N

$$D_N \leqq 4A \frac{\log N}{N}, \tag{25'}$$

also wegen (22)

$$\varrho_N(f) < \frac{K}{R-1} \frac{e^{c \log^2 N}}{R^N} \left(\frac{R+1}{R-1}\right)^{8A \log N}. \tag{22''}$$

Dies gilt insbesondere für quadratische Irrationalitäten.

Gilt

$$q_{i+1} < c q_i^r \quad \text{mit} \quad c > 0, \quad r > 1, \tag{26'}$$

dann ist

$$D_N \leqq CN^{-(1/r)} \quad \text{und} \quad D_N \geqq C_1 N^{-(1/r)} \tag{27}$$

für unendlich viele N.

Diese Abschätzung kann insbesondere angewendet werden, wenn α eine algebraische Zahl ist mit einem Grad $\geqq 3$ mit $r = 1 + \varepsilon\,(\varepsilon < 0)$; dann ist

$$D_N \leqq CN^{-1/(1+\varepsilon)}. \tag{27'}$$

Nach K. F. Roth ist nämlich stets

$$\left|\alpha - \frac{p}{q}\right| \geqq \frac{c}{q^{2+\varepsilon}}$$

Somit gilt für die i-ten Näherungsbrüche von α

$$\frac{1}{q_i q_{i+1}} > \left|\alpha - \frac{p_i}{q_i}\right| \geqq \frac{c}{q_i^{2+\varepsilon}}, \quad \text{also} \quad \frac{1}{q_{i+1}} \geqq \frac{c}{q_i^{1+\varepsilon}}.$$

Nehmen wir $\alpha = \pi$, dann ist nach K. Mahler [5]

$$\left|\pi - \frac{p_i}{q_i}\right| > \frac{1}{q_i^{42}},$$

also

$$q_{i+1} < q_i^{41},$$

also gilt (26′) mit $r = 41$.

Interessant ist der Fall $\alpha = \dfrac{1}{\pi}$, dann gilt ebenfalls

$$\left|\frac{1}{\pi} - \frac{p}{q}\right| > \frac{c_1}{q^{42}};$$

denn gäbe es p, q mit $\left|\dfrac{1}{\pi} - \dfrac{p}{q}\right| < \dfrac{c_1}{q^{42}}$ (o.B.d.A. $p < q$), dann wäre

$$\left|\frac{q}{p} - \pi\right| < \frac{\pi c_1}{pq^{41}} < \frac{\pi c_1}{p^{42}}.$$

Dann sind die ζ_h von der Gestalt e^{2ih} $(h = 0, 1, \ldots, N-1)$, und die Diskrepanz ist $\leqq CN^{-(1/41)}$.

Betrachtet man die Menge aller Folgen $(\varphi_1, \varphi_2, \ldots)$ mit dem zugehörigen Haarschen Maß, dann ist für fast alle Folgen für großes N nach Cassels

$$D_N < (1+\varepsilon)\sqrt{\frac{\log\log N}{N}}, \tag{27''}$$

und es gibt Folgen von natürlichen Zahlen N, so daß

$$D_N > (1-\varepsilon)\sqrt{\frac{\log\log N}{N}} \tag{27'''}$$

ist; also ist für unendlich viele N

$$\varrho_N(f) < \frac{K}{R-1}\,\frac{e^{16(1+\varepsilon)\sqrt{N\log\log N}\log N}}{R^N}\left(\frac{R+1}{R-1}\right)^{(1+\varepsilon)2\sqrt{N\,\mathrm{lglg}\,N}} \tag{28}$$

und

$$\varrho_N(f_0) > \frac{1}{R+1}\,\frac{e^{\alpha(1-\varepsilon)^2\log\log N}}{R^N}. \tag{28'}$$

Es ist daher für fast alle Folgen $\overline{\lim\limits_{N\to\infty}}\, R^N\varrho_N(f_0) = \infty$.

Bemerkung. Für jede Folge gibt es nach K. F. Roth unendlich viele N, so daß $D_N > c\dfrac{\sqrt{\log N}}{N}$ gilt. Für alle N ist $ND_N \geqq \dfrac{1}{2}$ [vgl. (35)]

Aus (28) folgt, daß für fast alle Folgen $\lim\limits_{N\to\infty}\varrho_N(f) = 0$ ist. Dieses Resultat ist schon lange bekannt und kann direkt sehr leicht hergeleitet werden (vgl. auch J. H. Curtiss [5]).

Aus (27'), (27''), (17), (18) folgt für $|z| = 1$ für fast alle Folgen

$$|\omega_N(z)| \leqq e^{O(\sqrt{N}\sqrt{\log\log N}\log N)}$$

und

$$\max_{|z|=1}|\omega_N(z)| \geqq e^{\Omega(\log\log^2 N)}, \quad \text{also} \quad \overline{\lim_{N\to\infty}}\,|\omega_N(z)| = \infty.$$

§ 3

Will man D_N für eine gegebene endliche Folge $\varphi_1, \ldots, \varphi_N$ abschätzen, so kann man so vorgehen: Es war $D_N = \sup\limits_t \dfrac{1}{N}\left|\sum\limits_{h=1}^{N}\chi_t(\varphi_h) - t\right|$, wobei das Supremum erstreckt wird über alle t mit $0 \leqq t < 1$. Es sei m eine natürliche Zahl $\geqq 1$ und man betrachte die m Intervalle j_l: $0 \leqq x < \dfrac{l}{m}$ $(l = 0, \ldots, m)$ und setze

$$\Delta_{l,m} = \left|\frac{1}{N}\sum_{h=1}^{N}\chi_{l/m}(\varphi_h) - \frac{l}{m}\right|.$$

Dabei ist $\chi(t)$ die charakteristische Funktion des Intervalls $0 \leqq \varphi < t$. Es ist $\Delta_{l,m} = \left|\frac{N'(j_l)}{N} - \lambda(j_l)\right|$, wobei $N'(j_l)$ die Anzahl der φ_h im Intervall j_l und $\lambda(j_l)$ die Länge des Intervalls j_l ist. Es kann also $\Delta_{l,m}$ für jedes l in endlich vielen Schritten bestimmt werden. Wir setzen nun $\operatorname{Max}_l \Delta_{l,m} = \Delta_m$. Dann ist

$$\Delta_m \leqq D_N \leqq \Delta_m + \frac{1}{m}. \tag{29}$$

Der Beweis von (29) ist, was die linke Seite betrifft, trivial. Um die rechte Seite zu beweisen nehmen wir ein Intervall $0 \leqq \varphi \leqq t$ und setzen $l = [tm]$. Es ist $0 \leqq l \leqq tm < l + 1$, und da $t \leqq 1$ ist, gilt $l \leqq m$. Damit ist

$$\frac{l}{m} \leqq t < \frac{l+1}{m}.$$

Es ist

$$\sum_h \chi_{l/m}(\varphi_h) \leqq \sum_h \chi_t(\varphi_h) \leqq \sum \chi_{(l+1)/m}(\varphi_h).$$

Daher ist

$$-\Delta_{l,m} - \frac{1}{m} \leqq \frac{1}{N}\sum \chi_t(\varphi_h) - t \leqq \Delta_{l+1,m} + \frac{1}{m},$$

also

$$\left|\frac{1}{N}\sum \chi_t(\varphi_h) - 1 - \Delta_m\right| \leqq \frac{1}{m},$$

und daraus folgt die Behauptung.

Für die Anwendung wird man $m \geqq N$ wählen müssen.

Es soll noch kurz der Fall behandelt werden, daß die φ_h nur bis auf einen Fehler $< \varepsilon$ bekannt sind; es gilt folgender Satz, der an sich bekannt ist:

Hilfssatz 7. *Sind* $(\varphi_1, \ldots, \varphi_N)$ *und* $(\psi_1, \ldots, \psi_N)$ *zwei Folgen und ist*

$$|\varphi_h - \psi_h| < \varepsilon \quad (h = 1, \ldots, N),$$

dann ist

$$|D - D^*| < \varepsilon. \tag{30}$$

Dabei wurde $D = D_N((\varphi_h))$, $D^* = D_N((\psi_h))$ *gesetzt.*

Beweis. Es sei $\langle 0, \alpha)$ zunächst ein Intervall I_α mit $\varepsilon < \alpha < 1 - \varepsilon$. Es sei $N_2(\alpha - \varepsilon)$ die Anzahl der ψ_h mit $0 \leqq \psi_h < \alpha - \varepsilon$. Da $0 \leqq \varphi_h < \psi_h + \varepsilon$ ist, gilt für die zugehörigen φ_h: $0 \leqq \varphi_h < \alpha$, d. h., wenn $N_1(\alpha)$ die Anzahl aller φ_h und I_α ist, daß $N_1(\alpha) \geqq N_2(\alpha - \varepsilon)$ gilt. Analog folgt aus $0 \leqq \varphi_h < \alpha$ die Ungleichung $0 \leqq \psi_h < \alpha + \varepsilon$, damit also

$$N_2(\alpha + \varepsilon) \geqq N_1(\alpha).$$

Nun folgt aus der Definition von D^*

$$N_2(\alpha \pm \varepsilon) = N(\alpha \pm \varepsilon) + \vartheta N \cdot D^* \quad \text{mit} \quad |\vartheta| \leqq 1,$$

d. h., wir erhalten

$$N(\varepsilon + D^*) \geqq N_1(\alpha) - N\alpha \geqq -N(\varepsilon + D^*)$$

oder

$$|N_1(\alpha) - N\alpha| \leqq N(\varepsilon + D^*).$$

Diese Ungleichung ist trivialerweise richtig, wenn α nicht der vorher eingeführten Bedingung genügt, da stets $N_1 \leqq N$ ist. Daraus folgt $D \leqq \varepsilon + D^*$. Vertauschen wir D mit D^*, so folgt (30). Wir wollen noch folgende Verallgemeinerung von Hilfssatz 7 beweisen:

Hilfssatz 8. *Es seien* $(\varphi_1, \ldots, \varphi_N)$, $(\psi_1, \ldots, \psi_N)$ *zwei Folgen mit*

$$|\varphi_h - \psi_h| < \varepsilon_h \quad (h = 1, \ldots, N).$$

Weiter sei $S(\varepsilon)$ *für* $\varepsilon > 0$ *die Anzahl der* $\varepsilon_h > \varepsilon$. *Dann gilt für die Diskrepanz* D, D^*

$$|D - D^*| < \varepsilon + \frac{S(\varepsilon)}{N}. \tag{31}$$

Wenn alle ε_h *gleich* ε *sind, so erhalten wir wieder* (30).

Beweis. Wir betrachten wie im Beweis von (30) Intervalle $I: \langle 0, \alpha)$ mit $\varepsilon < \alpha < 1 - \varepsilon$, und es sei $N_2(\alpha - \varepsilon)$ die Anzahl der ψ_h mit $0 \leqq \psi_h < \alpha - \varepsilon$. Für die zugehörigen φ_h mit $0 \leqq \varphi_h < \psi_h + \varepsilon_h$ gilt sicher, wenn $\varepsilon_h \leqq \varepsilon$ ist, daß sie in $\langle 0, \alpha)$ liegen. Ist also $N_1(\alpha)$ die Anzahl aller φ_h in I_α, dann ist $N_1(\alpha) \geqq N_2(\alpha - \varepsilon) - S$. Analog folgt $N_1(\alpha) < N_2(\alpha + \varepsilon) + S$. Schließt man nun weiter wie beim Beweis von (30), so ergibt sich

$$|N_1(\alpha) - N\alpha| \, \varepsilon \leqq N(\varepsilon + D^*) + S$$

und daraus folgt (31).

Daraus leiten wir nun folgenden Satz her:

Satz 4. *Es sei* (φ_j) *eine unendliche Folge mit der Folge der Diskrepanzen* D_N $(N = 1, \ldots)$. *Es sei* (ψ_j) *eine weitere unendliche Folge, welche im Einheitsintervall überall dicht liegt. Weiter sei* $\sigma(t) \geqq 1$ *monoton wachsend im stärkeren Sinne mit* $\lim_{t \to \infty} \sigma(t) = \infty$. *Dann können wir diese Folge so umordnen, daß für die neue Folge* $\vartheta_1, \vartheta_2, \ldots$ *die zugehörige*

Diskrepanz D_N^* *für jedes N die Ungleichung*

$$|D_N^* - D_N| \leqq \frac{\sigma(N)}{N} \tag{32}$$

erfüllt.

Beweis. Da die Folge (ψ_h) dicht ist im Einheitsintervall, gibt es sicher in ihr Elemente $\psi'_h = \psi_{j_h}$ $(h = 1, 2, \ldots)$, so daß

$$|\psi'_h - \varphi_h| < \frac{1}{\tau(h)} = \varepsilon_h$$

gilt, wobei $\tau(t)$ die inverse Funktion zu σ ist.

Ist $\varepsilon > 0$, so ist die Anzahl $S(\varepsilon)$ der $\varepsilon_h \geqq \varepsilon$, also der $\tau(h) \leqq \frac{1}{\varepsilon}$, sicher $\leqq \sigma\left(\frac{1}{\varepsilon}\right)$. Es gilt also nach (31) für die Diskrepanz D'_N der Folge (ψ'_h) mit $\varepsilon = \frac{1}{N}$

$$|D'_N - D_N| \leqq \frac{1}{N} + \frac{\sigma(N)}{N} \leqq \frac{2\sigma(N)}{N}. \tag{33}$$

Wir setzen nun $\vartheta_h = \psi'_h = \psi_{i_h}$, wenn h nicht von der Gestalt $[\tau(l)]$ ist. Alle übrigen Glieder der Folge (ψ_h) denken wir uns irgendwie durchnumeriert: $\lambda_1, \lambda_2, \ldots$ Setzen wir $\vartheta_h = \lambda_l$ für $h = [\tau(l)]$, dann ist für jedes α

$$\left|\frac{1}{N}\sum_{h=1}^{N} (\chi_\alpha(\vartheta_h) - \chi_\alpha(\psi'_h))\right| \leqq \frac{1}{N} + \frac{\sigma(N)}{N};$$

denn die Anzahl der $h = [\tau(l)] < N$ ist höchstens $\sigma(N)$. Es ist also

$$|D_N^* - D'_N| \leqq \frac{1 + \sigma(N)}{N},$$

d. h.

$$|D_N^* - D_N| < \frac{4\sigma(N)}{N}.$$

Wählt man statt σ die Funktion $\frac{\sigma}{4}$, so folgt die Behauptung.

Bemerkung. Aus (32) folgt: Die Folge (ψ_j) kann so umgeordnet werden, daß für die Diskrepanz der neuen Folge

$$|D_N^* - D_N| \leqq D_N \sigma(N) \tag{34}$$

gilt.

Beweis. Es ist stets für jede Folge (φ_j)

$$D_N \geqq \frac{1}{2N}. \tag{35}$$

Es sei φ_j irgendein Glied mit $j \leqq N$. Dann betrachten wir die Intervalle $\langle 0, \varphi_j - \varepsilon)$ und $\langle 0, \varphi_j + \varepsilon)$ mit $\varepsilon < \dfrac{1}{2N}$. (Der Fall, daß $\varphi_j = 0$ $(j \leqq N)$ ist, erledigt sich analog.)

Dann ist

$$N'(\varphi_j + \varepsilon) - N'(\varphi_j - \varepsilon) \geqq 1,$$

also

$$2ND \geqq |N'(\varphi_j + \varepsilon) - N(\varphi_j + \varepsilon) - N'(\varphi_j - \varepsilon) - N(\varphi_j - \varepsilon)| \geqq 1 - 2N\varepsilon,$$

und daraus folgt sofort (35).

Es erhebt sich die Frage, ob man erreichen kann, daß $c_1 D_N \leqq D_N^* \leqq c_2 D_N$ ist, wobei c_1, c_2 absolute Konstante sind, und zwar für alle N. (Für ein festes N ist dies leicht zu zeigen.)

Nehmen wir einerseits z. B. für die Folge (φ_j) eine Folge mit (25'), andererseits die Folge 0, 0, ..., für welche die Diskrepanzen D_N alle 1 sind, so folgt: Jede Folge (ψ_j), welche im Einheitsintervall überall dicht liegt, also auch jede gleichverteilte Folge, kann einerseits so umgeordnet werden, daß die neue Folge (ϑ_j') eine Diskrepanz $D_N' = O\left(\dfrac{\log N}{N}\right)$ besitzt (also gleichverteilt ist), andererseits so, daß für die Folge (ϑ_j'') die zugehörigen Diskrepanzen $D_N'' \geqq \dfrac{1}{2}$ für alle N sind, also (ϑ_j'') nicht gleichverteilt ist.

Bemerkung. Die Tatsache, daß jede in E dichte Folge so umgeordnet werden kann, daß sie dort gleichverteilt ist, stammt bereits von J. von Neumann. Aus (22) entnimmt man: Ist $f(z)$ analytisch in $|z| \leqq R$ $(R > 1)$ und ist die unendliche Folge $\varphi_1, \varphi_2, \ldots$ gleichverteilt, also $\lim\limits_{n\to\infty} D_N = 0$, dann gilt eine Abschätzung

$$\varrho_N(f, \varphi_h) \leqq \frac{C^{o(N)}}{R^N} \tag{36}$$

wobei C von R abhängt.

Daraus folgt natürlich der bekannte Satz, welcher in der Einleitung zitiert wurde:

Es ist $\lim\limits_{N\to\infty} \varrho_N(f, \varphi_h)$ genau dann und nur dann $=0$ für alle in $|z| \leqq 1$ analytischen Funktionen, wenn die Folge (φ_h) gleichverteilt ist. Dies folgt sofort aus (22') und (23'). Ist nämlich nicht $\lim D_N = 0$, dann gibt es eine Teilfolge $(D_{N'})$ mit $\lim D_{N'} = \varepsilon > 0$, und dann kann (23) angewendet werden mit $R = e^{\alpha\varepsilon^2/4} > 1$ mit $f_0(N) = \dfrac{1}{z - \zeta(N)}$, wo $|\omega_N(\zeta(N))| \leqq R^N$ mit passendem $\zeta(N)$ mit $|\zeta(N)| = R$. Es existiert eine Teilfolge (N_1) von N mit $\lim \zeta(N_1) = \zeta$, für genügend großes N_1 ist also

$$|\zeta(N_1) - \zeta| < (R - 1)^2.$$

Nun ist für jedes j wegen $|\zeta(N) - \zeta_j| \geqq R - 1$

$$|\zeta - \zeta_j| \leqq |\zeta - \zeta(N_1)| + |\zeta(N_1) - \zeta_j| \leqq |\zeta(N_1) - \zeta_j|\, R,$$

also

$$\omega_{N_1}(\zeta) \leqq \omega_{N_1}(\zeta(N_1))\, R^{N_1} \leqq R^{2N_1}.$$

Somit ist für $f_0(\zeta) = \dfrac{1}{z - \zeta}$

$$\varrho_{N_1}(f_0, \varphi_h) \geqq \frac{1}{R + 1}\left(e^{\alpha\left(D_{N_1}{}^2 - \frac{\varepsilon^2}{2}\right)}\right)^{N_1}, \tag{23''}$$

und es ist dann sogar $\lim\limits_{N_1 \to \infty} \varrho_{N_1} = \infty$.

Aus Satz 4 erkennt man aber, daß es bei Interpolation mittels einer unendlichen Folge auf die Reihenfolge ganz wesentlich ankommt, im Gegensatz zu endlichen Folgen. Wesentlich ist natürlich, daß die Folge φ_h überall dicht in E, also die Folge $(e^{2\pi i \varphi_h})$ dicht auf dem Einheitskreis sein muß, wenn die Interpolationspolynome für alle in $|z| \leqq R$ $(R > 1)$ analytischen Funktionen $f(z)$ gegen diese Funktion konvergieren soll. Ist die Folge ζ_h gleichverteilt, dann ist dies stets der Fall. Man kann aber, wie aus der Schlußbemerkung von § 3 hervorgeht, (φ_j) so umordnen, daß für die neue Folge (ϑ_j)

$$\varrho_N(f, \vartheta_h) \leqq \frac{e^{8\log^2 N}}{R^N} \cdot \left(\frac{R+1}{R-1}\right)^{8\log N} \frac{C}{R - 1}$$

ist. Man kann sie aber auch in eine „schlechte“ Folge (ϑ_h'') umordnen, daß es eine analytische Funktion f in $|z| \leqq R$ gibt mit

$$\varrho_N(f_0, \vartheta_h'') \geqq \frac{C_1^N}{R^N} \cdot \frac{1}{R + 1},$$

wobei C_1 eine absolute Konstante ist. (Man kann $C_1 = e^\beta$, $\beta = \frac{1}{4}\alpha$, $\alpha = \frac{1}{4} 10^{-8}$ wählen.)

Wenn also $1 < R < C$ ist, so ist das Verfahren divergent.

Literatur

[1] J. L. Walsh, Interpolation and approximation by rational functions in the complex domain, Amer. Math. Soc. Providence 1956 (2. Ed).

[2] Abschätzungen ohne Verwendung der Diskrepanz bei O. Kis, Hung. Acta Math. **7** (1956), 173—200.

[3] Annals of Math. **51** (1950), 105—119. Ein funktionentheoretischer Beweis bei T. Ganelius, Arkiv för Matematik 3 (1958), Nr. 1, 1—50.

[4] Vgl. etwa Cigler-Helmberg, Jahresbericht DMV **64** (1962), 1—50.

[5] J. H. Curtiss, Fourth Berkeley Symposium on Mathematical Statistics and Probability 1961, University of California Press, Berkeley and Los Angeles, Vol. II, pp. 79—93.

A. E. Ingham †

ON THE HIGH-INDICES THEOREM FOR BOREL SUMMABILITY

1. Introduction

In a volume dedicated to the memory of Landau, summability of series is an appropriate topic, for Landau was intensely interested in this subject and made important contributions to it. In particular, it was he who introduced the notion of one-sided Tauberian condition, thereby enriching the theory itself and opening the way to arithmetical applications. No such conditions will indeed appear in this paper, for we shall be concerned exclusively with a theorem of the 'high-indices' or 'gap' type, and it is characteristic of such theorems, in their most perfect forms, that they involve no explicit order condition (one-sided or two-sided) on the terms of the series.

Given an infinite series

$$\sum_{n=0}^{\infty} a_n, \tag{a}$$

let

$$A_n := \sum_{\nu=0}^{n} a_\nu, \quad A(x) := \sum_{n=0}^{\infty} A_n \frac{x^n}{n!}.$$

We say that (a) is B-summable, or summable by Borel's exponential method, to the sum A if

$$e^{-x}A(x) \to A \quad \text{as} \quad x \to \infty,$$

where it is assumed that A is finite, that x is a continuous real variable, and that the series for $A(x)$ is convergent for all x. A connected account of this method, and of the related integral method, or B′-summability, may be found in Hardy's book [3], Chapters VIII, IX, but very little of this will be needed here.

We shall say that (a) satisfies a gap condition $[p, h]$ ($p \geqq 0$, $h > 0$, fixed) if $a_n = 0$ when $n \neq n_r$, where $\{n_r\}$ is some sequence of integers for which

$$n_1 > 0, \quad n_{r+1} - n_r > h n_r^p \quad (r = 1, 2, \ldots).$$

We now state the

High-indices theorem. *If* (a) *is* B-*summable to* A *and satisfies a gap condition* $[\frac{1}{2}, h]$, *then* (a) *is convergent to* A.

Proofs of this long-conjectured theorem, without extra restrictions on $\{a_n\}$ or $\{n_r\}$, were published in 1965 by D. Gaier [1] and by V. I. Mel'nik [6]. An account of the somewhat intricate history of the problem is given in the introduction to Gaier's paper. Both authors note that, in view of known inclusion relations, the theorem carries with it the same theorem with B′ in place of B.

In this paper we present another proof, found independently. This proof is based on the use of a 'peak function' on lines suggested in Pitt's book [7], p. 41, but uses a more efficient peak function and works directly with B-summability.

In § 2 (theorem A) we first reduce the theorem to a known theorem of Pitt [7, p. 92], which asserts that the high-indices is true if its hypotheses are supplemented by the restriction $a_n = O(e^{K\sqrt{n}})$ for some fixed $K > 0$; and we complete the proof by an appeal to this theorem.

The reader who is content to quote Pitt's theorem need not read beyond § 2. In § 3, however, we develop the peak method into an independent proof of Pitt's theorem, and indeed of a stronger theorem (Theorem B) involving the weaker restriction $a_n = O(e^{\varepsilon n})$ for every fixed $\varepsilon > 0$. In § 4 we remove all such restrictions by combining theorems A and B. The resulting proof of the high-indices theorem is self-contained, and demands no knowledge beyond the elements of classical (real-variable) analysis, together with the simple fact that B-summation is a regular linear method: it sums every convergent series to its ordinary sum.

In § 5 we comment on the choice of a suitable peak function, and on other aspects of the peak method.

Gaier's proof of the high-indices theorem differs fundamentally from ours in using complex function theory, but structurally it may be compared with or § 2 in that it reduces the general theorem to a restricted theorem, taken as known. But his argument calls for a wider theorem than Pitt's, namely one involving the weaker restriction $a_n = O(e^{\varepsilon n})$ for every fixed $\varepsilon > 0$. Such a theorem was indeed known, but (as explained in Gaier's introduction) only through a still wider version, due to Meyer–König and Zeller, with 'every' replaced by 'some'. A proof requiring only Pitt's theorem is, however, contained implicitly in a second paper by Gaier [2], (2.9) and (3.3) ($\alpha = 1$). Substantially the same proof was found independently by G. Halász, of the University of Budapest.

Mel'nik's proof differs from those just described in using real-variable analysis, but resembles Gaier's first proof in reducing the general high-indices theorem to the theorem of Meyer–König and Zeller.

In the full account presented here, no independent knowledge of a theorem of high-indices type is called for. There is in fact an overlap between the two parts: the conclusion of Theorem A is stronger than the corresponding hypothesis of Theorem B.

Notation and conventions

Systematic notations involving the letters a, A (as above) and α (to be introduced in § 2.2) will be used also with other letters.

Limits, upper limits, bounds, etc. (such as A, Ω, M) occurring in hypotheses are assumed finite. The series defining $A(x)$ is assumed convergent for all x; statements about $A(x)$ and related functions are to be interpreted on this basis.

In the gap condition $[\frac{1}{2}, h]$ we may suppose $0 < h \leqq 1$ [by changing h to min $(h, 1)$]. We assume $n_1 > 0$ (rather than $n_1 \geqq 0$) for general convenience; this involves no loss of generality for the high-indices theorem [if $n_1 = 0$ change a_0 to 0 and renumber n_{r+1} as $n_r (r$ M $1)$].

$H, H', \ldots$ are positive constants depending only on h. Possible values will be given explicitly, on the assumption that $0 < h \leqq 1$.

$l := \log 2$. Note that $\frac{2}{3} < l < 1$ [e.g. from the power series for $\log\{(1+x)/(1-x)\}$ with $x = \frac{1}{3}$].

2. Reduction to a theorem of Pitt

2.1. We first construct a suitable peak function.

Lemma P. *For each integer $N \geqq 2$ we can construct a function $P_N(t)$ with the properties:*

(P 1) $$P_N(t) = \sum_{s=1}^{2N} p_s 2^{-st}, \quad p_s = p_s(N);$$

(P 2) $$\sum_{s=1}^{2N} |p_s| < 2^{4N-2};$$

(P 3,4,5) $$0 \leqq P_N(t) \leqq P_N(0) = 1 \quad (t \geqq -1);$$

(P 6) $$P_N(t) < \frac{5}{N^2 t^2} \quad (t \geqq -1;\ t \neq 0);$$

(P 7) $$P_N(t) > \tfrac{1}{4} \quad (-1/N \leqq t \leqq 1/N).$$

For $N = 0, 1, 2, \ldots$ let

$$F_N = F_N(T) := \left(\frac{\sin N\Theta}{\sin \Theta}\right)^2 \cos^2 \Theta = \frac{1 - \cos 2N\Theta}{2 \sin^2 \Theta} \cos^2 \Theta,$$

where

$$\sin \Theta = 1 - T \quad (-\tfrac{1}{2}\pi \leqq \Theta \leqq \tfrac{1}{2}\pi;\ 2 \geqq T \geqq 0)$$

(F_N being defined as N^2 when $\Theta = 0$). The correspondence $\Theta \leftrightarrow T$ is one-to-one, so this defines $F_N(T)$ for $0 \leqq T \leqq 2$. To extend the definition that, for $N \geqq 1$,

$$F_{N+1} - 2F_N + F_{N-1} = 2 \cos 2N\Theta \cos^2 \Theta = 2 \cos^2 \Theta - (4 \sin^2 \Theta)\, F_N,$$

and so

$$F_{N+1} = (-2 + 8T - 4T^2)\, F_N - F_{N-1} + (4T - 2T^2). \tag{1}$$

Since $F_0 = 0$ and $F_1 = 2T - T^2$, it follows by induction that

$$F_N(T) = \sum_{s=1}^{2N} f_s T^s, \quad f_s = f_s(N)$$

(the sum being empty when $N = 0$). We take this as definition of $F_N(T)$ for all T. For $N \geqq 1$ let

$$P_N(t) := N^{-2} F_N(T), \quad T := 2^{-t}.$$

Then we have (P1), with $p_s := N^{-2} f_s$. Also, if $F_N^* = \sum_{s=1}^{2N} |f_s|$, it follows from (1) that

$$F_{N+1}^* \leqq 14 F_N^* + F_{N-1}^* + 6 \quad (N \geqq 1),$$

whence $F_N^* < 15^N$ ($N \geqq 0$), since this obviously holds for $N = 0, 1$. This implies (P2) if $N \geqq 2$.

Now suppose $t > -1$, $t \neq 0$. Then $0 < T < 2$, $T \neq 1$; the original definition applies, and so (since $t = 0$ corresponds to $\Theta = 0$)

$$0 \leqq P_N(t) = \left(\frac{\sin N\Theta}{N \tan \Theta}\right)^2 < \frac{(N\Theta)^2}{N^2 \Theta^2} = 1 = P_N(0).$$

Also

$$N^2 P_N(t) \leqq \frac{\cos^2 \Theta}{\sin^2 \Theta} = \frac{2T - T^2}{(1 - T)^2} < \frac{2}{T^{-1} + T - 2} < \frac{2}{(lt)^2},$$

since $T^{-1} = e^{lt}$. If, in addition, $|t| \leqq 1/N$, then

$$|\sin \Theta| \leqq 2^{1/N} - 1 < \frac{1}{N}, \quad |\Theta| < \frac{\pi}{2} |\sin \Theta| < \frac{\pi}{2N} \leqq \frac{\pi}{4}$$

[where we have used the fact that $(1 + N^{-1})^N > 2$]; whence, since $(\sin\varphi)/\varphi \downarrow$ $(0 < \varphi \leqq \frac{1}{2}\pi)$ and $(\tan\varphi)/\varphi \uparrow (0 < \varphi < \frac{1}{2}\pi)$,

$$P_N(t) = \left(\frac{\sin N\Theta}{N\Theta} \cdot \frac{\Theta}{\tan\Theta}\right)^2 > \left(\frac{2}{\pi} \cdot \frac{\pi}{4}\right)^2 = \frac{1}{4}.$$

These results imply (P 3, 4, 5, 6, 7); the values $t = 0, -1$ (where applicable) may be included since $P_N(0) = 1$, $P_N(-1) = 0$.

In order to emphasize the real-variable character of our arguments we have avoided complex numbers in this proof, though their use would have enables us to evaluate the sum in (P 2) exactly (as in [5], Lemma B, 164). We have included (P 7) for completeness, but this will not be used in § 2.

2.2 Our main argument will be a "maximum term argument" applied, not to the series for $A(x)$, but to the same series with a_n in place of A_n. Let

$$\alpha_n := \frac{a_n}{n!}, \quad a(x) := \sum_{n=0}^{\infty} \alpha_n x^n.$$

Since $a_n = A_n - A_{n-1}$ $(n \geqq 1)$, our assumption that the series for $A(x)$ converges for all x implies that the same is true of the series for $a(x)$. Hence $|a_n x^n| \to 0$ as $n \to \infty$, for each fixed x, and we may define the *maximum term* $\mu(x)$ and *central index* $\nu(x)$ thus:

$$\mu(x) := \max_n |\alpha_n x^n|;$$

$$\nu(x) := \text{the least } n \text{ for which } |\alpha_n x^n| = \mu(x).$$

We begin with two lemmas connecting $\mu(x)$ and $\nu(x)$. In the proof of the first of these the maximal property of $\mu(x)$ will be reinforced by means of the peak function $P_N(t)$.

Lemma A 1. *If* (a) *satisfies a gap condition* $[\frac{1}{2}, h]$, *and*

$$|a(x)| \leqq 2M\, e^x \quad (x > 0), \tag{2}$$

then

$$\mu(x) \leqq M \exp\left(x + H\sqrt{\nu(x)}\right) \quad (x > 0).$$

We may suppose the a_n not all 0; otherwise $\mu(x) = \nu(x) = 0$ and there is nothing to prove. Suppose $x > 0$, and let $\mu := \mu(x)$, $\nu := \nu(x)$. Then $\mu > 0$, and $\nu = n_\varrho\ (\geqq 1)$ for some ϱ since $\alpha_n x^n = 0$ when $n \neq n_r$. Construct $P_N(t)$ for some $N \geqq 2$ and let

$$S := \sum_{n=0}^{\infty} \alpha_n x^n P_N\left(\frac{n - \nu}{\nu}\right).$$

By (P 1),

$$S = \sum_{n=0}^{\infty} \alpha_n x^n \sum_{s=1}^{2N} p_s 2^{s(\nu-n)/\nu} = \sum_{s=1}^{2N} p_s 2^s a(x2^{-s/\nu});$$

whence, by (2) and (P 2) (since $x2^{-s/\nu} < x$ and $2 = e^l$),

$$|S| \leqq \sum_{s=1}^{2N} |p_s| \, 2^{s+1} M \, e^x < \frac{1}{2} M \, e^{x+1/N}. \tag{3}$$

On the other hand, by (P 5, 6), $[\frac{1}{2}, h]$, and the definitions of μ and ν,

$$|S| \geqq \mu - \sum_{n \neq \nu} |\alpha_n x^n| \frac{5\nu^2}{N^2(n-\nu)^2} \geqq \mu - \mu \sum_{r \neq \varrho} \frac{5n_\varrho^2}{N^2(n_r - n_\varrho)^2}.$$

But, by $[\frac{1}{2}, h]$ (in which we suppose $0 < h \leqq 1$), we have

$$n_{i+1}^{1/2} > (n_i + hn_i^{1/2})^{1/2} > n_i^{1/2} + \tfrac{1}{3} h \quad (i \geqq 1)$$

(by squaring both sides), and so

$$|n_r - n_\varrho| = |(n_r^{1/2} - n_\varrho^{1/2})(n_r^{1/2} + n_\varrho^{1/2})| > \tfrac{1}{3} h \, |r - \varrho| \, n_\varrho^{1/2} \quad (r \neq \varrho).$$

Hence

$$|S| > \mu\left(1 - \sum_{r \neq \varrho} \frac{45\nu}{N^2 h^2 (r-\varrho)^2}\right) > \mu\left(1 - \frac{180\nu}{N^2 h^2}\right).$$

Now let

$$N := \left[\frac{20}{h}\sqrt{\nu}\right] > \frac{19}{h}\sqrt{\nu} \quad (\geqq 19).$$

Then

$$|S| > \mu(1 - \tfrac{1}{2}). \tag{4}$$

Combining (3) and (4), we obtain the result, with $H := 120l/h$.

Lemma A 2. *If the a_n ($n \geqq 1$) are not all 0, then, given $\varepsilon > 0$, we have*

$$y^{\nu(x)} < \varepsilon \mu(xy) \quad [y > 0; \; x > x_0(\varepsilon, \{a_n\})].$$

Let $\nu := \nu(x)$, $\mu' := \mu(x, y)$, and let k be the least $n \geqq 1$ with $a_n \neq 0$. Then, for $x > (\varepsilon|\alpha_k|)^{-1/k}$, $y > 0$, we have

$$\varepsilon^{-1} < |\alpha_k x^k| \leqq |\alpha_\nu x^\nu| = |\alpha_\nu (xy)^\nu| \, y^{-\nu} \leqq \mu' y^{-\nu}.$$

2.3. By combining the results of lemmas A 1 and A 2, and eliminating $\mu(\;)$ or $\nu(\frac{1}{2})$, we can obtain a functional inequality for the other function. The next lemma (which contains the kernel of the proof of theorem A) will enable us to replace this by an explicit inequality.

Lemma A 3. *Suppose that $\varphi(u) \geqq 0$ (and finite) for each $u > u_0$, and that, for some fixed $\delta > 0$,*

$$\varphi(u)\, v \leqq \sqrt{\varphi(u+v)} \quad (u > u_0,\ 0 < v \leqq \delta). \tag{5}$$

Then

$$\varphi(u) \leqq \delta^{-2} \quad (u > u_0).$$

For $q = 1, 2, \ldots$, we have

$$\varphi(u + v_1 + \cdots + v_q) \geqq (\varphi(u))^{2^q} v_1^2 v_2^{2^2} \cdots v_q^{2^q} \quad (u > u_0,\ 0 < v_r \leqq \delta). \tag*{$(6)_q$}$$

For $(6)_1$ is true by (5); and $(6)_{q+1}$ follows from $(6)_q$ (with $u + v_{q+1}$ in place of u) and (5) (with $v = v_{q+1}$).

In $(6)_q$ take $v_1 + \cdots + v_q = \delta$, and (to maximize the right-hand side subject to this condition)

$$v_r = \frac{2^r \delta}{Q}, \quad \text{where} \quad Q := 2 + 2^2 + \cdots + 2^q = 2^{q+1} - 2;$$

and let

$$S := \sum_{r=1}^{q} r 2^r = (q-1)\, 2^{q+1} + 2 > (q-1)\, Q.$$

Then $(6)_q$ gives, for $u > u_0$,

$$\varphi(u+\delta) \geqq (\varphi(u))^{2^q}\, 2^S \left(\frac{\delta}{Q}\right)^Q \geqq \varphi(u) \left(\frac{\varphi(u)\,\delta^2}{4^2}\right)^{1/2Q}, \tag{7}$$

since $2^q = 1 + \frac{1}{2}Q$ and $2^S > (2^{q-1})^Q > (Q/4)^Q$. It follows that $\varphi(u) \leqq 4^2\delta^{-2}$ for each $u > u_0$; otherwise the right-hand side of (7) tends to infinity when $q \to \infty$ (u fixed) while the left-hand side remains fixed an finite.

Thus we have proved that $\varphi(u) \leqq C\delta^{-2}$ ($u > u_0$), with $C = 16$. This would suffice for the application; but we note that, by (5) with $v = \delta$, such an inequality implies the same with C replaced by $C^{1/2}$. Applying this n times and making $n \to \infty$ (u fixed), we deduce the inequality with $C = 1$.

The example $\varphi(u) := \delta^{-2}$ shows that the result is best possible.

2.4. We now come to the main result of this section.

Theorem A. *If* (a) *satisfies a gap condition* $[\frac{1}{2}, h]$, *and*

$$e^{-x}\,|A(x)| \leqq M \quad (x > 0), \tag{8}$$

then

$$a_n = O(e^{H'\sqrt{n}}) \quad \textit{as} \quad n \to \infty. \tag{9}$$

We may suppose the a_n ($n \geqq 1$) not all 0. Then $M > 0$, and, since $a_n = A_n - A_{n-1}$ (with $A_{-1} := 0$), we have

$$|a(x)| = \left| A(x) - \int_0^x A(t)\,dt \right| < 2M\,e^x \quad (x > 0),$$

by (8). Hence, by lemmas A 2 ($\varepsilon = 1/M$) and A 1 (with xy in place of x),

$$\nu(x) \log y < xy + H\sqrt{\nu(xy)} \quad (x > x_0, y > 1),$$

where we may take $H \geqq 2(1 + e)$, $x_0 \geqq 4e$. Let

$$f(x) := \frac{1}{H^2} + \frac{\nu(x)}{H^2 x}.$$

Then $0 < f(x) < \infty$ for each $x > 0$; and, for $x > x_0$, $1 < y \leqq e$,

$$f(x) \log y < \frac{\log y}{H^2} + \frac{y}{H^2} + \frac{\sqrt{\nu(xy)}}{Hx} < \frac{1}{2H} + \frac{\sqrt{\nu(xy)}}{2H\sqrt{xy}} < \sqrt{f(xy)},$$

where we have used the inequalities $\log y + y \leqq \frac{1}{2}H$, $\sqrt{x} > 2\sqrt{y}$, $\frac{1}{2}\sqrt{X} + \frac{1}{2}\sqrt{Y} < \sqrt{X+Y}$ ($X, Y > 0$). Writing $x = e^u$, $y = e^v$, $x_0 = e^{u_0}$, we conclude that the function $\varphi(u) := f(e^u)$ satisfies the conditions of lemma A 3 with $\delta = 1$; whence $f(x) \leqq 1$ and so

$$\nu(x) < H^2 x \quad (x > x_0).$$

Hence, by lemma A 1 (again) and the definition of $\mu(x)$,

$$|a_n x^n / n!| \leqq \mu(x) < M \exp(x + H^2\sqrt{x}) \quad (x > x_0;\ n \geqq 0).$$

Taking $x = n$, we deduce that, for $n > x_0$,

$$|a_n| < M n^{-n} n! \exp(n + H^2\sqrt{n}) = O(\sqrt{n}\exp(H^2\sqrt{n}))$$

as $n \to \infty$, by Stirling's theorem [or by less precise relations, such as (E 3) below with $\zeta = 1$]. Hence the result, with (say) $H' := H^2 + 1$.

2.5. We can now complete the proof of the general high-indices theorem by an appeal to the restricted high-indices theorem of Pitt quoted in the introduction. The hypotheses of the general theorem imply the hypotheses, and therefore the conclusion (9), of theorem A. But, by Pitt's theorem, the hypotheses of the general theorem, supplement by the restriction (9), imply the desired convergence.

In the next two sections we shall show, however, that there is no need to assume a prior knowledge of Pitt's theorem. This we do by proving a similar, but wider, theorem independently.

3. A restricted high-indices theorem

3.1. To the peak properties of $P_N(t)$ we now add some properties of $x^n/n!$, regarded as a peak function in n $(= 0, 1, 2, \ldots)$.

Lemma E.

(E 1) $$e^{-u} \leqq 1 - u + \frac{1}{2}u^2 \quad (u \geqq 0);$$

(E 2) $$\sum_{n \leqq x-\xi} \frac{x^n}{n!} e^{\sigma(x-\xi-n)} < \exp\left(x - \frac{\xi^2}{2x}\right) \quad \left(x, \xi > 0;\ \sigma \leqq \frac{\xi}{x}\right);$$

(E 3) $$\sum_{|n-x|<\zeta} \frac{x^n}{n!} > \frac{\zeta\, e^x}{8\sqrt{x}} \quad (x \textit{ a positive integer};\ 0 < \zeta \leqq \sqrt{x}).$$

We have (E 1) by Taylor's theorem:

$$e^{-u} = 1 - u + \tfrac{1}{2} u^2 e^{-\Theta u}, \quad \text{where} \quad 0 < \Theta < 1.$$

For given x, ξ, denote the sum in (E 2) (possibly empty) by $E(\sigma)$. Taking $\varrho \geqq \max(\sigma, 0)$, we have

$$E(\sigma) \leqq E(\varrho) < e^{\varrho(x-\xi)} \exp(x\, e^{-\varrho}) \leqq e^{-\varrho\xi} \exp(x + \tfrac{1}{2} x\varrho^2),$$

by extending the sum $E(\varrho)$ to infinity and then applying (E 1) to $e^{-\varrho}$. Taking $\varrho = \xi/x$ (to minimize the right-hand side), we obtain (E 2).

Take a fixed integer x, and suppose first $x \geqq 3$. For integers $q \geqq 0$ let

$$E_q := \sum_{n=x-q}^{x+q} u_n, \quad \text{where} \quad u_n := \begin{cases} x^n/n! & (n \geqq 0), \\ 0 & (n < 0). \end{cases}$$

Since $u_{n+1} < u_n$ $(n \geqq x)$ and $u_{n-1} \leqq u_n$ $(n \leqq x)$, the mean value $E_q/(2q + 1)$ decreases as q increases. Hence, for any integer $Q > q$,

$$\frac{2Q + 1}{2q + 1} E_q > E_Q > \sum_{-\infty}^{\infty} \left(1 - \frac{(n - x)^2}{Q^2}\right) u_n = \left(1 - \frac{x}{Q^2}\right) e^x.$$

Given ζ with $0 < \zeta \leqq \sqrt{x}$, let q be the integer defined by $q < \zeta \leqq q + 1$; and let $Q := [2\sqrt{x}] + 1$. Then $Q > 2\sqrt{x} > q \geqq 0$, and $2Q + 1 < 6\sqrt{x}$ (since $3 \leqq \sqrt{3x} < 2\sqrt{x}$);

and so

$$\frac{E_q}{e^x} > \frac{2q+1}{2Q+1}\left(1 - \frac{x}{Q^2}\right) > \frac{\zeta}{6\sqrt{x}}\left(1 - \frac{1}{4}\right) = \frac{\zeta}{8\sqrt{x}}.$$

Thus (E 3) is true for $x \geqq 3$; and it holds also for $x = 1, 2$, since $u_x/e^x > 1/8$ in these cases.

3.2. The main theorem of this section will be proved by a "maximum term argument" of a different type from the one used in § 2. It will be convenient to state and prove the theorem as an independent item, without assuming any knowledge of § 2 beyond § 2.1.

Theorem B. *Suppose that* (a) *satisfies a gap condition* $[\frac{1}{2}, h]$, *and that*

$$a_n = O(e^{\varepsilon n}) \quad as \quad n \to \infty, \tag{10}$$

for every fixed $\varepsilon > 0$. *Then*:

(I) $\quad e^{-x}|A(x)| \leqq M \ (x \geqq 0) \Rightarrow |A_n| \leqq H_1 M \ (n \geqq 0);$

(II) $\quad \overline{\lim_{x\to\infty}}\, e^{-x}|A(x)| \leqq \Omega \Rightarrow \overline{\lim_{n\to\infty}}\, |A_n| \leqq H_2 \Omega.$

(I) We suppose first, instead of (10), that

$$A_n \to A \quad \text{as} \quad n \to \infty. \tag{11}$$

(This drastic assumption does not reduce our result to triviality, since the conclusion is a universal inequality and not an asymptotic relation.) We suppose further, in the first instance, that

$$|A_n| > |A| \quad \text{for some } n. \tag{12}$$

Then $|A_n|$ has a greatest value, say $|A_m|$, where m is taken to be the least n for which the maximum occurs. Since $a_n = 0$ when $n \neq n_r$, we must have $m = n_\mu$ ($\geqq 1$) for some μ, and $A_n = A_m$ for $n_\mu \leqq n < n_{\mu+1}$. Let

$$x := (n_\mu + \tfrac{1}{2} h n_\mu^{1/2}). \tag{13}$$

Then

$$n_\mu + \tfrac{1}{2} h n_\mu^{1/2} - 1 < x < n_{\mu+1} - \tfrac{1}{2} h n_\mu^{1/2} \tag{14}$$

by $[\frac{1}{2}, h]$ (in which we shall suppose $0 < h \leqq 1$). Take N (integral), η, ξ so that

$$N \geqq 2, \quad 0 < \eta \leqq \tfrac{1}{3} h \sqrt{x}, \quad \eta < \xi \leqq N\eta, \tag{15}$$

and let

$$S := \sum_{n=0}^{\infty} A_n \frac{x^n}{n!} P_N\left(\frac{n-x}{\xi}\right) = S_0 + S_1 + S_2,$$

where the summation conditions (besides $n \geqq 0$) are:

(S_0) $\quad |n-x| < \eta;$

(S_1) $\quad |n-x| \geqq \eta, \quad n > x - \xi;$

(S_2) $\quad n \leqq x - \xi.$

Denote the corresponding summations by $\sum_0, \sum_1, \sum_2$.

By (P 1), since $2 = e^l$,

$$S = \sum_{n=0}^{\infty} A_n \frac{x^n}{n!} \sum_{s=1}^{2N} p_s e^{ls(x-n)/\xi} = \sum_{s=1}^{2N} p_s e^{lsx/\xi} A(x e^{-ls/\xi}),$$

and so, by hypothesis,

$$|S| \leqq \sum_{s=1}^{2N} |p_s| e^{lsx/\xi} M \exp(x e^{-ls/\xi}) \leqq M \exp\left(4lN + x + x\frac{2l^2N^2}{\xi^2}\right), \tag{16}$$

by (E 1) ($u = ls/\xi$) and (P 2), since $s^2 \leqq 4N^2$ in each term.

In S_0 we have $(n-x)/\xi > -\eta/\xi > -1$; and also, by (15) and (13),

$$|n-x| < \eta \leqq \tfrac{1}{3} h (n_\mu + \tfrac{1}{2} hn_\mu)^{1/2} < \tfrac{1}{2} hn_\mu^{1/2}$$

(since $0 < h \leqq 1$), and so $n < n_{\mu+1}$ and $n > n_\mu - 1$ by (14). Thus we have $A_n = A_m$ throughout S_0, and so

$$|S_0| = \left| A_m \textstyle\sum_0 \dfrac{x^n}{n!} P_N\left(\dfrac{n-x}{\xi}\right)\right| \geqq |A_m| \dfrac{\xi e^x}{8N\sqrt{x}} \cdot \dfrac{1}{4}, \tag{17}$$

by restricting $\sum_0$ to $|n-x| < \xi/N$ ($\leqq \eta < \sqrt{x}$) [taking account of (P 3)] and then applying (P 7), (E 3) ($\zeta = \xi/N$).

In S_1 we still have $(n-x)/\xi > -1$, and so, by (P 3, 6), since $|A_n| \leqq |A_m|$ for all n,

$$|S_1| \leqq |A_m| \textstyle\sum_1 \dfrac{x^n}{n!} \dfrac{5\xi^2}{N^2\eta^2} \leqq |A_m| e^x \dfrac{5\xi^2}{N^2\eta^2}. \tag{18}$$

For S_2 (which may be empty) we have, in any existing term,

$$\left| P_N\left(\frac{n-x}{\xi}\right)\right| \leqq \sum_{s=1}^{2N} |p_s| e^{ls(x-n)/\xi} < e^{4lN + 2lN(x-n)/\xi}$$

by (P 1, 2); whence

$$|S_2| \leqq |A_m|\, e^{6lN} \sum\nolimits_2 \frac{x^n}{n!} e^{2lN(x-\xi-n)/\xi} \leqq |A_m| \exp\left(6lN + x - \frac{\xi^2}{2x}\right)$$

by (E 2), if $2lN/\xi \leqq \xi/x$. We now assume the stronger inequality

$$\xi^2/2x \geqq 7lN, \tag{19}$$

and deduce that

$$|S_2| \leqq |A_m|\, e^{x-lN}. \tag{20}$$

Inserting (16), (17), (18), (20) into the obvious inequality

$$|S_0| - |S_1| - |S_2| \leqq |S|,$$

dividing by e^x, and noting that $e^{lN} = 2^N > N$, we obtain

$$|A_m| \left(\frac{\xi}{32N\sqrt{x}} - \frac{5\xi^2}{N^2\eta^2} - \frac{1}{N}\right) \leqq M \exp\left(4lN + \frac{2l^2N^2x}{\xi^2}\right), \tag{21}$$

if (15) and (19) are satisfied.

Now let

$$\eta := \tfrac{1}{3} h \sqrt{x}, \quad \xi := 4\sqrt{Nx},$$

with N to be chosen later. Condition (19) is satisfied (since $8 > 7l$); and so are conditions (15) if $\sqrt{N} \geqq 12/h$. The coefficient of $|A_m|$ in (21) is

$$\frac{1}{8\sqrt{N}} - \frac{720}{Nh^2} - \frac{1}{N} \geqq \frac{1}{8\sqrt{N}} - \frac{721}{Nh^2} > \frac{1}{32\sqrt{N}}$$

if $N = N(h)$ is large enough. All conditions are certainly satisfied if $N := [2^{26}/h^4]$; and we then deduce from (21), since $|A_n| \leqq |A_m|$ (all n) and $e^{l^2} < e^l = 2$, that

$$|A_n| \leqq H_1 M \,(n \geqq 0), \quad \text{where} \quad H_1 := 32N^{1/2}\, 2^{(4+(1/8))N}.$$

This holds also if (a) satisfies (11) but not (12), since, by the regularity of B-summability, we then have, for each n,

$$|A_n| \leqq |A| = \lim_{x\to\infty} |e^{-x}A(x)| \leqq M \leqq H_1 M.$$

We have thus proved (I) in the special case (11). Now suppose only that (a) satisfies (10), together with the other hypotheses of (I). Let

$$b_n = b_n(\beta) := a_n\beta^n \quad (0 < \beta < 1;\ \beta \text{ fixed}).$$

The series (b) is convergent, by (10); so $B_n > B$ as $n \to \infty$, for some $B = B(\beta)$. Also (b) satisfies $[\frac{1}{2}, h]$. Further, $b(x) = a(\beta x)$, $A'(x) - A(x) = a'(x)$, etc.; whence

$$e^{-x}B(x) = e^{-x}A(\beta x) + (1-\beta)\int_0^x e^{-u}A(\beta u)\,du, \tag{22}$$

since the two sides are equal to $b_0 = a_0$ when $x = 0$ and have equal derivatives $e^{-x}b'(x) = e^{-x}\beta a'(\beta x)$ for each x. Hence

$$e^{-x}|B(x)| \leqq e^{-x}M\,e^{\beta x} + (1-\beta)\int_0^x e^{-u}M\,e^{\beta u}\,du = M \quad (x \geqq 0).$$

Thus (b) satisfies all conditions of the special case, with the same M. Hence $|B_n| \leqq H_1 M$ $(n \geqq 0)$; from which our result follows, since $B_n = B_n(\beta) \to A_n$ when $\beta \to 1-$, for each fixed $n \geqq 0$.

(II) Consider the series (c), where

$$c_n = c_n(\beta) := a_n(1-\beta^n) = a_n - b_n \quad (\tfrac{1}{2} < \beta < 1;\ \beta \text{ fixed}).$$

Since $C(x) = A(x) - B(x)$, we have, by (22),

$$e^{-x}C(x) = e^{-x}(A(x) - A(\beta x)) - \int_0^x e^{-u}(1-\beta)\,A(\beta u)\,du. \tag{23}$$

Given $\varepsilon > 0$, take $U = U(\varepsilon) > 0$ so that

$$e^{-v}|A(v)| < \Omega + \varepsilon \quad (v > \tfrac{1}{2}U). \tag{24}$$

With U thus fixed, A(v) is continuous, therefore uniformly continuous and bounded, in $0 \leqq v \leqq U$. We can therefore choose our $\beta = \beta(\varepsilon, U) = \beta(\varepsilon)$ so near to 1 that

$$|A(v) - A(\beta v)| < \varepsilon, \quad (1-\beta)\,|A(v)| < \varepsilon \quad (0 \leqq v \leqq U).$$

Combined with (24) this implies (since $\frac{1}{2} < \beta < 1$) that

$$\left.\begin{aligned} &|A(v) - A(\beta v)| < (\Omega + \varepsilon)(e^{v} + e^{\beta v}), &(25)\\ &(1-\beta)|\,A(v)| < (1-\beta)(\Omega+\varepsilon)\,e^{v} + \varepsilon, &(26)\end{aligned}\right\} \quad (v \geqq 0).$$

Substituting into (23), we obtain

$$e^{-x}|C(x)| \leqq (\Omega + \varepsilon)(1 + e^{(\beta-1)x} + 1 - e^{(\beta-1)x}) + \varepsilon = 2\Omega + 3\varepsilon$$

for all $x \geqq 0$. But (c) satisfies $[\frac{1}{2}, h]$. Hence, by (I),

$$|C_n| \leqq H_1(2\Omega + 3\varepsilon) \quad (n \geqq 0). \tag{27}$$

Now $C_n = A_n - B_n$; and, by (10), the series (*b*) is convergent, so that $B_n \to B$ as $n \to \infty$, for some B depending on ε. Also, by (22), (24) ($v = \beta x$), (26) ($v = \beta u$),

$$e^{-x}|B(x)| < (\Omega + \varepsilon) + \varepsilon \quad (x > U),$$

and so, by regularity of B-summability

$$\lim_{n\to\infty} |B_n| = \lim_{x\to\infty} |e^{-x}B(x)| \leqq \Omega + 2\varepsilon. \tag{28}$$

From (27) and (28) we deduce, since $A_n = C_n + B_n$, that

$$\overline{\lim_{n\to\infty}} |A_n| \leqq H_1(2\Omega + 3\varepsilon) + (\Omega + 2\varepsilon).$$

Since $\varepsilon > 0$ is arbitrary and A_n is independent of ε, our result follows, with $H_2 := 2H_1 + 1$.

4. The general high-indices theorem

We reach our final conclusion by combining the results of §§ 2 and 3.

Theorem B.

(i) *The hypothesis* (10) *may be omitted from theorem* B.

(ii) (*High-indices theorem*). *If* (*a*) *satisfies a gap condition* $[\frac{1}{2}, h]$ *and is* B-*summable to A, then it is convergent to A.*

(i) The hypotheses, other than (10), of theorem B, (I) or (II), imply the hypotheses, and therefore the conclusion (9), of theorem A. But (9) in turn implies (10), which is thus a redundant hypothesis.

(ii) By changing a_k to $a_k - A$ for (say) $k = n_1$ we may suppose $A = 0$. The result then follows from the case $\Omega = 0$ of the extended Theorem B (II).

5. Concluding remarks

These remarks may be read in conjunction with similar comments on Abel summability [5, § 4]. In that context we used functions $H_A(x)$ and $H_B(x)$ with peaks at $x = 0$. The analogue here of $H_B(x)$ is $P_N(t) = N^{-2}H_B(1 + t)$ with peak at $t = 0$. The analogue of $H_A(x)$ would be $p_N(t) := (2^{1-t} - 2^{-2t})^N$; and the function used by Pitt [7, 41] has a somewhat similar structure when allowance is made for differences of notation. As explained in the earlier account, $P_N(t)$ has a sharper peak than $p_N(t)$

for a given N: an analogue of (P 6) for $p_N(t)$ would have only N (not N^2) in the denominator. With Abel summability, so long as we are not concerned with explicit O-estimates, the blunter function suffices, even for the high-indices theorem (with gap condition [1, h]) [3, 172–174; 5, Theorem A]. With B-summability, however, it seems that the sharper $P_N(t)$ must be used if arguments like those of §§ 2 and 3 are to succeed. The use of $p_N(t)$ in lemma A 1 would give a weaker result that could not be developed; and in theorem B it would leave no room for a possible N.

For reasons indicated in the earlier account, the use of $p_N(t)$ may be expected to be similar in scope to the method of repeated differentiation, though simpler in execution. It is a familiar fact that, in the context of B-summability, these methods achieve only a limited degree of "peak-sharpening" in one step, so that repeated application is necessary (see [4] and [7, 41]). This is no obstacle if we are working with a summation method in which the factor $x^n/n!$ has been replaced by e^{-cu^2}; for "peak-sharpening" then amounts only to a suitable increase in c. But the transfer to such a method presupposes a heavy restriction on the order of a_n, and is hardly practicable for the high-indices theorem. We therefore work directly with B-summability, and are virtually compelled to use the sharper function $P_N(t)$.

The peak method has been applied to other aspects of B-summability by Miss R. Deb in a Cambridge dissertation (1963). Since this work is in process of publication, we make no further comment on it here.

References

[1] D. Gaier, Der allgemeine Lückenumkehrsatz für das Borel-Verfahren, Math. Z. **88** (1965), 410—417.

[2] D. Gaier, On the coefficients and the growth of gap power series, J. SIAM Numer. Anal. **3** (1966), 248—265.

[3] G. H. Hardy, Divergent series, Oxford 1949.

[4] G. H. Hardy and J. E. Littlewood, Theorems concerning the summability of series by Borel's exponential method, Rend. Circ. Mat. Palermo **41** (1916), 36—53.

[5] A. E. Ingham, On Tauberian theorems, Proc. London Math. Soc. (3) **14A** (1965), 157—173.

[6] V. I. Mel'nik, The Tauberian theorem of "large exponents" for the method of Borel (Russian), Mat. Sb. (N.S.) **68 (110)** (1965), 17—25.

[7] H. R. Pitt, Tauberian theorems, Oxford 1958.

V. Jarník in Praha

BEMERKUNGEN ZU LANDAUSCHEN METHODEN IN DER GITTERPUNKTLEHRE

§ 1. Einleitung

Im folgenden sei stets

$$Q(u) = Q(u_1, \ldots, u_r) = \sum_{i,k=1}^{r} \alpha_{ik} u_i u_k \quad (\alpha_{ik} = \alpha_{ki})$$

eine positiv-definite quadratische Form in r Veränderlichen mit der Determinante D (die Buchstaben r, Q, α_{ik}, D behalten immer diese Bedeutung). Für $x > 0$ sei $A(x) = A(x; Q) = \sum_{Q(m) \leqq x} 1$ die Anzahl der Gitterpunkte (d. h. der Punkte $(m) = (m_1, \ldots, m_r)$ mit ganzen m_i) im Ellipsoid $Q(u) \leqq x$; das Volumen dieses Ellipsoids ist $V(x) = \pi^{r/2} x^{r/2} D^{-(1/2)} / \Gamma\left(\frac{r}{2} + 1\right)$, und $P(x) = A(x) - V(x)$ ist der „Gitterrest", dessen Verhalten für $x \to +\infty$ Gegenstand zahlreicher Untersuchungen gewesen ist. Man setze noch für $\varrho > 0$

$$\left.\begin{aligned}
&A_0(x) = A(x), \quad A_\varrho(x) = \frac{1}{\Gamma(\varrho)} \int_0^x A(y)(x-y)^{\varrho-1}\, dy, \\
&V_0(x) = V(x), \\
&V_\varrho(x) = \frac{1}{\Gamma(\varrho)} \int_0^x V(y)(x-y)^{\varrho-1}\, dy = \pi^{r/2} x^{r/2+\varrho} D^{-(1/2)} / \Gamma\left(\frac{r}{2} + \varrho + 1\right), \\
&P_0(x) = P(x), \quad P_\varrho(x) = \frac{1}{\Gamma(\varrho)} \int_0^x P(y)(x-y)^{\varrho-1}\, dy,
\end{aligned}\right\} \tag{1}$$

so daß

$$P_\varrho(x) = A_\varrho(x) - V_\varrho(x) \quad \text{für} \quad \varrho \geqq 0 \tag{2}$$

gilt. Für $\varrho > 0$ ist nach (1)

$$A_\varrho(x) = \frac{1}{\Gamma(\varrho)} \int_0^x \sum_{Q(m) \leqq y} (x-y)^{\varrho-1}\, dy = \frac{1}{\Gamma(\varrho)} \sum_{Q(m) \leqq x} \int_{Q(m)}^x (x-y)^{\varrho-1}\, dy,$$

d. h.

$$A_\varrho(x) = \frac{1}{\Gamma(\varrho + 1)} \sum_{Q(m) \leqq x} (x - Q(m))^\varrho, \tag{3}$$

und dies gilt offenbar auch für $\varrho = 0$. Daraus folgt für $\varrho \geqq 0$

$$\int_0^x A_\varrho(y)\, dy = \frac{1}{\Gamma(\varrho + 1)} \int_0^x \sum_{Q(m) \leqq y} (y - Q(m))^\varrho\, dy = \frac{1}{\Gamma(\varrho + 2)} \sum_{Q(m) \leqq x} (x - Q(m))^{\varrho + 1},$$

d. h.

$$A_{\varrho+1}(x) = \int_0^x A_\varrho(y)\, dy \quad \text{für} \quad \varrho \geqq 0. \tag{4}$$

Dieselbe Formel gilt offenbar für V_ϱ und daher wegen (2) auch für P_ϱ. Für ganze $\varrho > 0$ kann man also A_ϱ auch durch (4) definieren; wir werden aber im folgenden alle reellen $\varrho \geqq 0$ zulassen.

Als ich im Herbst 1923 nach Göttingen kam, um drei Semester lang unter Leitung von Edmund Landau zu arbeiten, endete die etwa zwölfjährige Periode, während der sich Landau intensiv mit Gitterpunktproblemen beschäftigte; zugleich erreichten zu jener Zeit seine diesbezüglichen Methoden ihren Höhepunkt in Wirkungskraft und Einfachheit. Die vorliegende Note schließt sich sehr eng an die im Jahre 1924 entstandenen Landauschen Arbeiten [4] bis [7] an; es ist aber vielleicht gerade bei dieser Gelegenheit nicht unangemessen, sich wieder einmal die Vollkommenheit der Landauschen Methoden zu vergegenwärtigen.

Um das Ziel dieser Note zu erläutern, wähle ich das Kreisproblem, d. h. $r = 2$, $Q(u) = u_1^2 + u_2^2$. Die Funktion $P(x)$ $\left(\text{genauer gesagt } \lim_{h \to 0} \frac{1}{2}(P(x + h) + P(x - h))\right)$ läßt sich mit Hilfe Besselscher Funktionen durch eine Reihe darstellen, die sich aber wegen „schlechter" Konvergenz zur Abschätzung von P nicht eignet. Dagegen sind die analogen Reihenentwicklungen von $P_1(x)$, $P_2(x)$, … für $x > 0$ absolut und gleichmäßig konvergent und lassen bequeme Abschätzungen zu. Daher läßt sich das O-Ω-Problem für diese Funktionen vollständig lösen; es ist

$$P_\varrho(x) = O(x^{1/4 + \varrho/2}), \quad P_\varrho(x) = \Omega(x^{1/4 + \varrho/2}) \quad \text{für} \quad \varrho = 1, 2, \ldots. \tag{5}$$

Durch geeignete Differenzenbildung erhält man – von P_1 ausgehend – auch Auskunft über $P = P_0$:

$$P(x) = O(x^{1/3}), \quad P(x) = \Omega(x^{1/4}). \tag{6}$$

Mit Hilfe von scharfsinnigen und methodisch sehr wichtigen Methoden ist es gelungen, (6) zu verschärfen: in der O-Formel kann man $\frac{1}{3}$ durch eine kleinere Zahl ersetzen, und die Ω-Formel kann man z. B. zu $P(x) = \Omega(x^{1/4} \log^{1/4} x)$ verschärfen.

Man kennt aber nicht einmal den „wahren Exponenten", d. h. die untere Grenze der α mit $P(x) = O(x^\alpha)$.

Analoges gilt für $r = 3$, und auch für größere r findet man, daß die P_ϱ für große ϱ (nämlich für $\varrho > r/2 - \frac{1}{2}$) leicht behandelt werden können. Man könnte also vermuten, daß die P_ϱ um so leichter zu handhaben sind, je größer ϱ ist; z. B. wenn man für eine Form Q den wahren Exponenten für $P = P_0$ kennt, daß man dann um so leichter den wahren Exponenten für P_ϱ mit $\varrho > 0$ bestimmen kann. Aber es sieht anders aus, mindestens für ganze α_{ik}:

Satz 1. *Es seien die α_{ik} ganz, $\varrho \geqq 0$. Dann gelten für $P_\varrho(x)$ folgende Abschätzungen:*

$$O(x^{r/2-1}),\quad \Omega(x^{r/2-1}) \quad \textit{für} \quad \varrho < \frac{r}{2} - 2, \tag{7}$$

$$O(x^{r/2-1}\log x),\quad \Omega(x^{r/2-1}) \quad \textit{für} \quad \varrho = \frac{r}{2} - 2, \tag{8}$$

$$\left.\begin{array}{l} O(\mathrm{Min}\,(x^{\varrho+r/2-r/(r+1-2\varrho)}, x^{\varrho/2+r/4})), \\ \Omega(\mathrm{Max}\,(x^{\varrho/2+(r-1)/4}, x^{r/2-1})) \quad \textit{für} \quad \dfrac{r}{2} - 2 < \varrho < \dfrac{r}{2} - \dfrac{1}{2}, \end{array}\right\} \tag{9}$$

$$O(x^{\varrho/2+(r-1)/4}\log x),\quad \Omega(x^{\varrho/2+(r-1)/4}) \quad \textit{für} \quad \varrho = \frac{r}{2} - \frac{1}{2}, \tag{10}$$

$$O(x^{\varrho/2+(r-1)/4}),\quad \Omega(x^{\varrho/2+(r-1)/4}) \quad \textit{für} \quad \varrho > \frac{r}{2} - \frac{1}{2}, \tag{11}$$

mit der Ausnahme, daß (8) *für $r = 4$, $\varrho = r/2 - 2 = 0$ durch*

$$O(x\log^2 x),\quad \Omega(x) \quad \textit{für} \quad r = 4, \varrho = 0 \tag{12}$$

zu ersetzen ist.

Ich habe schon bemerkt, daß für $\varrho = 0$ und $r = 2, 3$ schärfere Resultate als (9) bekannt sind. Dasselbe gilt im Falle $\varrho = 0$, $r = 4$: Man kann in (12) $\log^2 x$ durch $\log x$, ja durch eine noch niedrigere Potenz von $\log x$ ersetzen, während andererseits die Abschätzung $O(x)$ z. B. für die vierdimensionale Kugel falsch ist.

Man analysiere ein wenig den Sinn des Satzes 1. Für $\varrho \geqq r/2 - \frac{1}{2}$ gibt Satz 1 den wahren Exponenten von P_ϱ an (und für $\varrho > r/2 - \frac{1}{2}$ sogar die vollständige Lösung des O-Ω-Problems). Analoges gilt im Intervall $0 \leqq \varrho \leqq r/2 - 2$, das für $r = 4$ auf den Punkt $\varrho = 0$ zusammenschrumpft und für $r < 4$ überhaupt wegfällt. Für $r/2 - 2 < \varrho < r/2 - \frac{1}{2}$ (und $\varrho \geqq 0$) löst unser Satz die Frage nach dem wahren Exponenten nicht; der Leser wird sehen, daß dieses Problem ebenso schwierig wie das Kreisproblem zu sein scheint. Man könnte analog zu den für $\varrho = 0$, $r = 2, 3, 4$

bekannten Verschärfungen versuchen, (8), (9), (10) zu verschärfen; ich gehe aber nicht darauf ein.

Man betrachte noch die Exponenten in (9). Die beiden Ω-Exponenten sind einander gleich für $\varrho = r/2 - \frac{3}{2}$; für größere (kleinere) ϱ ist der erste (zweite) größer, d. h. schärfer. Dabei kommen Werte $\varrho < r/2 - \frac{3}{2}$ nur für $r > 3$ in Betracht. Analog ist von den beiden O-Exponenten der erste schärfer (d. h. kleiner) als der zweite genau dann, wenn

$$(2\varrho)^2 - 2\varrho - r(r-3) > 0 \tag{13}$$

gilt. Ist $r = 2$, so gilt (13) für alle ϱ; für $r > 2$ hat aber die linke Seite von (13) eine positive Nullstelle ϱ_0 mit $r/2 - 2 < \varrho_0 < r/2 - \frac{1}{2}$. Daher ist der erste (zweite) Exponent schärfer als der andere, wenn $\varrho > \varrho_0$ ($\varrho < \varrho_0$) ist. Eine Verschärfung der Ω-Abschätzung für $r/2 - \frac{3}{2} \leqq \varrho \leqq r/2 - \frac{1}{2}$ ($\varrho \geqq 0$) im Falle der r-dimensionalen Kugel findet man in [13].

§ 2. Anwendung der Besselschen Funktionen

In diesem Paragraphen sind alle Zahlen reell. Wir brauchen eine Identität (Satz 2) mit Besselschen Funktionen

$$J_\nu(x) = \left(\frac{x}{2}\right)^\nu \sum_{k=0}^{\infty} \left(-\frac{1}{4}x^2\right)^k (k!\,\Gamma(\nu + k + 1))^{-1} \tag{14}$$

(wir werden sie nur für $x > 0$ und reelles ν brauchen). Diese Identität wurde von Landau in [2], [3] (oder [1], S. 11–29 und S. 258–264) auf komplexem Wege bewiesen.[1]) In [6] (oder [1], S. 112–147) hat Landau zur Herleitung derartiger Identitäten eine einfache Methode im reellen Gebiet entwickelt und auf Gitterpunktprobleme in der Ebene angewandt. Ich gebe hier einen Beweis im Reellen, der die Landausche Methode aus [6] imitiert.

Bekanntlich ist für $x \to +\infty$

$$J_\nu(x) = 2^{1/2}\pi^{-1/2}x^{-1/2}\cos\left(x - \frac{1}{2}\nu\pi - \frac{1}{4}\pi\right) + O(x^{-3/2}). \tag{15}$$

Durch gliedweises Differenzieren bekommt man leicht für $x > 0$, $A > 0$

$$\frac{d}{dx}(A^{-\nu}x^{\nu/2}J_\nu(Ax^{1/2})) = \frac{1}{2}A^{-(\nu-1)}x^{(\nu-1)/2}J_{\nu-1}(Ax^{1/2}). \tag{16}$$

---

[1]) Landau setzt ϱ ganz voraus, das ist aber unerheblich.

Nach Liouville gilt weiter: Es sei

$$A = \int_0^1 \varphi(u)\, u^{p_1 + \cdots + p_r - 1}\, du \tag{17}$$

konvergent, $p_i > 0$. Der Bereich M sei durch $x_1 > 0, \ldots, x_r > 0$, $x_1 + \cdots + x_r < 1$ gegeben. Dann ist

$$\int \cdots \int_M \varphi(x_1 + \cdots + x_r)\, x_1^{p_1 - 1} \cdots x_r^{p_r - 1}\, dx_1 \cdots dx_r = \frac{\Gamma(p_1) \cdots \Gamma(p_r)}{\Gamma(p_1 + \cdots + p_r)} A. \tag{18}$$

Hilfssatz 1. *Es sei* $A > 0$, $B > 0$. *Die Funktion* $f(u_1, \ldots, u_r)$ *habe folgende Eigenschaften:*

1. f *ist stetig im r-dimensionalen Raum* R^r.
2. $f(u_1, \ldots, u_r) = 0$ *für* Max $(|u_1|, \ldots, |u_r|) \geqq A$.
3. *Für jedes* $i\,(i = 1, \ldots, r)$ *und jede Wahl von* $u_1, \ldots, u_{i-1}, u_{i+1}, \ldots, u_r$ *ist die Schwankung der (als Funktion von* u_i *betrachteten) Funktion* $f(u_1, \ldots, u_r)$ *höchstens gleich* B.

Dann ist

$$\sum_{m_1, \ldots, m_r = -\infty}^{+\infty} f(m_1, \ldots, m_r) = \sum_{a_1 = -\infty}^{+\infty} \cdots \sum_{a_r = -\infty}^{+\infty} \int \cdots \int_{R^r} f(u_1, \ldots, u_r) \times \prod_{j=1}^{r} \cos(2\pi a_j u_j)\, du_1 \cdots du_r. \tag{19}$$

Bemerkung. Man beachte, daß es genügt, links über $|m_i| < A$ zu summieren. Rechts ist es wesentlich, daß es sich um eine *iterierte* Reihe handelt.

Beweis. Der Fall $r = 1$ ist wohlbekannt. Induktion von $r - 1$ auf r: Man setze

$$f_1(u_2, \ldots, u_r) = \sum_{m_1 = -\infty}^{+\infty} f(m_1, u_2, \ldots, u_r), \tag{20}$$

also

$$f_1(u_2, \ldots, u_r) = \sum_{a_1 = -\infty}^{+\infty} F_{a_1}(u_2, \ldots, u_r), \tag{21}$$

wobei

$$F_{a_1}(u_2, \ldots, u_r) = \int_{-\infty}^{+\infty} f(u_1, \ldots, u_r) \cos 2\pi a_1 u_1\, du_1 \tag{22}$$

ist. Nun hat F_{a_1} offenbar die Eigenschaften 1, 2, 3 mit $r - 1$, A, $2AB$ statt r, A, B. Also ist nach Induktionsvoraussetzung

$$\sum_{m_2,\ldots,m_r=-\infty}^{+\infty} F_{a_1}(m_2, \ldots, m_r) = \sum_{a_2=-\infty}^{+\infty} \cdots \sum_{a_r=-\infty}^{+\infty} \int_{R^{r-1}} \cdots \int F_{a_1}(u_2, \ldots, u_r)$$

$$\times \prod_{j=2}^{r} \cos(2\pi a_j u_j)\, du_2 \cdots du_r. \tag{23}$$

Daher ist nach (20), (21)

$$\sum_{m_1,\ldots,m_r=-\infty}^{+\infty} f(m_1, \ldots, m_r) = \sum_{\mathrm{Max}(|m_2|,\ldots,|m_r|)<A} f_1(m_2, \ldots, m_r)$$

$$= \sum_{\mathrm{Max}(|m_2|,\ldots,|m_r|)<A} \sum_{a_1=-\infty}^{+\infty} F_{a_1}(m_2, \ldots, m_r) = \sum_{a_1=-\infty}^{+\infty} \sum_{m_2,\ldots,m_r=-\infty}^{+\infty} F_{a_1}(m_2, \ldots, m_r);$$

daraus und aus (23), (22) folgt die Behauptung.

Hilfssatz 2. *Für $\varrho > 0$, $x > 0$ ist*

$$A_\varrho(x) = \frac{1}{\Gamma(\varrho+1)} \sum_{a_1=-\infty}^{+\infty} \cdots \sum_{a_r=-\infty}^{+\infty} \int_{Q(u)\leq x} \cdots \int (x - Q(u))^\varrho \prod_{j=1}^{r} \cos(2\pi a_j u_j)\, du_1 \cdots du_r$$

$$= \sum_{a_1=-\infty}^{+\infty} \cdots \sum_{a_r=-\infty}^{+\infty} 2^{-r} \sum_{\pm} K(\pm a_1, \pm a_2, \ldots, \pm a_r), \tag{24}$$

wobei

$$K(h_1, \ldots, h_r) = \frac{1}{\Gamma(\varrho+1)} \int_{Q(u)\leq x} \cdots \int (x - Q(u))^\varrho \cos(2\pi(h_1 u_1 + \cdots + h_r u_r))\, du_1 \cdots du_r \tag{25}$$

ist und $\sum_{\pm}$ über die 2^r Kombinationen der Vorzeichen $\pm$ erstreckt wird (auch wenn einige $a_j = 0$ sind).

Beweis. Man wende Hilfssatz 1 auf die Funktion $(\mathrm{Max}(x - Q(u), 0))^\varrho$ an und beachte, daß $\cos\alpha_1 \cos\alpha_2 \cdots \cos\alpha_r = 2^{-r} \sum_{\pm} \cos(\pm\alpha_1 \pm \cdots \pm \alpha_r)$ ist (dies folgt sofort durch Induktion nach r).

Wir berechnen jetzt $K(h_1, \ldots, h_r)$. Es sei (α) die Matrix der Koeffizienten α_{ik} von $Q(u)$. Durch eine Substitution $u_i = \sum_k \gamma_{ik} v_k$ mit der regulären Matrix (γ) geht $Q(u)$ in eine positiv-definite Form $Q'(v)$ mit der Matrix $\overline{(\gamma)}\,(\alpha)\,(\gamma)$ über; dabei bedeutet $\overline{(\gamma)}$ die zu (γ) transponierte Matrix. Man bezeichne mit (A) bzw. (Γ) die zu (α) bzw. (γ) inverse Matrix, mit (ε) die Einheitsmatrix. Mit Q_1 bezeichne man die sogenannte zu Q inverse Form, d. h. die (bekanntlich positiv-definite) Form mit der Matrix (A). Es gibt eine reguläre Matrix (γ) mit der Determinante $D^{-1/2}$, welche die Form $Q(u)$ in $v_1^2 + \cdots + v_r^2$ überführt. Sind daneben noch reelle Zahlen $h_1, \ldots, h_r$ mit

$|h_1| + \cdots + |h_r| > 0$ gegeben, so kann man noch erreichen, daß

$$h_1 u_1 + \cdots + h_r u_r = C v_1 \tag{26}$$

ist mit positivem $C = C(h_1, \ldots, h_r)$ (denn $v_1^2 + \cdots + v_r^2$ ist invariant gegenüber orthogonalen Transformationen). Wir berechnen C. Es ist $\overline{(\gamma)}\,(\alpha)\,(\gamma) = (\varepsilon)$, also $(\Gamma)\,(A)\,\overline{(\Gamma)} = (\varepsilon)$, insbesondere

$$\varepsilon_{11} = 1 = \sum_{i,k} \Gamma_{1i} A_{ik} \Gamma_{1k} = Q_1(\Gamma_{11}, \Gamma_{12}, \ldots, \Gamma_{1r}),$$

aber

$$h_1 u_1 + \cdots + h_r u_r = C v_1 = \sum_{s=1}^{r} C \Gamma_{1s} u_s, \quad \text{also} \quad h_s = C \Gamma_{1s},$$

$$C^2 = C^2 Q_1(\Gamma_{11}, \ldots, \Gamma_{1r}) = Q_1(h_1, \ldots, h_r). \tag{27}$$

Hilfssatz 3. *Es sei Q_1 die zu Q inverse Form. Für $\varrho \geqq 0$, $x > 0$ und reelle $h_1, \ldots, h_r$ mit $|h_1| + \cdots + |h_r| > 0$ ist*

$$A = \frac{1}{\Gamma(\varrho + 1)} \int \cdots \int_{Q(u) \leqq x} (x - Q(u))^{\varrho} \, du_1 \cdots du_r = \frac{x^{\varrho + r/2} \pi^{r/2}}{D^{1/2} \Gamma\left(\varrho + \frac{1}{2} r + 1\right)}, \tag{28}$$

$$B = \frac{1}{\Gamma(\varrho + 1)} \int \cdots \int_{Q(u) \leqq x} (x - Q(u))^{\varrho} \cos 2\pi (h_1 u_1 + \cdots + h_r u_r) \, du_1 \cdots du_r$$

$$= \frac{x^{\varrho/2 + r/4}}{\pi^{\varrho} D^{1/2} (Q_1(h))^{\varrho/2 + r/4}} J_{\varrho + r/2}(2\pi x^{1/2} Q_1^{1/2}(h)); \tag{29}$$

dabei schreiben wir freilich $Q_1(h) = Q_1(h_1, \ldots, h_r)$.

Beweis. Schreibt man $x^{1/2} u_j$ statt u_j, so erhält man

$$B = \frac{x^{\varrho + r/2}}{\Gamma(\varrho + 1)} \int \cdots \int_{Q(u) \leqq 1} (1 - Q(u))^{\varrho} \cos (2\pi x^{1/2} (h_1 u_1 + \cdots + h_r u_r)) \, du_1 \cdots du_r$$

(hierin ist auch A für $h_1 = \cdots = h_r = 0$ enthalten). Durch die oben beschriebene Substitution erhält man nach (26), (27)

$$B = \frac{x^{\varrho + r/2}}{D^{1/2} \Gamma(\varrho + 1)} \int \cdots \int_{v_1^2 + \cdots + v_r^2 \leqq 1} (1 - v_1^2 - \cdots - v_r^2)^{\varrho} \cos F v_1 \, dv_1 \cdots dv_r, \tag{30}$$

wobei zur Abkürzung $2\pi x^{1/2} Q_1^{1/2}(h) = F$ gesetzt wurde. Daher ist

$$B = \frac{x^{\varrho + r/2}}{D^{1/2}\Gamma(\varrho + 1)} \sum_{k=0}^{\infty} \frac{(-1)^k F^{2k}}{(2k)!} B_k \tag{31}$$

mit

$$B_k = 2^r \int_{\substack{v_1^2 + \cdots + v_r^2 \leqq 1 \\ v_j > 0}} \cdots \int (1 - v_1^2 - \cdots - v_r^2)^{\varrho} v_1^{2k} \, dv_1 \cdots dv_r$$

$$= \int_{\substack{x_1 + \cdots + x_r \leqq 1 \\ x_j > 0}} \cdots \int (1 - x_1 - \cdots - x_r)^{\varrho} x_1^{k-1/2} x_2^{-1/2} \cdots x_r^{-1/2} \, dx_1 \cdots dx_r .$$

Nach (17), (18) ist also

$$B_k = \pi^{r/2 - 1/2} \frac{\Gamma\left(k + \frac{1}{2}\right)}{\Gamma\left(k + \frac{r}{2}\right)} \int_0^1 (1 - u)^{\varrho} u^{k + r/2 - 1} \, du$$

$$= 2^{-k} \cdot 1 \cdot 3 \cdot 5 \cdots (2k - 1) \, \pi^{r/2} \frac{\Gamma(\varrho + 1)}{\Gamma\left(k + \varrho + \frac{r}{2} + 1\right)}. \tag{32}$$

Wegen $A = x^{\varrho + r/2} D^{-1/2} B_0 / \Gamma(\varrho + 1)$ folgt daraus (28), und durch Einsetzen von B_k aus (32) in (31) ergibt sich auch (29).[1])

Hilfssatz 4. *Es gibt ein* $c = c(Q, \lambda) > 0$, *so daß für jedes* $K \geqq 1$

$$\left.\begin{aligned} &\sum_{Q_1(h) < K}' (Q_1(h))^{-\lambda} < c K^{r/2 - \lambda} && \text{für } \lambda < \frac{r}{2}, \\ &\sum_{Q_1(h) < K}' (Q_1(h))^{-\lambda} < c \log (K + 1) && \text{für } \lambda = \frac{r}{2}, \\ &\sum_{Q_1(h) \geqq K} (Q_1(h))^{-\lambda} < c K^{r/2 - \lambda} && \text{für } \lambda > \frac{r}{2} \end{aligned}\right\} \tag{33}$$

gilt. Dabei bedeutet $\sum'$, daß das Glied mit $h_1 = \cdots = h_r = 0$ wegzulassen ist. Analoges gilt später in ähnlichen Fällen.

Beweis. Klar.

Im folgenden seien $0 < \lambda_1 < \lambda_2 < \lambda_3 < \cdots$ diejenigen Werte, die $Q_1(h)$ für ganze $h_1, \ldots, h_r$ annimmt; a_n sei die Anzahl der Darstellungen von λ_n durch Q_1, d. h.

---

[1]) Spezialfälle von (29) kommen in der Literatur seit Bessel vor; vgl. die Fußnoten in [1], S. 15, 16, 118 und den Schluß der Einleitung von [6]. Den allgemeinen Fall kann man unschwer aus [11] oder [12] ablesen.

$a_n = \sum_{Q_1(h) = \lambda_n} 1$. Weiter schreiben wir oft (m) statt $(m_1, \ldots, m_r)$,

$$\sum_{(m)=-\infty}^{+\infty} = \sum_{m_1,\ldots,m_r=-\infty}^{+\infty}, \quad \sum_{(m)=1}^{k} = \sum_{m_1,\ldots,m_r=1}^{k} \quad \text{usw.}$$

Satz 2. *Für $\varrho > r/2 - \frac{1}{2}$, $x > 0$ ist*

$$P_\varrho(x) = A_\varrho(x) - \frac{\pi^{r/2} x^{\varrho + r/2}}{D^{1/2}\Gamma\left(\varrho + \frac{r}{2} + 1\right)}$$

$$= \frac{x^{r/4+\varrho/2}}{\pi^\varrho D^{1/2}} \sum_{(h)=-\infty}^{+\infty}{}' (Q_1(h))^{-\varrho/2-r/4} J_{\varrho+r/2}(2\pi x^{1/2} Q_1^{1/2}(h)), \tag{34}$$

$$P_\varrho(x) = \frac{x^{r/4+\varrho/2}}{\pi^\varrho D^{1/2}} \sum_{n=1}^{\infty} a_n \lambda_n^{-\varrho/2-r/4} J_{\varrho+r/2}(2\pi x^{1/2}\lambda_n^{1/2}), \tag{35}$$

$$P_\varrho(x) = \frac{x^{(r-1)/4+\varrho/2}}{\pi^{\varrho+1} D^{1/2}} \sum_{n=1}^{\infty} a_n \lambda_n^{-\varrho/2-r/4-1/4} \cos\left(2\pi x^{1/2}\lambda_n^{1/2} - \varrho\frac{\pi}{2} - (r+1)\frac{\pi}{4}\right) \tag{36}$$

$$+ O(x^{(r-3)/4+\varrho/2}).$$

Beweis. Man setze in (24) für die $K(\pm a_1, \ldots, \pm a_r)$ die im Hilfssatz 3 gegebenen Werte ein. Man benutze (15) und Hilfssatz 4 (es ist $\varrho/2 + r/4 + \frac{1}{4} > r/2$). Man sieht, daß man die Reihe in (24) beliebig umgruppieren kann; daraus folgen unmittelbar die Behauptungen.

Weiter werden die Beweise nach Landaus Muster geführt. Ist f eine Funktion einer Veränderlichen, $z \neq 0$, so definiere man die Funktionen $\Delta_{k,z}f$ für $k = 1, 2, \ldots$ wie folgt:

$$\Delta_{1,z}f(x) = \Delta_z f(x) = f(x+z) - f(x), \quad \Delta_{k+1,z}f(x) = \Delta_{k,z}f(x+z) - \Delta_{k,z}f(x). \tag{37}$$

Wenn f im abgeschlossenen Intervall mit den Endpunkten $x, x + kz$ k-mal differenzierbar ist, so gibt es bekanntlich ein ξ zwischen $x, x + kz$, so daß

$$z^{-k}\Delta_{k,z}f(x) = f^{(k)}(\xi) \tag{38}$$

gilt. Da $A_\varrho(x)$ ($\varrho \geqq 0$) eine nichtabnehmende Funktion ist, gilt für $0 < kz < x$

$$(-z)^{-k}\Delta_{k,-z}A_{\varrho+k}(x) \leqq A_\varrho(x) \leqq z^{-k}\Delta_{k,z}A_{\varrho+k}(x). \tag{39}$$

(Denn z. B. ist

$$(-z)^{-k}\Delta_{k,-z}A_{\varrho+k}(x) = z^{-k}\int_{x-z}^{x}\left(\int_{x_1-z}^{x_1}\cdots\left(\int_{x_{k-1}-z}^{x_{k-1}} A_\varrho(x_k)\,dx_k\right)\cdots dx_2\right)dx_1.\Big)$$

Satz 3. *Es ist*

$$P_\varrho(x) = O(x^{(r-1)/4+\varrho/2}) \qquad \textit{für} \quad \varrho > \frac{r}{2} - \frac{1}{2}, \tag{40}$$

$$P_\varrho(x) = O(x^\varrho \log x) = O(x^{(r-1)/4+\varrho/2} \log x) \quad \textit{für} \quad \varrho = \frac{r}{2} - \frac{1}{2},^{1)} \tag{41}$$

$$P_\varrho(x) = O(x^{\varrho + r/2 - r/(r+1-2\varrho)}) \qquad \textit{für} \quad 0 \leqq \varrho < \frac{r}{2} - \frac{1}{2}. \tag{42}$$

Beweis. (40) folgt sofort aus (36). Ist $0 \leqq \varrho \leqq r/2 - \frac{1}{2}$, so wähle man ein ganzes k mit $\varrho + k > r/2 - \frac{1}{2}$ und eine Funktion $z(x)$ mit $0 < z(x) = o(x)$ und bilde die beiden Ausdrücke

$$(\pm z(x))^{-k} \Delta_{k, \pm z(x)} A_{\varrho+k}(x), \tag{43}$$

indem man von jedem Glied der Reihe (34) (mit $\varrho + k$ statt ϱ) die k-te Differenz bildet. Es ist erstens (ich schreibe z statt $z(x)$) nach (38)

$$(\pm z)^{-k} \Delta_{k, \pm z} \frac{\pi^{r/2} x^{r/2+\varrho+k}}{D^{1/2} \Gamma\left(\varrho + \frac{r}{2} + k + 1\right)} = \frac{\pi^{r/2} \xi^{r/2+\varrho}}{D^{1/2} \Gamma\left(\varrho + \frac{r}{2} + 1\right)}$$
$$= \frac{\pi^{r/2} x^{r/2+\varrho}}{D^{1/2} \Gamma\left(\varrho + \frac{r}{2} + 1\right)} + O(z x^{r/2+\varrho-1}).$$

Zweitens hat man für

$$L = (\pm z)^{-k} \Delta_{k, \pm z} x^{r/4+\varrho/2+k/2} (Q_1(h))^{-\varrho/2-k/2-r/4} J_{\varrho+k+r/2}(2\pi x^{1/2} Q_1^{1/2}(h))$$

zwei Abschätzungen: Nach (15) ist (für $x \to +\infty$)

$$L = O(z^{-k} x^{r/4+\varrho/2+k/2-1/4} (Q_1(h))^{-\varrho/2-k/2-r/4-1/4}), \tag{44}$$

und nach (38), (16), (15) ist

$$L = O(x^{r/4+\varrho/2-1/4} (Q_1(h))^{-\varrho/2-r/4-1/4}). \tag{45}$$

Man wähle nun ein $K = K(x) \geqq 1$. Für $0 < Q_1(h) < K$ benutze man (45), für $Q_1(h) \geqq K$ die Abschätzung (44). Nach (33) erhält man für (43) folgende Darstellung:

$$\frac{\pi^{r/2} x^{r/2+\varrho}}{D^{1/2} \Gamma\left(\frac{r}{2} + \varrho + 1\right)} + O(z x^{r/2+\varrho-1})$$
$$+ O(z^{-k} x^{r/4+\varrho/2+k/2-1/4} K^{r/4-\varrho/2-k/2-1/4}) + O(x^{r/4+\varrho/2-1/4} K^{r/4-\varrho/2-1/4}) \tag{46}$$

---

1) Für $r = 1$, $\varrho = 0$ ist natürlich trivialerweise $P_0(x) = O(1)$.

mit der Ausnahme, daß für $\varrho = r/2 - \frac{1}{2}$ im letzten Glied K^0 durch $\log(K+1)$ ersetzt werden muß. Man erhält die beste Abschätzung, wenn man $K = xz^{-2}$ und dann $z = x^{1-r/(r+1-2\varrho)}$ wählt. Diese Wahl ist zulässig, da $z(x) \leqq x^{1-r/(r+1)} \leqq x^{1/2} = o(x)$, $K \geqq 1$ ist. Setzt man z, K in (46) ein und beachtet (39), so erhält man (41), (42).

Satz 4. *Für* $\varrho \geqq 0$ *ist* $P_\varrho(x) = \Omega(x^{(r-1)/4+\varrho/2})$. *Schärfer: Es gibt ein* $c = c(Q, \varrho) > 0$, *ein* $n_0 = n_0(Q, \varrho)$ *und eine von* Q *und* ϱ *abhängige Folge* $x_1, x_2, \ldots$ *mit*

$$\left.\begin{aligned} &x_n = x_{n,\varrho} = \frac{n^2}{4\lambda_1} + O(n) \quad (n \to \infty), \\ &P_\varrho(x_{2n}) > c x_{2n}^{(r-1)/4+\varrho/2}, \quad P_\varrho(x_{2n+1}) < -c x_{2n+1}^{(r-1)/4+\varrho/2} \quad \textit{für} \quad n > n_0. \end{aligned}\right\} \tag{47}$$

Beweis. Ist $s > r/2 + 1$, so ist

$$\sum_{n=2}^{\infty} a_n \lambda_n^{-s} \leqq \lambda_2^{-s} \sum_{n=2}^{\infty} a_n (\lambda_2/\lambda_n)^{r/2+1} = o(a_1 \lambda_1^{-s}) \tag{48}$$

für $s \to +\infty$ (die Konvergenz der Reihe folgt aus (33)). Man definiere y_n durch

$$2\pi\lambda_1^{1/2} y_n^{1/2} - \left(\varrho + \frac{r}{2} + \frac{1}{2}\right)\frac{\pi}{2} = n\pi,$$

also $y_n = \frac{1}{4} n^2 \lambda_1^{-1} + O(n)$. Ist ϱ hinreichend groß, so folgt aus (36) und (48)

$$(-1)^n P_\varrho(y_n) > \frac{1}{2} \cdot \frac{y_n^{(r-1)/4+\varrho/2}}{\pi^{\varrho+1} D^{1/2}} \cdot \frac{a_1}{\lambda_1^{\varrho/2+r/4+1/4}} + O(y_n^{(r-3)/4+\varrho/2})$$

für $n \to \infty$. Also gilt (47) für hinreichend große ϱ (mit $x_n = y_n$). Gilt aber die Behauptung für ein $\varrho \geqq 1$, so gilt sie auch mit $\varrho - 1$ statt ϱ. In der Tat, aus (47) folgt für große n

$$(-1)^n \int_{x_{n-1}}^{x_n} P_{\varrho-1}(x)\, dx > c x_n^{(r-1)/4+\varrho/2}.$$

Da die Länge des Integrationsintervalls $O(n) = O(x_{n-1}^{1/2})$ ist, gibt es ein ξ_n mit $x_{n-1} < \xi_n < x_n$, $(-1)^n P_{\varrho-1}(\xi_n) > c' \xi_n^{(r-3)/4+\varrho/2}$, wobei $c' > 0$ von n unabhängig ist und $\xi_n = \frac{1}{4} n^2 \lambda_1^{-1} + O(n)$ gilt.

Bemerkung. Die Anwendbarkeit der Landauschen Methode aus [6] ist nicht auf Ellipsoide beschränkt. Landau selbst hat mit ihrer Hilfe einen Beweis eines allgemeinen van der Corputschen O-Satzes gegeben; der Satz gibt die Abschätzung $P(x) = O(x^{1/3})$ für sehr allgemeine ebene Bereiche. Man kann aber mit derselben Methode auch eine Ω-Abschätzung (mit dem Ergebnis $P(x) = \Omega(x^{(r-1)/4})$) für ziemlich allgemeine konvexe r-dimensionale Bereiche beweisen. Dies wird dadurch ermöglicht, daß die entsprechende Verallgemeinerung der $K(h_1, \ldots, h_r)$ aus Hilfssatz 2 asymptotische Eigenschaften besitzt, die denjenigen der Besselschen Funktionen ähnlich sind. Vgl. [9] für $r = 2$ und [10] für allgemeines r.

§ 3. Anwendung von Thetafunktionen

In diesem Paragraphen werden auch imaginäre Zahlen auftreten. Mit $c_1, c_2, \ldots$, aber öfter unterschiedslos mit c bezeichne ich positive Zahlen, die nur von Q und ϱ abhängen. Statt $c_1 < f(x) < c_2$ schreibe ich also oft $c < f(x) < c$ usw. (a, b) bedeute den größten gemeinsamen Teiler der ganzen Zahlen a, b. Wir leiten jetzt eine andere Formel für $P_\varrho(x)$ ab. Dazu brauchen wir die bekannte Formel: Ist $a > 0, \varrho > 0$, so ist

$$\int_{a-i\infty}^{a+i\infty} e^{\lambda s} s^{-\varrho-1}\, ds = \begin{cases} 0 & \text{für } \lambda \leqq 0, \\ 2\pi i \lambda^\varrho / \Gamma(\varrho + 1) & \text{für } \lambda > 0. \end{cases} \tag{49}$$

Dabei wird hier und im folgenden derjenige Zweig von s^α (α reell) genommen, der für $s > 0$ positiv ist.

Beweis. In dem (schwierigeren) Fall $\lambda > 0$ läßt sich das Integral wegen $\varrho > 0$ leicht auf das Hankelsche Schlingenintegral zurückführen. Man findet übrigens einen vollständigen Beweis z. B. in [1], S. 248.

Nach (3) ist also für $\varrho > 0$, $x > 0$, $a > 0$

$$A_\varrho(x) = \frac{1}{\Gamma(\varrho + 1)} \sum_{Q(m) \leqq x} (x - Q(m))^\varrho = \frac{1}{2\pi i} \sum_{(m)=-\infty}^{+\infty} \int_{a-i\infty}^{a+i\infty} \frac{\exp((x - Q(m))\, s)}{s^{\varrho+1}}\, ds$$

(ich schreibe gelegentlich $\exp(x)$ statt e^x). Hier kann man offenbar die Integration mit der Summation vertauschen. Setzt man also

$$\Theta(s) = \sum_{(m)=-\infty}^{+\infty} e^{-Q(m)s}, \tag{50}$$

so folgt

Hilfssatz 5. *Für* $a > 0$, $x > 0$, $\varrho > 0$ *ist*

$$A_\varrho(x) = \frac{1}{2\pi i} \int_{a-i\infty}^{a+i\infty} \Theta(s)\, e^{xs} s^{-\varrho-1}\, ds. \tag{51}$$

Bisher war in § 2, § 3 Q eine beliebige positiv-definite quadratische Form. Von nun an setzen wir voraus, daß die α_{jl} ganz sind.

Hilfssatz 6. *Es seien die* α_{jl} *ganz,* $(h, k) = 1$, $k > 0$; $m_1, \ldots, m_r$ *ganz. Man setze*

$$S_{h,k,(m)} = \sum_{(a)=1}^{k} \exp\left(-\frac{2\pi i h}{k} Q(a) - 2\pi i \frac{a_1 m_1 + \cdots + a_r m_r}{k}\right). \tag{52}$$

Dann gilt $S_{h,1,(m)} = 1$; $|S_{h,k,(m)}| < ck^{r/2}$; *ist* $\mathrm{Re}\, s > 0$, *so gilt weiter*

$$\Theta(s) = \pi^{r/2} D^{-1/2} k^{-r} \left(s - 2\pi i \frac{h}{k}\right)^{-r/2} \sum_{(m)=-\infty}^{+\infty} S_{h,k,(m)} \exp\left(\frac{-\pi^2 Q_1(m)}{k^2(s - 2\pi i h/k)}\right), \tag{53}$$

wobei Q_1 *die zu* Q *inverse Form ist.*

Beweis. Vgl. etwa [8], Hilfssatz 3 A und 4; oder (in einer etwas anderen Bezeichnung) die Beweise der Formeln (11), (12) in [7] oder [1], S. 148–154. In [7], [1] wird insbesondere $\Theta(s) = \sum \exp(-\pi Q(m)\, s)$ gesetzt (die dortigen α_j sind bei uns gleich Null); Landaus k heißt bei uns r.

Satz 5. *Q habe ganze Koeffizienten* α_{jl}. *Dann gilt*

(I) $\quad P_\varrho(x) = \Omega(x^{r/2-1}) \quad$ *für* $\varrho \geqq 0$,

(II) $\quad P_\varrho(x) = O(x^{r/2-1}) \quad$ *für* $0 \leqq \varrho < \frac{r}{2} - 2$,

(III) $\quad P_\varrho(x) = O(x^{r/2-1} \log x) \quad$ *für* $0 < \varrho = \frac{r}{2} - 2$,

(IV) $\quad P_\varrho(x) = O(x^{r/2-1} \log^2 x) \quad$ *für* $r = 4$, $0 = \varrho = \frac{r}{2} - 2$,

(V) $\quad P_\varrho(x) = O(x^{r/4+\varrho/2}) \quad$ *für* $\varrho > \frac{r}{2} - 2$, $\varrho \geqq 0$, $r \geqq 3$.

Vorbemerkung. In (II) bis (V) ist automatisch $r \geqq 3$. Der Fall $\varrho = 0$, $r = 3$ von (V) wurde in Satz 3 bewiesen. Satz 1 aus § 1 folgt offenbar aus den Sätzen 3, 4, 5.

Beweis. Wir beweisen zunächst (II) bis (V). Von nun an sei $x > 2$. Man setze $z = z(x) = x^{-\varrho}$. Nach (39) (mit $k = 1$) genügt es, die beiden Zahlen

$$I = \pm x^\varrho \int_x^{x \pm z} A_\varrho(y)\, dy = \pm x^\varrho (A_{\varrho+1}(x \pm z) - A_{\varrho+1}(x)) \tag{54}$$

mit der gewünschten Präzision abzuschätzen. Es sei noch bemerkt, daß man für $\varrho > 0$ direkt A_ϱ untersuchen könnte; um aber den (bekannten) Fall $\varrho = 0$ mit einzubeziehen, rechne ich mit den Differenzen (54). Für $A_{\varrho+1}$ wende ich Hilfssatz 5 mit $a = 1/x$ an und erhalte

$$I = \pm \frac{x^\varrho}{2\pi} \int_{-\infty}^{+\infty} \Theta(s)\, e^{xs}(e^{\pm zs} - 1)\, s^{-\varrho-2}\, dt \tag{55}$$

mit $s = 1/x + it$.

Ich konstruiere nun die sogenannte zu $x^{1/2}$ gehörige Fareyreihe, d. h. die Menge aller Brüche h/k mit $0 < k \leqq x^{1/2}$, $h \gtreqless 0$, $(h, k) = 1$; ich nenne sie kurz Fareybrüche. Sind h/k, h'/k' zwei benachbarte Fareybrüche, so heißt der Bruch $(h + h')/(k + k')$ ihre Mediante. Also ist $x^{1/2} < k + k' \leqq 2x^{1/2}$. Weiter ist bekanntlich $hk' - h'k = \pm 1$, also

$$\frac{1}{2kx^{1/2}} \leqq \left|\frac{h + h'}{k + k'} - \frac{h}{k}\right| < \frac{1}{kx^{1/2}}. \tag{56}$$

Zu jedem Fareybruch h/k konstruiere ich das Intervall (α, β), wobei α, β die beiden zu h/k benachbarten Medianten sind, und bezeichne mit $B_{h,k}$ das Intervall $(2\pi\alpha, 2\pi\beta)$. Nach (56) ist

$$B_{h,k} = (2\pi h/k - \lambda_1 k^{-1}x^{-1/2}, 2\pi h/k + \lambda_2 k^{-1}x^{-1/2}) \quad \text{mit} \quad \pi \leqq \lambda_j < 2\pi. \tag{57}$$

Es sei

$$I_{h,k} = \pm \frac{x^\varrho}{2\pi} \int_{B_{h,k}} \Theta(s)\, e^{xs}(e^{\pm zs} - 1)\, s^{-\varrho-2}\, dt. \tag{58}$$

Ich berechne $I_{0,1}$ möglichst genau und schätze die übrigen $I_{h,k}$ ab; dabei kann ich mich auf $h > 0$ beschränken, da $I_{h,k}$, $I_{-h,k}$ konjugiert-komplex sind.

Für alle reellen t und $s = 1/x + it$ ist offenbar $|e^{xs}| = e$, $|e^{\pm zs} - 1| < c \operatorname{Min}(|zs|, 1)$. Für $t \in B_{h,k}$ ist nach (57) und wegen $k \leqq x^{1/2}$

$$\operatorname{Re} \frac{1}{k^2(s - 2\pi ih/k)} = \frac{x}{k^2(1 + x^2(t - 2\pi h/k)^2)} > c. \tag{59}$$

Also ist auf $B_{h,k}$

$$\sum_{(m)=-\infty}^{+\infty} \left|\exp\left(\frac{-\pi^2 Q_1(m)}{k^2(s - 2\pi ih/k)}\right)\right| < \sum_{(m)=-\infty}^{+\infty} \exp(-c(|m_1| + \cdots + |m_r|)) < c, \tag{60}$$

und auf $B_{0,1}$ ist

$$\sum_{(m)=-\infty}^{+\infty}{}' \left|\exp\left(\frac{-\pi^2 Q_1(m)}{s}\right)\right|$$

$$< c \sum_{(m)=-\infty}^{+\infty}{}' \exp\left(-\frac{cx}{1 + x^2t^2}(|m_1| + \cdots + |m_r|)\right) < c \exp\left(\frac{-cx}{1 + x^2t^2}\right). \tag{61}$$

Nach (61) und Hilfssatz 6 ist auf $B_{0,1}$

$$\Theta(s) = \frac{\pi^{r/2}}{D^{1/2}s^{r/2}}(1 + \psi(s)) \quad \text{mit} \quad |\psi(s)| < c \exp\left(\frac{-cx}{1 + x^2t^2}\right).$$

Da für $\alpha > 0$ die Funktion $\xi^{\alpha} e^{-c\xi}$ für $\xi > 0$ beschränkt ist, erhält man

$$\left| \pm \frac{x^{\varrho}}{2\pi} \int_{B_{0,1}} \frac{\pi^{r/2}(e^{\pm zs} - 1)\, e^{xs}}{D^{1/2} s^{r/2+\varrho+2}} \psi(s)\, dt \right| < cx^{\varrho} \int_{-2\pi/x^{1/2}}^{2\pi/x^{1/2}} \frac{z}{|s^{r/2+\varrho+1}|} \exp\left(\frac{-cx}{1 + x^2 t^2}\right) dt$$

$$= c \int_{-2\pi/x^{1/2}}^{2\pi/x^{1/2}} \frac{x^{r/2+\varrho+1}}{(1 + x^2 t^2)^{r/4+\varrho/2+1/2}} \exp\left(\frac{-cx}{1 + x^2 t^2}\right) dt < cx^{-1/2} \cdot x^{r/4+\varrho/2+1/2}.$$

Man schätze nun den Fehler ab, den man begeht, wenn man

$$\pm \frac{x^{\varrho}}{2\pi} \int_{B_{0,1}} \frac{\pi^{r/2} e^{xs}(e^{\pm zs} - 1)}{D^{1/2} s^{r/2+\varrho+2}}\, dt$$

durch

$$\pm \frac{x^{\varrho}}{2\pi i} \int_{1/x - i\infty}^{1/x + i\infty} \frac{\pi^{r/2} e^{xs}(e^{\pm zs} - 1)}{D^{1/2} s^{r/2+\varrho+2}}\, ds$$

$$= \pm \frac{\pi^{r/2} x^{\varrho}}{D^{1/2}\Gamma\left(\frac{r}{2} + \varrho + 2\right)} \left((x \pm x^{-\varrho})^{\varrho+r/2+1} - x^{\varrho+r/2+1}\right)$$

$$= \frac{\pi^{r/2} x^{\varrho+r/2}}{D^{1/2}\Gamma\left(\frac{r}{2} + \varrho + 1\right)} + O(x^{\varrho} \cdot x^{-2\varrho} \cdot x^{\varrho+r/2-1})$$

ersetzt (Formel (49) und Taylorsche Formel). Dieser Fehler ist absolut kleiner als

$$cx^{\varrho} \int_{\pi x^{-1/2}}^{+\infty} x^{-\varrho} t^{-r/2-\varrho-1}\, dt = cx^{r/4+\varrho/2}.$$

Also ist

$$I_{0,1} = \frac{\pi^{r/2} x^{\varrho+r/2}}{D^{1/2}\Gamma\left(\varrho + \frac{r}{2} + 1\right)} + O(x^{r/2-1}) + O(x^{r/4+\varrho/2}). \tag{62}$$

Es genügt, noch die Summe der $I_{h,k}$ mit $h > 0$ abzuschätzen. Auf $B_{h,k}$ mit $h > 0$ ist offenbar $cx^{-1/2} < ch/k < t < ch/k$, und (58), Hilfssatz 6 und (60) ergeben (man beachte $r \geqq 3$)

$$|I_{h,k}| \leqq ck^{-r/2}(k/h)^{\varrho+2}\, x^{\varrho}\, \mathrm{Min}\,(hk^{-1}x^{-\varrho}, 1) \int_{-\infty}^{+\infty} |s - 2\pi i h/k|^{-r/2}\, dt$$

$$= ck^{\varrho+1-r/2} h^{-\varrho-1}\, \mathrm{Min}\,(1, kh^{-1}x^{\varrho}) \int_{-\infty}^{+\infty} x^{r/2}(1 + x^2 t^2)^{-r/4}\, dt.$$

Die Summation über h, k ergibt höchstens

$$W = cx^{r/2-1} \sum_{0<k\leqq x^{1/2}} k^{\varrho+1-r/2} \sum_{h=1}^{\infty} h^{-\varrho-1}\, \mathrm{Min}\,(1, kh^{-1}x^{\varrho}). \tag{63}$$

Ist $\varrho > 0$, so ersetze man das Min durch Eins, und man erhält

$$W = O(x^{r/2-1}) \quad \text{für} \quad 0 < \varrho < \frac{r}{2} - 2,$$

$$W = O(x^{r/2-1} \log x) \quad \text{für} \quad 0 < \varrho = \frac{r}{2} - 2,$$

$$W = O(x^{r/4+\varrho/2}) \quad \text{für} \quad \varrho > 0, \varrho > \frac{r}{2} - 2.$$

Für $\varrho = 0$ hat die innere Summe in (63) die Größenordnung $\log(k+1)$ und man erhält

$$W = O(x^{r/2-1}) \quad \text{für} \quad \varrho = 0, r > 4,$$

$$W = O\left(x \sum_{0<k\leqq x^{1/2}} \frac{\log(k+1)}{k}\right) = O(x \log^2 x) \quad \text{für} \quad \varrho = 0, r = 4.$$

Daraus und aus (62) folgt (II) bis (V). Man beachte, daß sich für $r = 3$, $\varrho = 0$ nur $W = O(x^{3/4} \log x)$ ergibt, ein Resultat, das schwächer als das Resultat des Satzes 3 ist.

Beweis von (I). Für natürliches n sei α_n die Anzahl der Lösungen von $Q(m) = n$. Dann ist $\sum_{n\leqq N} \alpha_n = A_0(N)$ asymptotisch gleich dem Volumen $\pi^{r/2}D^{-1/2}N^{r/2}/\Gamma\left(\frac{r}{2}+1\right)$; also ist

$$\alpha_n = \Omega(n^{r/2-1}). \tag{64}$$

Nun ist $P_0(x) = A_0(x) - \pi^{r/2}D^{-1/2}x^{r/2}/\Gamma\left(\frac{r}{2}+1\right)$. Hier ist das zweite Glied rechts stetig, während das erste für $x = n$ einen Sprung von der Größe α_n erleidet. Nach (64) ist also $P_0(x) = \Omega(x^{r/2-1})$.

Nun sei $\varrho > 0$. Setzt man $K = \pi^{r/2}D^{-1/2}/\Gamma\left(\frac{r}{2}+\varrho+1\right)$, so ist

$$P_\varrho(x) = \sum_{m\leqq x} \alpha_m (x-m)^\varrho/\Gamma(\varrho+1) - Kx^{r/2+\varrho}. \tag{65}$$

Man setze nun $P_\varrho(x) = o(x^{r/2-1})$ voraus; daraus werden wir einen Widerspruch ableiten. Man wähle ein ganzes $k \geqq \varrho + 1$, und für jedes natürliche n bilde man die k-te Differenz

$$\Delta_{k,1/(k+2)} P_\varrho\left(n + \frac{1}{k+2}\right) = D_n. \tag{66}$$

Man beachte, daß in D_n die Werte von P_ϱ in den Punkten $n + \frac{1}{k+2}, n + \frac{2}{k+2}, \ldots, n + \frac{k+1}{k+2}$ vorkommen. Nach der Voraussetzung ist $D_n = o(n^{r/2-1})$. Nun sind alle

Glieder rechts in (65) für $n < x < n + 1$ unbeschränkt differenzierbar ($m \leqq x$ bedeutet für diese x dasselbe wie $m \leqq n$). Also gibt es ein x_n mit

$$n + \frac{1}{k+2} < x_n < n + \frac{k+1}{k+2}, \tag{67}$$

$$(k+2)^k D_n = P_\varrho^{(k)}(x_n) = \varrho(\varrho - 1) \cdots (\varrho - k + 1) \sum_{m \leqq n} \alpha_m (x_n - m)^{\varrho - k} / \Gamma(\varrho + 1)$$
$$- \left(\frac{r}{2} + \varrho\right)\left(\frac{r}{2} + \varrho - 1\right) \cdots \left(\frac{r}{2} + \varrho - k + 1\right) K x_n^{r/2 + \varrho - k}. \tag{68}$$

Dieser Ausdruck ist also gleich $o(n^{r/2-1})$. Es sei erstens ϱ nicht ganz, also $k > \varrho + 1$, und aus (68) folgt

$$\sum_{m \leqq n} \alpha_m (x_n - m)^{\varrho - k} = o(n^{r/2-1}),$$

also (wegen (67)) erst recht

$$\alpha_n (x_n - n)^{\varrho - k} = o(n^{r/2-1}), \quad \alpha_n = o(n^{r/2-1});$$

das steht im Widerspruch zu (64).

Es sei zweitens ϱ ganz, und man setze $k = \varrho + 1$, so daß die erste Summe in (68) rechts verschwindet, während $r/2 + \varrho - k + 1 > 0$ ist. Also würde aus (68) $x_n^{r/2-1} = o(n^{r/2-1})$ folgen, was wiederum einen Widerspruch liefert.

Literatur

[1] Ausgewählte Abhandlungen zur Gitterpunktlehre von Edmund Landau, herausgegeben von Arnold Walfisz; VEB Deutscher Verlag der Wissenschaften, Berlin 1962.

[2] E. Landau, Zur analytischen Zahlentheorie der definiten quadratischen Formen (Über die Gitterpunkte in einem mehrdimensionalen Ellipsoid), Sitzungsber. d. Kgl. Preuß. Akad. d. Wiss. **31** (1915), 458–476.

[3] E. Landau, Über eine Aufgabe aus der Theorie der quadratischen Formen, Sitzungsber. d. Kaiserl. Akad. d. Wiss. in Wien, Math.-Naturwiss. Kl. Abt. IIa, **124** (1915), 445–468.

[4] E. Landau, Über die Anzahl der Gitterpunkte in gewissen Bereichen (IV. Abhandlung), Nachr. d. Ges. d. Wiss. zu Göttingen, Math.-Phys. Kl., 1924, 137–150.

[5] E. Landau, Über die Gitterpunkte in einem Kreise (V. Mitteilung), Nachr. d. Ges. d. Wiss. zu Göttingen, Math.-Phys. Kl., 1924, 135–136.

[6] E. Landau, Die Bedeutungslosigkeit der Pfeiffer'schen Methode für die analytische Zahlentheorie, Monatshefte für Math. u. Physik **34** (1925), 1–36.

[7] E. Landau, Über die Gitterpunkte in mehrdimensionalen Ellipsoiden, Math. Z. **21** (1924), 126–132.

[8] V. Jarník, Über die Mittelwertsätze der Gitterpunktlehre, 5. Abh., Časopis pro pěst. mat. a fys. **69** (1940), 148–174.

[9] V. Jarník, Sur les points à coordonnées entières dans le plan, Bulletin international de l'Acad. Tchèque des Sci. **25** (1925), 341—352.

[10] S. Krupička, О целых точках в болеемерных выпуклых телах, Czechoslovak Math. Journal **7** (82) (1957), 524—550. Deutscher Auszug daselbst 550—552.

[11] D. G. Kendall, On the number of lattice points inside a random oval, Quarterly J. of Math., Oxford Ser. **19** (1948), 1—26.

[12] S. Bochner and K. Chandrasekharan, Summations over lattice points in $\varkappa$-space. Quarterly J. of Math., Oxford Ser. **19** (1948), 238—248.

[13] K. Chandrasekharan and R. Narasimhan, Hecke's functional equation and the average order of arithmetical functions, Acta Arithmetica **6** (1961), 487—503.

S. Knapowski † und P. Turán in Budapest

ÜBER EINIGE FRAGEN DER VERGLEICHENDEN PRIMZAHLTHEORIE

1. Von Tschebyscheff [1] stammt die Behauptung, daß „es mehr Primzahlen $\equiv 3 \pmod 4$ als $\equiv 1 \pmod 4$ gibt" in dem Sinne, daß

$$\lim_{x \to +\infty} \sum_{p > 2} (-1)^{(p-1)/2} e^{-p/x} = -\infty \tag{1.1}$$

ist. Diese 1853 aufgestellte Behauptung ist bis heute unbewiesen; Landau [2] und Hardy-Littlewood [3] hatten im Jahre 1917 wenigstens gezeigt, daß die Gültigkeit von (1.1) äquivalent ist mit der Behauptung

$$L(s, \chi_1) \neq 0 \quad \text{für} \quad \sigma > \frac{1}{2}, \tag{1.2}$$

wobei $s = \sigma + it$ ist und χ_1 den Nicht-Hauptcharakter mod 4 bedeutet. Daß die heuristische Behauptung von Tschebyscheff im „natürlichen" Sinne, d. h.

$$\pi(x, 4, 1) - \pi(x, 4, 3) \to -\infty \tag{1.3}$$

falsch ist, hatten Hardy und Littlewood [3] gezeigt; vielmehr hatten sie bewiesen, daß

$$\overline{\lim_{x \to +\infty}} \{\pi(x, 4, 1) - \pi(x, 4, 3)\} = \pm\infty \tag{1.4}$$

ist. Noch allgemeiner hatten sie l.c. bewiesen, daß

$$\overline{\lim_{x \to +\infty}} \sum_{p \leqq x} \chi(p, k) = \pm\infty \tag{1.5}$$

gilt, wobei $\chi \neq \chi_0$ einen beliebigen Charakter mod k bedeutet; Landau [4] hat dieses Resultat auf beliebige algebraische Zahlkörper mit verschiedenen Bedeutungen des Klassenbegriffs ausgedehnt.

2. Diese Sätze gaben Anlaß zu vielen natürlichen Fragen, deren Gesamtheit wir vergleichende Primzahltheorie nannten (siehe Knapowski-Turán [5]). Die bisher berührten Fragen lassen sich grob in vier Gruppen einteilen:

a) Vergleich der Primzahlen zweier verschiedener Restklassen mod k, d. h. Untersuchung von

$$\pi(x, k, l_1) - \pi(x, k, l_2)$$

und analogen Funktionen.

b) Vergleich von $\pi(x, k, l)$ mit seinem „rechten“ Anteil, d. h. Untersuchung von

$$\pi(x, k, l) - \frac{\pi(x)}{\varphi(k)}$$

und analogen Funktionen.

Die obigen Funktionen sind — wenigstens im Fall, daß die Riemann-Piltzsche Vermutung wahr ist — etwa von der Größenordnung $O(\sqrt{x})$. Nun haben wir gefunden, daß man eine solche Diskrepanz „in viel konzentrierterer Form“ erreichen kann. Etwas genauer gesprochen kann man in vielen Fällen die Existenz von „nicht allzu kleinen“ Intervallen von der Form (U_1, U_2) mit

$$U_2\, e^{-\log^{\vartheta} U_2} < U_1 < U_2, \quad 0 < \vartheta < 1$$

sichern, so daß

$$\sum_{\substack{U_1 < p < U_2 \\ p \equiv l_1 (\mathrm{mod}\, k)}} 1 - \sum_{\substack{U_1 < p < U_2 \\ p \equiv l_2 (\mathrm{mod}\, k)}} 1$$

in beiden Richtungen etwa die Größenordnung $O(\sqrt{U_2})$ erreicht. Diese Sätze nannten wir

c) Akkumulationssätze.

Zu der vierten Gruppe der Probleme führt die natürliche Verallgemeinerung des klassischen Satzes von Littlewood [6], nach welchem für eine geeignete Folge

$$x_1 < x_2 < \cdots \to +\infty$$

die Ungleichung

$$\pi(x_\nu) - \mathrm{Li}\, x_\nu > 0 \qquad \left(\mathrm{Li}\, x = \int_2^x \frac{dv}{\log v}\right)$$

gilt, d. h. also

d) Untersuchung von

$$\Delta = (x, k, l) \overset{\mathrm{def}}{=} \pi(x, k, l) - \frac{1}{\varphi(k)} \mathrm{Li}\, x \tag{2.1}$$

und analogen Funktionen (Vergleich der Restglieder, Vorzeichenverteilung usw.).

3. Mit den Gruppen a) und b) wollen wir uns jetzt nicht beschäftigen. Hinsichtlich der Gruppe c) bezogen sich unsere bisherigen Resultate auf die Fälle, daß l_1 und l_2 „von demselben quadratischen Charakter" sind, d. h. die Kongruenzen

$$x^2 \equiv l_1 (\mathrm{mod}\, k), \quad x^2 \equiv l_2 (\mathrm{mod}\, k), \quad (l_1, k) = (l_2, k) = 1$$

gleichzeitig lösbar oder unlösbar sind. Kritisch ist aber auch hier der Fall, daß l_1 und l_2 von *entgegengesetztem* quadratischen Charakter sind. Wir behaupten nun den folgenden

Satz 1.[1]) *Für $T > c_1$ gibt es Zahlen U_1, U_2, U_3, U_4, so daß*

$$\left.\begin{aligned} \log_3 T &\leqq U_2\, e^{-\log^{9/10} U_2} < U_1 < U_2 \leqq T, \\ \log_3 T &\leqq U_4\, e^{-\log^{9/10} U_0} < U_3 < U_4 \leqq T \end{aligned}\right\} \tag{3.1}$$

ist und die Ungleichungen

$$Z_1 \overset{\mathrm{def}}{=} \sum_{\substack{U_1 < p < U_2 \\ p \equiv 1 (\mathrm{mod}\, 4)}} \log p - \sum_{\substack{U_1 < p < U_2 \\ p \equiv 3 (\mathrm{mod}\, 4)}} \log p > \sqrt{U_2}, \tag{3.2}$$

$$Z_2 \overset{\mathrm{def}}{=} \sum_{\substack{U_3 < p < U_4 \\ p \equiv 1 (\mathrm{mod}\, 4)}} \log p - \sum_{\substack{U_3 < p < U_4 \\ p \equiv 3 (\mathrm{mod}\, 4)}} \log p < -\sqrt{U_4} \tag{3.3}$$

gelten.

Den Beweis dieses Satzes werden wir an anderer Stelle publizieren; die Pointe des Satzes ist natürlich die Ungleichung (3.2). Hier möchten wir nur eine, vielleicht nicht uninteressante Folgerung daraus ziehen. Die Frage ist – offenbar ein Spezialfall einer viel allgemeineren –, ob es unendlich viele *konsekutive* Primzahlen $p_\nu, p_{\nu+1}$ gibt, die beide $\equiv 1$ (mod 4) sind. Der Satz (1.4) von Hardy-Littlewood hat als unmittelbare Folgerung, daß *unendlich viele* solche Paare $(p_\nu, p_{\nu+1})$ existieren (obwohl es scheint, daß dieses Korollar früher nie bemerkt wurde). Die Ungleichung (3.2) gestattet, etwas mehr zu behaupten. Wenn nämlich in Z_1 nach jeder Primzahl $p_\nu \equiv 1$ (mod 4) eine Primzahl $p_{\nu+1} \equiv 3$ (mod 4) folgte, so wäre

$$Z_1 < \sum_{\substack{U_1 < p < U_2 \\ p \equiv 1 (\mathrm{mod}\, 4)}} \{\log p - \log (p+2)\} + \log U_2 < \log U_2, \tag{3.4}$$

was (3.2) widerspricht, also gilt

Korollar 1. *Für $T > c_1$ gibt es konsekutive Primzahlen $p_\nu, p_{\nu+1}$ im Intervall* $(\log_3 T, T)$, *so daß*

$$p_\nu \equiv p_{\nu+1} \equiv 1 (\mathrm{mod}\, 4)$$

ist.

---

[1]) $c_1, c_2, \ldots$ bezeichnen immer positive, numerisch bestimmbare Konstanten, und $\log_\nu T$ bzw. $e_\nu(x)$ bedeuten ν-fach iterierter Logarithmus bzw. iterierte Exponentialfunktion. p ist immer eine rationale Primzahl.

Bei der analogen Frage mit $p_\nu \equiv p_{\nu+1} \equiv 3 \pmod 4$ kann man natürlich viel mehr behaupten. Wie wir gezeigt hatten (siehe [7]), gibt es für $T > c_2$ Zahlen x_1, x_2, so daß[1])

$$T e^{-\log^{11/12} T} \leqq x_1 < x_2 \leqq T$$

und

$$\sum_{\substack{x_1 < n < x_2 \\ n \equiv 1 (\mathrm{mod}\, 4)}} \Lambda(n) - \sum_{\substack{x_1 < n < x_2 \\ n \equiv 3 (\mathrm{mod}\, 4)}} \Lambda(n) < -\sqrt{T} e^{-\log^{11/12} T}$$

gilt. Daraus folgt aber für $T > c_3$

$$\sum_{\substack{x_1 < p < x_2 \\ p \equiv 1 (\mathrm{mod}\, 4)}} \log p - \sum_{\substack{x_1 < p < x_2 \\ p \equiv 3 (\mathrm{mod}\, 4)}} \log p < -\frac{1}{2} \sqrt{T} e^{-\log^{11/12} T},$$

und dann können wir wie in (3.4) verfahren; somit erhalten wir

Korollar 8. *Für $T > c_3$ gibt es konsekutive Primzahlen $p_\mu, p_{\mu+1}$ im Intervall $(T e^{-\log^{11/12} T}, T)$, so daß*

$$p_\mu \equiv p_{\mu+1} \equiv 3 (\mathrm{mod}\ 4)$$

ist.

Soweit uns bekannt ist, gibt es keinen *elementaren* Beweis für die Tatsache, daß unendlich viele *konsekutive* Primzahlen $(p_\nu, p_{\nu+1})$ bzw. $(p_\mu, p_{\mu+1})$ mit

$$p_\nu \equiv p_{\nu+1} \equiv 1 (\mathrm{mod}\ 4), \quad p_\mu \equiv p_{\mu+1} \equiv 3 (\mathrm{mod}\ 4)$$

existieren. Es wäre sehr interessant, einen solchen zu finden, auch für die richtige Beurteilung der Wirksamkeit arithmetischer und analytischer Beweisführungen. Zur Zeit gibt es aber keinen Beweis dafür, daß unendlich viele λ-Indizes mit

$$p_\lambda \equiv p_{\lambda+1} \equiv p_{\lambda+2} \equiv 1 (\mathrm{mod}\ 4)$$

existieren. Daß unendlich oft

$$p_j \equiv 1 (\mathrm{mod}\ 4), \quad p_{j+1} \equiv 3 (\mathrm{mod}\ 4)$$

und

$$p_j \equiv 3 (\mathrm{mod}\ 4), \quad p_{j+1} \equiv 1 (\mathrm{mod}\ 4)$$

vorkommt, ist natürlich trivial.

Eine andere Bedeutung von (3.2) liegt in der folgenden Tatsache begründet. Wie wir in [9] bewiesen haben, folgt aus (1.2) die Ungleichung

$$\sum_{p \equiv 1 (\mathrm{mod}\, 4)} \log p \, e^{-c_4 \log^2 (p/x)} - \sum_{p \equiv 3 (\mathrm{mod}\, 4)} \log p \, e^{-c_4 \log^2 (p/x)} < -c_5 \sqrt{x}, \quad x > c_6, \tag{3.5}$$

[1]) Die andere Ungleichung dieses Satzes hatten wir *hier* als belanglos weggelassen.

wobei in beiden Summen p das Intervall

$$\left(x\,e^{-10\sqrt{\frac{\log x}{c_4}}},\, x\,e^{10\sqrt{\frac{\log x}{c_4}}}\right)$$

durchläuft; hier kann man mit numerischen Mitteln zeigen, daß c_4 und c_5 ziemlich groß, c_6 nicht allzu groß gewählt werden können.[1]) (3.2) bedeutet eine „Anhäufung" der Primzahlen $\equiv 1 \pmod 4$ gegenüber denen $\equiv 3 \pmod 4$ im Intervall (U_1, U_2); wählt man also $U_1 > c_6$, so könnte man annehmen, daß $x = x_0$ in (U_1, U_2) mit Benutzung von Rechenmaschinen so gewählt werden kann, daß (3.5) *nicht* erfüllt ist. Diese Untersuchung scheint uns mit modernen Rechenmaschinen durchführbar zu sein; allerdings scheint uns die Gegenbehauptung zur Riemann-Piltzschen Vermutung im Fall von $L(s, \chi_1)$ leichter zugänglich zu sein als für $\zeta(s)$ selbst.

4. Was die Gruppe d) der Probleme anbelangt, so scheint uns schon für $k = 4$, $l \not\equiv 1 \pmod 4$ die Methode von Littlewood nicht zu funktionieren. Somit ist diese Methode, obwohl sie die Existenz von

$$y_1' < y_2' < \cdots \to +\infty$$

mit

$$\pi(y_\nu', 4, 1) > \frac{1}{2} \operatorname{Li} y_\nu' \tag{4.1}$$

ergibt, *nicht* einmal fähig, die Existenz von

$$y_1'' < y_2'' < \cdots \to +\infty$$

mit

$$\pi(y_\nu'', 4, 3) > \frac{1}{2} \operatorname{Li} y_\nu''$$

zu erbringen (obwohl die alte Methode von Landau [8] dies ohne irgendwelche Lokalisation leisten kann), was schon die vielen neuartigen Schwierigkeiten dieser Fragen andeutet. Im folgenden werden wir (4.1) in einem viel allgemeineren Rahmen zu einem Akkumulationssatz verfeinern, nämlich für die Lösungszahlen ν_p der Kongruenzen

$$f(x) \equiv 0 \pmod p, \tag{4.2}$$

wobei $f(x) = x^n + \cdots + a_0$ ein festes Polynom mit ganzrationalen Koeffizienten und Diskriminante Δ bedeutet, welches irreduzibel über dem rationalen Zahlkörper R ist.[2]) Es sei α eine Nullstelle von $f(x)$, weiter sei

$$K = R(\alpha)$$

---

[1]) Zum Beispiel kann $c_4 = 37$ gewählt werden; durch sorgfältigere Durchführung der Abschätzungen lassen sich die Konstanten gewiß verbessern.

[2]) Der Fall (4.1) ergibt sich für $f(x) = x^2 + 1$.

und $\zeta_K(s)$ die zu K gehörige Dedekindsche Zetafunktion. Wir setzen voraus, daß $\delta = \delta(K)$ mit $0 < \delta(K) \leqq \dfrac{1}{\pi^6 \cdot 7!}$ so existiere, daß $\zeta_K(s)$ im Bereich

$$\frac{1}{2} \leqq \sigma < 1, \quad |t| \leqq \delta(K) \tag{4.3}$$

nicht verschwindet. Dann gilt der folgende

Satz 2. *Ist*

$$T > \max\left\{e_2\left(\frac{1}{\delta(K)}\right), e_5((1 + |\Delta|)^{c_7 n})\right\} \tag{4.4}$$

so existieren Zahlen Y_1, Y_2, Y_3, Y_4 *mit*

$$\left.\begin{aligned} \log_3 T &\leqq Y_2 \exp(-8(\log Y_2)^{5/6}) < Y_1 < Y_2 \leqq T, \\ \log_3 T &\leqq Y_4 \exp(-8(\log Y_4)^{5/6}) < Y_3 < Y_4 \leqq T, \end{aligned}\right\} \tag{4.5}$$

so daß die Ungleichungen

$$\sum_{Y_1 \leqq p \leqq Y_2} \nu_p > \int_{Y_1}^{Y_2} \frac{du}{\log u} + \frac{\sqrt{Y_2}}{\log Y_2}, \tag{4.6}$$

$$\sum_{Y_3 \leqq p \leqq Y_4} \nu_p < \int_{Y_3}^{Y_4} \frac{du}{\log u} - \frac{\sqrt{Y_4}}{\log Y_4} \tag{4.7}$$

gelten.

5. Wir skizzieren jetzt den Beweis des Satzes 2. Es sei für $\sigma > 1$

$$-\frac{\zeta_K'(s)}{\zeta_K(s)} = \sum_{m=1}^{\infty} \frac{g(m)}{m^s}. \tag{5.1}$$

Es sei r eine ganze Zahl $\geqq 4$, $b \geqq 100$, welche wir erst später genauer bestimmen werden; wir gehen aus von dem Integral

$$J_r \overset{\text{def}}{=} \frac{1}{2\pi i} \int_{(2)} e^{r(s+b)^2} \left\{-\frac{\zeta_K'}{\zeta_K}(s) - \zeta(s)\right\} ds. \tag{5.2}$$

Wenn man Dirichlet-Reihen einführt, folgt sofort

$$J_r = \frac{e^{b^2 r}}{2\sqrt{\pi r}} \sum_{m=1}^{\infty} \{g(m) - 1\}\, e^{-(1/4r)(\log m - 2br)^2}.$$

Da man leicht die Existenz eines Weges H im vertikalen Streifen $\frac{1}{3} \leqq \sigma \leqq 5/12$ zeigen kann, welcher abwechselnd aus horizontalen und vertikalen Strecken besteht, und wobei für jedes ganze m im Horizontalstreifen $m \leqq t < m + 1$ genau eine horizontale Strecke von H liegt, und auf welchem die Ungleichung

$$\left|\frac{\zeta_K'}{\zeta_K}(s)\right| \leqq c_8 n^2 \log^2 \{|\Delta|(|t| + 2)\}$$

gilt, so gibt die Anwendung des Cauchyschen Integralsatzes nach Routineabschätzungen (ϱ bedeute die nichttrivialen Nullstellen von $\zeta_K(s)$)

$$\left|\sum_{m=1}^{\infty} (g(m) - 1)\, e^{-(\log m - 2br)^2/4r} - 2\sqrt{\pi r} \sum_{\varrho \text{ rechts von } H} e^{r(\varrho^2 + 2b\varrho)}\right|$$

$$< c_9 \sqrt{r}\, n^2 \log^2(1 + |\Delta|)\, e^{r(25/144 + (5/6)b)}. \tag{5.3}$$

Wenn man noch in Betracht zieht, daß für reelle y die Anzahl der Nullstellen von $\zeta_K(s)$ für $0 < \sigma < 1$, $y \leqq t < y + 1$ nicht größer als

$$c_{10} n \log \{|\Delta|(2 + |y|)\} \tag{5.4}$$

sein kann, so ist, wenn man $\varrho = \sigma_\varrho + it_\varrho$ setzt,

$$\left|\sum_{\substack{|t_\varrho| > 2\sqrt{b}+1 \\ \varrho \text{ rechts von } H}} e^{r(\varrho^2 + 2b\varrho)}\right| < \sum_{\substack{|t_\varrho| < 2\sqrt{b}+1 \\ \varrho \text{ rechts von } H}} e^{(2b+1)r - rt_\varrho^2}$$

$$< c_{11} n \log(|\Delta| + 1) \left\{1 + \sum_{m > 2\sqrt{b}+1} \log m\, e^{-m^2 r/2}\right\}$$

$$\leqq c_{12} n \log(|\Delta| + 1).$$

Daraus und aus (5.3) folgt

$$\left|\sum_{m=1}^{\infty} (g(m) - 1)\, e^{-(\log m - 2br)^2/4r} - 2\sqrt{\pi r}\, Z(r)\right|$$

$$< c_{13} \sqrt{r}\, n^2 \log^2(|\Delta| + 1)\, e^{r(25/144 + (5/6)b)} \tag{5.5}$$

mit

$$Z(r) \overset{\text{def}}{=} \operatorname{Re} \sum_{\substack{|t_\varrho| \leqq 2\sqrt{b}+1 \\ \varrho \text{ rechts von } H}} e^{r(\varrho^2 + 2b\varrho)}. \tag{5.6}$$

6. Jetzt benötigen wir zwei Hilfssätze.

Lemma 1. *Für die reelle Zahlenfolge $\alpha_1, \alpha_2, \ldots$ seien $U, V > 0$ und $\gamma > 1$ so beschaffen, daß*

$$|\alpha_\nu| \geqq U \quad \textit{und} \quad \sum_\nu \frac{1}{1 + |\alpha_\nu|^\gamma} \leqq V$$

ist. Dann gibt es für jede reelle Folge $\beta_1, \beta_2, \ldots$ *und* $D > \frac{1}{U}$, *in jedem reellen Intervall I der Länge D ein* ξ *derart, daß für jedes* $\gamma = 1, 2, \ldots$ *die Ungleichung*

$$\frac{1}{24V}\frac{1}{1+|\alpha_\nu|^\gamma} \leqq \alpha_\nu\xi + \beta_\nu - [\alpha_\nu\xi + \beta_\nu]$$

$$\leqq 1 - \frac{1}{24V}\frac{1}{1+|\alpha_\nu|^\gamma}$$

erfüllt ist.

Für den Beweis siehe Knapowski-Turán [9].

Lemma 2. *Für die komplexen Zahlen* $z_1, z_2, \ldots, z_n$ *mit*

$$|z_1| \geqq |z_2| \geqq \cdots \geqq |z_n| \quad \text{und} \quad n \leqq N \tag{6.1}$$

gebe es ein $0 < \varkappa \leqq \frac{\pi}{2}$ *derart, daß*

$$\varkappa \leqq |\mathrm{arc}\, z_j| \leqq \pi, \quad j = 1, 2, \ldots, n \tag{6.2}$$

ist, weiter sei für die komplexen Zahlen $d_1, d_2, \ldots, d_n$

$$\min_\lambda \mathrm{Re} \sum_{j=1}^{\lambda} d_j \geqq A > 0.$$

Dann gibt es für jedes positive m ganzrationale Zahlen ν_1 *und* ν_2 *mit*

$$m \leqq \nu_1, \quad \nu_2 \leqq m + N\left(3 + \frac{\pi}{\varkappa}\right),$$

so daß

$$\mathrm{Re} \sum_{j=1}^{n} d_j z_j^{\nu_1} \geqq \left\{\frac{N}{8e\left(m + N\left(3 + \frac{\pi}{x}\right)\right)}\right\}^{2N} \frac{A}{3N} |z_1|^{\nu_1}$$

und

$$\mathrm{Re} \sum_{j=1}^{n} d_j z_j^{\nu_2} \leqq -\left\{\frac{N}{8e\left(m + N\left(3 + \frac{\pi}{x}\right)\right)}\right\}^{2N} \frac{A}{3N} |z_1|^{\nu_2}$$

gilt.

Für den Beweis siehe Turán [10].

7. Nun kehren wir zur Abschätzung von $Z(r)$ in (5.5) zurück. Zuerst bestimmen wir $b = b_0$ durch Lemma 1. Für die α_ν nehme man die Zahlen $\frac{t_\varrho}{\pi}$, für die β_γ die Zahlen

$\frac{1}{2\pi}\operatorname{Im}(\varrho^2)$; es sei weiter

$$U = \frac{\delta(K)}{\pi}, \quad \gamma = \frac{11}{10},$$

dann ist wegen (5.4)

$$V = c_{14} n \log(|\Delta| + 1).$$

Mit

$$D = \log^{1/6} T$$

ist die Bedingung $D > \frac{1}{U}$ wegen (4.4) erfüllt. Wenn man also als I das Intervall

$$\log^{1/3} T \leqq b \leqq \log^{1/3} T + \log^{1/6} T \tag{7.1}$$

wählt, so erhält man mit Lemma 1 ein $b = b_0$ im Intervall (7.1), so daß für *jedes* ϱ die Ungleichung

$$\frac{c_{15}\log^{-11/60} T}{n\log(|\Delta|+1)} \leqq \frac{t_\varrho}{\pi} b_0 + \frac{1}{2\pi}\operatorname{Im}(\varrho^2) - \left[\frac{t_\varrho}{\pi} b_0 + \frac{1}{2\pi}\operatorname{Im}(\varrho^2)\right]$$

$$\leqq 1 - \frac{c_{15}\log^{-11/60} T}{n\log(|\Delta|+1)} \tag{7.2}$$

gilt. Das bedeutet aber, wenn man als z_j die Zahlen $e^{\varrho^2 + 2b_0\varrho}$ wählt, daß die Einschränkung (6.1) mit

$$\varkappa = \log^{-1/5} T \tag{7.3}$$

erfüllt ist. Nun können wir $Z(r)$ in (5.6) durch Lemma 2 mit

$$N = \log^{1/6} T(\log_2 T)^3, \tag{7.4}$$

$$m = \frac{\log T}{2b_0} \tag{7.5}$$

abschätzen. Es ergibt sich die Existenz von ganzen Zahlen r_1 und r_2 mit

$$\frac{\log T}{2b_0} \leqq r_1, \quad r_2 \leqq \frac{\log T}{2b_0} + \log^{11/30} T(\log_2 T)^4, \tag{7.6}$$

so daß

$$Z(r_1) > |z_1|^{r_1} e^{-\log^{1/6}T\,(\log_2 T)^6} \tag{7.7}$$

und

$$Z(r_2) < -|z_1|^{r_1} e^{-\log^{1/6}T\,(\log_2 T)^6} \tag{7.8}$$

gilt.

Es sei zuerst $\varrho_0 = \beta_0 + i\gamma_0$ eine beliebige, nichttriviale Nullstelle von $\zeta_K(s)$ mit $\beta_0 \geqq \frac{1}{2}$. Dann ist

$$|z_1|^{r_1} \geqq |e^{\varrho_0^2 + 2b_0\varrho_0}|^{r_1} > (e^{2b_0 r_1})^{\beta_0} e^{-\gamma_0^2 r_1} > T^{\beta_0} e^{-\gamma_0^2 \log^{2/3} T},$$

und dasselbe gilt auch für $|z_1|^{r_2}$. Zusammen mit (7.7) und (7.8) ergibt das

$$\left.\begin{aligned} Z(r_1) &> T^{\beta_0} e^{-(\gamma_0^2+1)\log^{2/3} T}, \\ Z(r_2) &< -T^{\beta_0} e^{-(\gamma_0^2+1)\log^{2/3} T}. \end{aligned}\right\} \tag{7.9}$$

Wenn man für $j = 1, 2$

$$e^{2b_0 r_j} = x_j$$

setzt, so ergeben (5.5) und (7.9) die Abschätzungen

$$\left.\begin{aligned} \sum_{m=1}^{\infty} \{g(m) - 1\} e^{-\frac{1}{4r_1}\log^2 \frac{m}{x_1}} &> 3T^{\beta_0} e^{-(\gamma_0^2+1)\log^{2/3} T}, \\ \sum_{m=1}^{\infty} \{g(m) - 1\} e^{-\frac{1}{4r_2}\log^2 \frac{m}{x_2}} &< -3T^{\beta_0} e^{-(\gamma_0^2+1)\log^{2/3} T} \end{aligned}\right\} \tag{7.10}$$

mit

$$T \leqq x_1, \quad x_2 \leqq T e^{\log^{3/4} T}. \tag{7.11}$$

8. Nun schreiben wir mit

$$A(y) \overset{\text{def}}{=} \sum_{2 \leqq m \leqq y} \frac{g(m) - 1}{\log m} \tag{8.1}$$

und

$$\varphi_j(y) = e^{-\frac{1}{4r_j}\log^2 \frac{y}{x_j}} \log y, \quad j = 1, 2, \tag{8.2}$$

die linke Seite von (7.10) in der Form

$$\int_1^{\infty} e^{-\frac{1}{4r_j}\log^2 \frac{y}{x_j}} \log y \, dA(y) = -\int_1^{\infty} A(y)\, \varphi_j'(y)\, dy.$$

Wenn wir

$$\xi_1 = x_1 e^{-3\sqrt{r_1 \log x_1}}, \quad \xi_2 = x_1 e^{3\sqrt{r_1 \log x_1}} \tag{8.3}$$

setzen, so folgt leicht, mit Rücksicht auf (4.4),

$$\left|\int_1^{\xi_1}\right| + \left|\int_{\xi_2}^{\infty}\right| < c_{16}.$$

Wenn auch die Ungleichung

$$|\gamma_0| < \frac{1}{2}\log^{1/12} T \tag{8.4}$$

erfüllt ist, so folgt (es genügt, nur den Fall $j = 1$ auszuführen und $\varphi(y)$ an Stelle von $\varphi_1(y)$ zu schreiben)

$$-\int_{\xi_1}^{\xi_2} A(y)\,\varphi'(y)\,dy > 2T^{\beta_0} e^{-(2/3)\log^{5/6} T}. \tag{8.5}$$

Aus

$$\varphi'(y) = e^{-\frac{1}{4r_1}\log^2\frac{y}{x_1}} \frac{1}{y}\left\{1 - \frac{1}{2r_1}\log\frac{y}{x_1}\log y\right\}$$

ergibt sich offenbar ein $y = y_0$ mit

$$x_1 < y_0 < 2x_1,$$

so daß $\varphi(y)$ von 1 bis y_0 wächst und dann abnimmt. Es gilt also

$$-\int_{\xi_1}^{\xi_2} A(y)\,\varphi'(y)\,dy \leqq \max_{y_0 \leqq y \leqq \xi_2} A(y) \int_{y_0}^{\xi_2} |\varphi'(y)|\,dy - \min_{\xi_1 \leqq y \leqq y_0} A(y) \int_{\xi_1}^{y_0} |\varphi'(y)|\,dy. \tag{8.6}$$

Da ferner

$$\int_{\xi_1}^{y_0} |\varphi'(y)|\,dy - \int_{y_0}^{\xi_2} |\varphi'(y)|\,dy = \int_{\xi_1}^{\xi_2} \varphi'(y)\,dy = O\left(\frac{\log x_1}{x_1^2}\right)$$

gilt, folgt aus (8.5) und (8.6)

$$\left\{\max_{y_0 \leqq y \leqq \xi_2} A(y) - \min_{\xi_1 \leqq y \leqq y_0} A(y)\right\} \int_{\xi_1}^{y_0} \varphi'(y)\,dy > T^{\beta_0} e^{-(2/3)\log^{5/6} T}. \tag{8.7}$$

Weiter gilt nach (7.11)

$$\int_{\xi_1}^{y_0} \varphi'(y)\,dy < \varphi(y_0) < \log 2x_1 < 2\log T,$$

also haben wir für geeignete

$$\xi_1 \leqq U_1 < U_2 \leqq \xi_2 \tag{8.8}$$

die Ungleichung

$$\sum_{U_1 \leqq m \leqq U_2} \frac{g(m)}{\log m} - \int_{U_1}^{U_2} \frac{dv}{\log v} > T^{\beta_0} e^{-(3/4)\log^{5/6} T}. \tag{8.9}$$

Wir bemerken, daß aus (8.3), (7.11) und (7.6)

$$\left.\begin{aligned} U_2 &\leqq \xi_2 \leqq T e^{4\log^{5/6} T}, \\ U_1 &\geqq \xi_1 \geqq T e^{-4\log^{5/6} T} \end{aligned}\right\} \tag{8.10}$$

folgt. Da der Beitrag von $m = p^\alpha$ mit $\alpha \geqq 3$ offenbar höchstens gleich $nI^{2/5}$ ist, folgt aus (8.9) die Ungleichung

$$\sum_{\substack{U_1 \leqq p^\alpha \leqq U_2 \\ \alpha = 1,2}} \frac{g(p^\alpha)}{\log(p^\alpha)} - \int_{U_1}^{U_2} \frac{dv}{\log v} > T^{\beta_0} e^{-(4/5)\log^{5/6} T}. \tag{8.11}$$

Analog folgt für geeignete

$$T e^{-4\log^{5/6} T} \leqq U_3 < U_4 \leqq T e^{4\log^{5/6} T} \tag{8.12}$$

die Ungleichung

$$\sum_{\substack{U_3 \leqq p^\alpha \leqq U_4 \\ \alpha = 1,2}} \frac{g(p^\alpha)}{\log p^\alpha} - \int_{U_3}^{U_4} \frac{dv}{\log v} < -T^{\beta_0} e^{-(4/5)\log^{5/6} T}. \tag{8.13}$$

9. Jetzt sind wir in der Lage, die folgenden beiden Fälle zu behandeln:

Fall 1. *Es gebe eine Nullstelle $\varrho^* = \sigma^* + it^*$ von $\zeta_K(s)$ im Gebiet*

$$\sigma \geqq \frac{1}{2} + 20 \log^{-1/6} T, \quad |t| \leqq \frac{1}{2} \log^{1/12} T. \tag{9.1}$$

Dann ist (8.4) erfüllt und somit (8.11)–(8.13) mit $\varrho_0 = \varrho^*$ anwendbar. Nun folgt offenbar

$$\sum_{U_1 \leqq p \leqq U_2} \frac{g(p)}{\log p} - \int_{U_1}^{U_2} \frac{dv}{\log v} > \sqrt{T} e^{10\log^{5/6} T}. \tag{9.2}$$

Nach (8.10) und (4.4) gilt $U_1 > |\Delta|$, also nach dem klassischen Satz von Dedekind

$$\frac{g(p)}{\log p} = \nu_p.$$

Im Falle 1 folgt also die Behauptung (4.6) aus (9.2); (4.7) kann man im Falle 1 auf ähnliche Weise erledigen. Wie man aus (8.10) und (9.2) sieht, konnte man die Behauptung in diesem Fall viel schärfer fassen (und auch die rechte Seite von (4.4) viel kleiner wählen).

Fall 2. *$\zeta_K(s)$ verschwindet nicht im Gebiet*

$$\sigma \geqq \frac{1}{2} + 20 \log^{-1/6} T, \quad |t| \leqq \frac{1}{2} \log^{1/12} T. \tag{9.3}$$

Da die Behandlung dieses Falles ziemlich umfangreich ist und die Grundgedanken – sie gehen auf Littlewood, Ingham und Skewes zurück –, obwohl in einer etwas anderen Form, an unsere Arbeit [11] erinnern, sehen wir hier davon ab.

Literatur

[1] P. L. Tschebyscheff, Lettre de M. le professeur Tchebychev à M. Fuss sur un nouveau théorème rélatif aux nombres premiers contenus dans les formes $4n + 1$ et $4n - 3$, Bull. de la Classe phys.-math. de l'Acad. Imp. des Sciences St. Petersburg **11** (1853), 208. Oder Oeuvres Bd. 1, S. 697.

[2] E. Landau, Über einige ältere Vermutungen und Behauptungen in der Primzahltheorie, Math. Z. **1**, Nr. 1 (1918), 1–24.

[3] G. H. Hardy und J. E. Littlewood, Contributions to the theory of the Riemann zeta function and the theory of the distribution of primes, Acta Math. **41** (1917), 119–196.

[4] E. Landau, Über Ideale und Primideale in Idealklassen, Math. Z. **2**, Nr. 1–2 (1918), 52–154.

[5] Die 14 Abhandlungen sind in zwei Serien in Acta Math. Hung., Bd. XIII, XIV bzw. in Acta Arithmetica B. IX, X, XI und XII publiziert. Diejenigen, die zu dieser Abhandlung in Beziehung stehen, werden wir ausführlich zitieren.

[6] J. E. Littlewood, Sur la distribution des nombres premiers, Comptes Rendus Paris **158** (1914), 1869–1872.

[7] S. Knapowski und P. Turán, Further developments in the comparative prime number theory I, Acta Arithm. IX (1964), 23–40.

[8] E. Landau, Über einen Satz von Tschebyscheff, Math. Ann. **61**, Heft 4 (1905), 527–550.

[9] S. Knapowski und P. Turán, Further developments in the comparative prime number theory II, Acta Arithm. X (1964), 293–313.

[10] P. Turán, On some further one-sided theorems of new type in the theory of diophantine approximation, Acta Math. Hung. XII (1961), 455–468.

[11] S. Knapowski und P. Turán, Comparative prime number theory II, Acta Math. Hung. XIII, Nr. 3–4 (1962), 313–342.

J. Kubilius in Vilnius

ON LOCAL THEOREMS FOR ADDITIVE NUMBER-THEORETIC FUNCTIONS

In the present paper the distribution of values of additive number-theoretic functions is considered. A function $f(m)$ defined for all positive integers $m = 1, 2, \ldots$ is called additive if $f(mn) = f(m) + f(n)$ provided $(m, n) = 1$. The theory of integral limit laws for these functions has been developed by many authors. As to local laws which are generally speaking deeper very little is known. In this case it is a matter of finding an asymptotic expression for the number $N_n(a)$ of positive integers $m \leqq n$ for which $f(m)$ assumes a given value a. Theorems of such kind are the asymptotic law of prime numbers as well as the asymptotic laws of positive integers having a given number of prime divisors.

It is comparatively simple to prove local theorems for additive functions $f(m)$ assuming only integral values and satisfying the condition $f(p) = 0$ for all primes p. The first result of this kind is due to A. Rényi [10] (see also [4, 5, 11]). Let $\omega(m)$ denote the number of distinct prime factors of m and let $\Omega(m)$ denote the number of all prime factors of m, where each factor is counted according to its multiplicity. Rényi proved that for every non-negative integer a the set of positive integers m with property $f(m) = \Omega(m) - \omega(m) = a$ has asymptotic density d_a:

$$\frac{N_n(a)}{n} \to d_a \tag{1}$$

as $n \to \infty$. He also proposed a method for counting the value of d_a. H. Delange [1] and I. Kátai [6] strengthened Rényi's result by estimating the rate of convergence in (1).

The author of this paper succeded [7] in generalizing Rényi's theorem to any integral-valued additive function $f(m)$ satisfying the condition $f(p) = 0$ for all primes p. It turned out that this condition could be broken for any set of primes with the property

$$\sum_{f(p) \neq 0} \frac{1}{p} < \infty,$$

where the summation is extended over all primes p with $f(p) \neq 0$. Our condition seems to be natural because, in view of Erdös-Wintner theorem [2], it is necessary for the existence of the integral limit law for an additive function assuming only integral values. The remainder in this more general theorem was also estimated by the author [7] and S. A. Fainleib [3].

It occurs natural to one to replace the condition $f(p) = 0$ by the condition $f(p) = b \neq 0$ and to suppose that $f(p^\alpha)$ are multiples of b for all positive integer power of primes p^α. It is sufficient to restrict our attention to the case $b = 1$. A local theorem for functions of this kind was obtained by A. Wintner [17, 18]. He proved that if $f(m)$ is a positive integral-valued additive function and $f(p) = 1$ for all primes, then for any fixed positive integer a

$$N_n(a) \sim \frac{n\,(\ln\ln n)^{a-1}}{(a-1)!\,\ln n}.$$

L. G. Sathe [14] (see also [15, 13]) showed more precise results for the functions $\omega(m)$ and $\Omega(m)$. If $1 \leqq a \leqq c_1 \ln\ln n$, where $c_1 < 2$ is a constant, then

$$N_n(a) = \frac{nw(\eta)\,(\ln\ln n)^{a-1}}{(a-1)!\,\ln n}\left(1 + O\left(\frac{1}{\ln\ln n}\right)\right).$$

Here $\eta = a(\ln\ln n)^{-1}$ and $w(\eta)$ equals

$$\frac{1}{\Gamma(1+\eta)}\prod_p \left(1 - \frac{1}{p}\right)^\eta \left(1 + \frac{\eta}{p-1}\right)$$

or

$$\frac{1}{\Gamma(1+\eta)}\prod_p \left(1 - \frac{1}{p}\right)^\eta \left(1 - \frac{\eta}{p}\right)^{-1}$$

for $\omega(m)$ or $\Omega(m)$ respectively, Γ is the gamma-function. The constant in the symbol O depends on the constant c_1 only.

The aim of this paper is to prove some analogues of the classical probability-theoretic local laws for additive number-theoretic functions $f(m)$ assuming only integral values and satisfying the condition $f(p) = 1$ for all primes p with the exception of primes from a rare set. More precisely, we suppose that

$$\sum_{f(p)\neq 1} \frac{\ln p}{p} < \infty. \tag{C}$$

For the sake of brevity we introduce some notations. We put

$$\lambda = \sqrt{\ln\ln n}, \quad y = \frac{a - \lambda^2}{\lambda}, \quad \varphi(u) = \frac{1}{\sqrt{2\pi}} e^{-(1/2)u^2},$$

$$\mu(z) = e^z - 1 - z, \quad \Lambda(z) = \frac{h(z)}{\Gamma(e^z)}, \quad h(z) = \prod_p \psi_p(z),$$

$$\psi_p(z) = \left(1 - \frac{1}{p}\right)^{e^z} \sum_{\alpha=0}^{\infty} \frac{e^{zf(p^\alpha)}}{p^\alpha}.$$

$c_2, c_3, \dots$ are suitably chosen positive constants. They are either absolute or depending only on the function $f(m)$ throughout the whole paper, except in theorem 2, where they are dependent on r as well. B denotes a number (not always the same) which is bounded by a constant. In what follows we consider only the principal value of logarithms and powers. We shall suppose n to be sufficiently large.

We shall prove that for any function $f(m)$ satisfying (C) and any real τ

$$\sum_{m=1}^{n} e^{i\tau f(m)} = n\Lambda(i\tau)(\ln n)^{e^{i\tau}-1} + \frac{Bn\lambda}{\sqrt{\ln n}}, \tag{2}$$

which implies

$$N_n(a) = \frac{n}{2\pi}\int_{-\pi}^{\pi} \Lambda(i\tau)\, e^{\lambda^2\mu(i\tau) - i\lambda\tau y}\, d\tau + \frac{Bn\lambda}{\sqrt{\ln n}}. \tag{3}$$

Thus we have to evaluate this integral. The exactness of this evaluation depends on the analytical properties of the function $h(i\tau)$ which in turn depends on the properties of $f(m)$. Under various assumptions concerning the function $f(m)$ we obtain three theorems.

Theorem 1. *If the additive function $f(m)$ assumes only integral values and satisfies the condition* (C) *as well as the conditions*

$$\sum_{f(p)\neq 1} \frac{|f(p)|}{p} < \infty, \quad \sum_{p}\sum_{\alpha=2}^{\infty} \frac{|f(p^\alpha)|}{p^\alpha} < \infty, \tag{D_1}$$

then uniformly for all integers a and $n > 30$

$$N_n(a) = \frac{n\varphi(y)}{\lambda} + \frac{Bn}{\lambda^2}. \tag{*}$$

If $|a - \lambda^2| > (1+\delta)\lambda\sqrt{2\ln\lambda}$ or $|a-\lambda^2| < (1-\delta)\lambda\sqrt{2\ln\lambda}$, where δ is any fixed positive number, then the principal term in (*) is less or greater than the remainder correspondingly. If besides the condition (C) the function $f(m)$ satisfies a stronger condition than (D_1) we can enlarge the region of the validity of the local theorem. Let us suppose that there exists an integer $r \geqq 2$ such that

$$\sum_{f(p)\neq 1} \frac{|f^r(p)|}{p} < \infty, \quad \sum_{p}\sum_{\alpha=2}^{\infty} \frac{|f^r(p^\alpha)|}{p^\alpha} < \infty. \tag{D_r}$$

Then for any sufficiently small τ we can write the integrand of (3) in the form

$$\Lambda\left(\frac{i\tau}{\lambda}\right) e^{\lambda^2\mu(i\tau/\lambda)} = \sum_{k=0}^{r-1} \frac{P_k(i\tau)}{\lambda^k} e^{-(1/2)\tau^2} + \frac{B|\tau|^r}{\lambda^r}(1+\tau^2)^r e^{-(1/4)\tau^2},$$

where $P_k(i\tau)$ is a polynomial of the degree $3k$ with coefficients depending only on the function $f(m)$. The first two of these polynomials are

$$P_0(i\tau) = 1,$$

$$P_1(i\tau) = \frac{1}{6}(i\tau)^3 + i\tau\left\{-\gamma + \sum_p \left[\ln\left(1 - \frac{1}{p}\right) + \left(1 - \frac{1}{p}\right)\sum_{\alpha=1}^{\infty}\frac{f(p^\alpha)}{p^\alpha}\right]\right\},$$

where $\gamma = -\Gamma'(1)$ is Euler's constant. In this case the following more precise theorem holds.

Theorem 2. *If the additive number-theoretic function $f(m)$ assumes only integer values and satisfies the conditions* (C) *and* (D_r), *then uniformly for all integers a and $n > 30$*

$$N_n(a) = n\sum_{k=0}^{r-1}\frac{P_k(-\varphi)}{\lambda^{k+1}} + \frac{Bn}{\lambda^{r+1}},$$

where $P_k(-\varphi)$ is obtained from $P_k(-i\tau)$ by replacing all powers $(i\tau)^l$ $(l = 0, 1, \ldots)$ by $\varphi^{(l)}(y)$.

If $|a - \lambda^2| < (1 - \delta)\,\lambda\sqrt{2r\ln\lambda}$ or $|a - \lambda^2| > (1 + \delta)\,\lambda\sqrt{2r\ln\lambda}$, then the principal term of the last formula is greater or less than the remainder respectively.

Further restriction on the function $f(m)$ makes it possible to enlarge the region of the validity of the local theorem to $a - \lambda^2 = o(\lambda^2)$.

Theorem 3. *Let the additive number-theoretic function $f(m)$ assuming only integral values satisfy the condition* (C) *and there exists a positive constant c such that for all real x, $|x| \leqq c$, the series*

$$\sum_{f(p)\neq 1}\frac{e^{xf(p)}}{p}, \quad \sum_p\sum_{\alpha=2}^{\infty}\frac{e^{xf(p^\alpha)}}{p^\alpha} \tag{A}$$

converge. Then for all integers a satisfying $a = \lambda^2 + o(\lambda^2)$

$$N_n(a) = \frac{n}{\sqrt{2\pi}\,\lambda\ln n}\left(\frac{e\lambda^2}{a}\right)^a\left\{1 + B\left(|\xi| + \frac{\ln^3\lambda}{\lambda^2}\right)\right\},$$

where $\xi = a\lambda^{-2} - 1$.

At first the formula (2) is proved. It is deduced from the following lemma.

Lemma. *Let $g(m)$ be a complex-valued multiplicative function which is not identical with zero. If $|g(m)| \leqq 1$ for all positive integers m and for some $\varkappa$*

$$\sum_{g(p)\neq\varkappa}\frac{\ln p}{p} < \infty, \tag{4}$$

then

$$\sum_{m \leqq x} g(m) = \frac{x(\ln x)^{\varkappa-1}}{\Gamma(\varkappa)} \prod_p \left(1 - \frac{1}{p}\right)^{\varkappa} \sum_{\alpha=0}^{\infty} \frac{g(p^\alpha)}{p^\alpha} + Bx \sqrt{\frac{\ln\ln x}{\ln x}}$$

uniformly for all functions $g(m)$ *and* $x > 3$.

Proof. Our method of proof makes use of the standard tools of analytic number theory (see e.g. [11, 12, 15, 16]). Another proof can be obtained by application of Wirsing's method [19] (see also [8, 9].)

Consider the Dirichlet series

$$Z(s) = \sum_{m=1}^{\infty} \frac{g(m)}{m^s},$$

where $s = \sigma + it$ is a complex variable. This series is uniformly convergent for $\sigma \geqq 1 + \delta$, where δ is any fixed positive number. Hence $Z(s)$ is regular in the half-plane $\sigma > 1$. From the absolute convergence of the series for $\sigma > 1$ we deduce

$$Z(s) = \prod_p \sum_{\alpha=0}^{\infty} \frac{g(p^\alpha)}{p^{\alpha s}}.$$

Now let us put

$$H(s) = \prod_p \chi_p(s),$$

where

$$\chi_p(s) = \left(1 - \frac{1}{p^s}\right)^{\varkappa} \sum_{\alpha=0}^{\infty} \frac{g(p^\alpha)}{p^{\alpha s}} = 1 + \frac{g(p) - \varkappa}{p^s} + \omega_p(s),$$

$$\omega_p(s) = \left(1 - \frac{1}{p^s}\right)^{\varkappa} - 1 + \frac{\varkappa}{p^s} + \frac{g(p)}{p^s}\left[\left(1 - \frac{1}{p^s}\right)^{\varkappa} - 1\right] + \left(1 - \frac{1}{p^s}\right)^{\varkappa} \sum_{\alpha=2}^{\infty} \frac{g(p^\alpha)}{p^{\alpha s}}.$$

In virtue of the evident estimate $\omega_p(s) = Bp^{-2\sigma}$ and the condition (4) we have that $H(s)$ is regular for $\sigma > 1$, continuous for $\sigma \geqq 1$ and $H(s) = B$ for $\sigma \geqq 1$. Further, denoting the zeta-function of Riemann as usually by $\zeta(s)$, we deduce from the equality

$$Z(s) = H(s)\zeta^{\varkappa}(s)$$

that $Z(s)$ is continuous for $\sigma \geqq 1$ with a possible exception of the point $s = 1$.

For $\sigma \geqq 1$ we have by a simple calculation that

$$\chi_p'(s) = \left(1 - \frac{1}{p^s}\right)^{\varkappa-1} \left(\frac{\varkappa - g(p)}{p^s} + \sum_{\alpha=2}^{\infty} \frac{(\alpha - 1 + \varkappa)\, g(p^{\alpha-1}) - \alpha g(p^\alpha)}{p^{\alpha s}}\right) \ln p = Bw_p$$

and

$$\chi_p(s) = \chi_p(1) + B\,|s - 1|\, w_p,$$

where $w_p = p^{-1} \ln p$ if $g(p) = \varkappa$ and $w_p = p^{-2} \ln p$ if $g(p) = \varkappa$. Using the condition (4), hence we deduce that $H(1 + it)$ as a function of t has the derivative

$$H'(1 + it) = B \tag{5}$$

and that

$$H(s) = H(1) + B\,|s - 1| \tag{6}$$

for $\sigma \geqq 1$.

We shall also need the estimates

$$\zeta^{\pm 1}(s) = B \ln(|t| + 1), \quad \zeta'(s) = B \ln(|t| + 1) \tag{7}$$

for $\sigma \geqq 1$, $|s - 1| \geqq 1$. Still less exact estimates are sufficient for the proof.

By the well-known formula for Dirichlet series, putting

$$T(x) = \sum_{m \leqq x} g(m) \ln \frac{x}{m},$$

we have

$$T(x) = \frac{1}{2\pi i} \int_{2-i\infty}^{2+i\infty} \frac{x^s Z(s)}{s^2}\, ds. \tag{8}$$

Let L_j and L_j' denote the following lines:

$$L_1: \ s = 1 + it, \quad -\infty < t \leqq -1,$$

$$L_2: \ s = 1 + it, \quad -1 < t \leqq \varrho,$$

$$L_3: \ s = \varrho\, e^{i\varphi}, \quad -\frac{\pi}{2} < \varphi < \frac{\pi}{2},$$

$$L_4: \ s = 1 + it, \quad \varrho \leqq t < 1,$$

$$L_5: \ s = 1 + it, \quad 1 \leqq t < \infty,$$

$$L_1': \ s = \sigma - i, \quad -\infty < \sigma < 1,$$

$$L_5': \ s = \sigma + i, \quad 1 > \sigma > -\infty,$$

where $\varrho = \ln^{-1} x$. By the above properties of $\zeta(s)$ and $H(s)$ we may replace the contour of the integral in (8) by the contour $L = L_1 \cup L_2 \cup L_3 \cup L_4 \cup L_5$. Let us put

$$\frac{Z(s)}{s^2} = \frac{H(1)}{(s-1)^{\varkappa}} + K(s).$$

Then we have

$$T(x) = \frac{1}{2\pi i}\left(\int_{L_1 \cup L_5} \frac{x^s Z(s)}{s^2}\, ds + \int_{L_2 \cup L_4} x^s K(s)\, ds + \int_{L_3} x^s K(s)\, ds \right.$$
$$\left. - \int_{L_2' \cup L_5'} \frac{x^s\, ds}{(s-1)^{\varkappa}} + H(1)\int_{L'} \frac{x^s\, ds}{(s-1)^{\varkappa}}\right)$$
$$= I_1 + I_2 + I_3 - I_4 + I_5,$$

where $L' = L_1' \cup L_2 \cup L_3 \cup L_4 \cup L_5'$. By (5), (7) and the estimate $H(s) = B$ we have that

$$(Z(s)\, s^{-2})' = Bt^{-2} \ln^3 (|t| + 1)$$

on the contour $L_1 \cup L_5$. Applying partial integration in such a manner that x is chosen as the factor to be integrated, we obtain

$$I_1 = \frac{Bx}{\ln x}.$$

In virtue of the identities

$$K(s) = H(s)\frac{\zeta^{\varkappa}(s)(s-1)^{\varkappa} - 1}{s^2(s-1)^{\varkappa}} + \frac{H(s) - H(1)}{s^2(s-1)^{\varkappa}} - \frac{H(1)(s+1)}{s^2(s-1)^{\varkappa-1}},$$

$$K'(s) = \varkappa H(s)\frac{\zeta^{\varkappa-1}(s)\zeta'(s)(s-1)^{\varkappa+1} + 1}{s^2(s-1)^{\varkappa+1}} - \varkappa\frac{H(s) - H(1)}{s^2(s-1)^{\varkappa+1}}$$
$$+ \frac{\varkappa H(1)(s+1)}{s^2(s-1)^{\varkappa}} + s^{-2}\zeta^{\varkappa}(s)\, H'(s) - 2s^{-3}\zeta^{\varkappa}(s)\, H(s),$$

the well-known properties of $\zeta(s)$ and (5), (6) we have that $K'(s) = B|s-1|^{-\operatorname{Re}\varkappa}$ on the contour $L_2 \cup L_4$ and $K(s) = B|S-1|^{1-\operatorname{Re}\varkappa}$ on the contour L_3. Applying again partial integration we obtain

$$I_2 = \frac{Bx}{\ln x}\ln\ln x.$$

The trivial estimation gives

$$I_3 = \frac{Bx}{\ln x}, \quad I_4 = \frac{Bx}{\ln x}.$$

Finally, from the representation of the Γ-function by the Hankel contour integral, we conclude that

$$I_5 = \frac{H(1)\, x}{\Gamma(\varkappa)}(\ln x)^{\varkappa-1}.$$

Thus collecting all estimates we have

$$T(x) = \frac{H(1)\,x}{\Gamma(\varkappa)}(\ln x)^{\varkappa-1} + \frac{Bx}{\ln x}\ln\ln x. \tag{9}$$

Hence we can obtain the estimate for the sum

$$S(x) = \sum_{m \leqq x} g(m).$$

Let us put $\Delta = (\ln\ln x/\ln x)^{1/2}$. From the identity

$$T(x) = \int_1^x \frac{S(y)}{y}\,dy$$

it follows that

$$S(x) = \frac{T(x + x\Delta) - T(x)}{\ln(1+\Delta)} - \frac{1}{\ln(1+\Delta)}\int_x^{x+x\Delta} \frac{S(y) - S(x)}{y}\,dy.$$

In view of the estimate $|S(y) - S(x)| \leqq x\Delta + 1$ for $x \leqq y \leqq x + x\Delta$ we deduce from (9)

$$S(x) = \frac{xH(1)}{\Gamma(\varkappa)}(\ln x)^{\varkappa-1}\,\frac{(1+\Delta)\left(1 + \dfrac{\ln(1+\Delta)}{\ln x}\right)^{\varkappa-1} - 1}{\ln(1+\Delta)} + Bx\Delta$$

$$= \frac{\varkappa H(1)}{\Gamma(\varkappa)}(\ln x)^{\varkappa-1} + Bx\Delta.$$

If the function $g(m)$ satisfies some stronger condition than (4) we may get more exact results. In particular, if $g(p) \neq \varkappa$ only for a finite set of primes, then the function $H(s)$ is regular in the half-plane $\sigma > \frac{1}{2}$. In this case we could transfer the path of integration to the left of the line $\sigma = 1$. Using well-known estimates for $\zeta(s)$, we could obtain the estimate $B\,x(\ln x)^{\mathrm{Re}\,\varkappa-2}$ for the remainder term.

Let $f(m)$ satisfy the condition (C). Put $\varepsilon = c_2\lambda^{-1}(\ln\lambda)^{1/2}$. Since $\mathrm{Re}\,\mu(i\tau) = \cos\tau - 1 \leqq -\frac{1}{12}\tau^2$ for $|\tau| \leqq \pi$ and $\Lambda(i\tau) = B$ we have that the part of the integral in (3) along the intervals $\varepsilon < |\tau| \leqq \pi$ equals

$$B\int_\varepsilon^\pi e^{-(1/12)\lambda^2\tau^2}\,d\tau = \frac{B}{\varepsilon\lambda^2}e^{-(1/12)\varepsilon^2\lambda^2}.$$

Consequently,

$$N_n(a) = \frac{n}{2\pi}\int_{-\varepsilon}^{\varepsilon} \Lambda(i\tau)\,e^{\lambda^2\mu(i\tau) - i\lambda\tau y}\,d\tau + Bn\lambda^{-1-(1/12)c_2^2}. \tag{10}$$

Proof of the theorem 1. Let us put $c_2 = 5$ and consider the function $\Lambda(i\tau)$ for $|\tau| \leqq \varepsilon$. Denote

$$F_p = \sum_{\alpha=2}^{\infty} \frac{|f(p^\alpha)|}{p^\alpha}.$$

In view of the estimates

$$e^{i\tau} \ln\left(1 - \frac{1}{p}\right) = -\frac{e^{i\tau} - 1}{p} + \ln\left(1 - \frac{1}{p}\right) + \frac{B|\tau|}{p^2},$$

$$\sum_{\alpha=0}^{\infty} \frac{e^{i\tau f(p^\alpha)}}{p^\alpha} = \frac{1}{1 - \frac{1}{p}} + \frac{e^{i\tau f(p)} - 1}{p} + B|\tau|\, F_p$$

and the conditions (C) and (D_1) it follows

$$\ln \psi_p(i\tau) = \frac{e^{i\tau f(p)} - e^{i\tau}}{p} + B|\tau| \left(\frac{1}{p^2} + \frac{f^2(p)}{p^2} + F_p\right)$$

and $h(i\tau) = 1 + B|\tau|$. Taking into account the estimate $\Gamma^{-1}(e^{i\tau}) = 1 + B|\tau|$, we have $\Lambda(i\tau) = 1 + B|\tau|$. Further,

$$e^{\lambda^2 \mu(i\tau)} = \exp\left(-\tfrac{1}{2}\lambda^2\tau^2 + B\lambda^2|\tau|^3\right) = e^{-(1/2)\lambda^2\tau^2}\left(1 + B\lambda^2|\tau|^3\right).$$

Substituting these estimates into (10), we get

$$\begin{aligned} &N_n(a) \\ &= \frac{n}{2\pi\lambda}\left(\int_{-\infty}^{\infty} e^{-(1/2)u^2 - iuy}\, du + B \int_{\varepsilon\lambda}^{\infty} e^{-(1/2)u^2}\, du\right) + \frac{Bn}{\lambda^2}\int_0^{\varepsilon\lambda} u(1 + u^2)\, e^{-(1/2)u^2}\, du + \frac{Bn}{\lambda^3} \\ &= \frac{n\varphi(y)}{\lambda} + \frac{Bn}{\lambda^2}. \end{aligned}$$

Proof of the theorem 2. The following notations are used throughout the proof:

$$E_{pk} = \frac{f^k(p)}{p} \quad (k = 1, \dots, r - 1), \quad E_{pr} = \frac{|f^r(p)|}{p}, \quad E = \sum_{f(p) \neq 1} E_{pr},$$

$$F_{pk} = \sum_{\alpha=2}^{\infty} \frac{f^k(p^\alpha)}{p^\alpha} \quad (k = 1, \dots, r - 1), \quad F_{pr} = \sum_{\alpha=2}^{\infty} \frac{|f^r(p^\alpha)|}{p^\alpha}, \quad F = \sum_p F_{pr},$$

$$\eta = 10 \max(1, E, F), \quad c_2 = \sqrt{12r}.$$

The condition (D_r) allows us to represent the function $h(i\tau)$ for any sufficiently small τ in the form

$$\ln h(i\tau) = \sum_{k=1}^{r-1} \beta_k (i\tau)^k + R\tau^r. \tag{11}$$

For this aim we note that

$$\ln \psi_p(i\tau) = (e^{i\tau} - 1) \ln\left(1 - \frac{1}{p}\right)$$
$$+ \ln\left[1 + \left(1 - \frac{1}{p}\right) \sum_{k=1}^{r-1} (E_{pk} + F_{pk}) \frac{(i\tau)^k}{k!} + \frac{\Theta}{r!}(E_{pr} + F_{pr})\tau^r\right]$$

and in case $f(p) = 1$

$$\ln \psi_p(i\tau) = (e^{i\tau} - 1)\left[\ln\left(1 - \frac{1}{p}\right) + \frac{1}{p}\right] + \left[\ln\left(1 + \frac{e^{i\tau} - 1}{p}\right) - \frac{e^{i\tau} - 1}{p}\right]$$
$$+ \ln\left[1 - \frac{e^{i\tau} - 1}{p^2\left(1 + \dfrac{e^{i\tau} - 1}{p}\right)} + \frac{1 - \dfrac{1}{p}}{1 + \dfrac{e^{i\tau} - 1}{p}} \left(\sum_{k=1}^{r-1} F_{pk} \frac{(i\tau)^k}{k!} + \frac{\Theta}{r!} F_{pr}\tau^r\right)\right],$$

where $|\Theta| \leqq 1$.

We introduce the function

$$\tilde{h}(z) = \prod_p \tilde{\varphi}_p(z),$$

where

$$\ln \tilde{\psi}_p(z) = -(e^z - 1) \ln\left(1 - \frac{1}{p}\right) - \ln\left[1 - (E_{pr} + F_{pr})(e^z - 1)\right],$$

if $f(p) \neq 1$, and

$$\ln \tilde{\psi}_p(z) = -(e^z - 1)\left[\ln\left(1 - \frac{1}{p}\right) + \frac{1}{p}\right] - \left[\ln\left(1 - \frac{e^z - 1}{p}\right) + \frac{e^z - 1}{p}\right]$$
$$- \ln\left[1 - \frac{e^z - 1}{p^2\left(1 - \dfrac{e^z - 1}{p}\right)} - \frac{F_{pr}(e^z - 1)}{1 - \dfrac{e^z - 1}{p}}\right],$$

if $f(p) = 1$. For $|z| \leqq \frac{1}{2}$, $0 \leqq x \leqq \frac{1}{2}$ we have

$$|e^z - 1| \leqq 2|z|, \quad 0 \leqq -\ln(1 - x) \leqq 2x, \quad 0 \leqq -x - \ln(1 - x) \leqq 2x^2.$$

Consequently, for $|z| \leqq \eta^{-1}$

$$|\ln \tilde{\psi}_p(z)| \leqq \frac{4|z|}{p} - \ln\left[1 - 2|z|(E_{pr} + F_{pr})\right] < \frac{1}{p} + E_{pr} + F_{pr},$$

if $f(p) \neq 1$, and

$$|\ln \tilde{\varphi}_p(z)| \leqq \frac{4|z|}{p^2} + \sum_{k=2}^{\infty} \frac{(2|z|)^k}{kp^k} - \ln\left[1 - \frac{2|z|}{p^2\left(1 - \frac{2|z|}{p}\right)} - \frac{2|z|\, F_{pr}}{1 - \frac{2|z|}{p}}\right]$$

$$< \frac{1}{p^2} + F_{pr},$$

if $f(p) = 1$. Hence by the conditions (C) and (D_r) we conclude that for $|z| \leqq \eta^{-1}$

$$|\ln \tilde{h}(z)| \leqq c_3. \tag{12}$$

We represent the function $\ln \tilde{h}(z)$ in the form

$$\ln \tilde{h}(z) = \sum_{k=1}^{r-1} \tilde{\beta}_k z^k + \tilde{R} z^r.$$

Now we have to estimate the coefficients $\tilde{\beta}_k$ and $\tilde{R}$. By Cauchy's theorem we have

$$\tilde{\beta}_k = \frac{1}{2\pi i} \int_{|w| = \eta^{-1}} \frac{\ln \tilde{h}(w)}{w^{k+1}}\, dw,$$

which by (12) implies

$$\tilde{\beta}_k \leqq c_3 \eta^k \quad (k = 1, \dots, r-1).$$

Analogously from the formula

$$\tilde{R} = \frac{1}{2\pi i} \int_{|w| = \eta^{-1}} \frac{\ln \tilde{h}(w)}{w^r (w - z)}\, dw$$

we obtain

$$|\tilde{R}| \leqq 2c_3 \eta^r$$

for $|z| \leqq \frac{1}{2}\eta^{-1}$. By the evident inequalities $|\beta_k| \leqq \tilde{\beta}_k\, (k = 1, \dots, r-1)$, $|R| \leqq |\tilde{R}|$ we conclude

$$|\beta_k| \leqq c_3 \eta^k \quad (k = 1, \dots, r-1), \quad |R| \leqq 2c_3 \eta^r \tag{13}$$

for $|\tau| \leqq \frac{1}{2}\eta^{-1}$.

By Cauchy's theorem and the well-known properties of the Γ-function we deduce that

$$\ln \Gamma(e^{i\tau}) = \sum_{k=1}^{r-1} \gamma_k (i\tau)^k + R_1 \tau^r, \tag{14}$$

where

$$|\gamma_k| \leqq c_4 \eta^k \quad (k = 1, \dots, r-1), \quad |R_1| \leqq 2c_4 \eta^r \tag{15}$$

for $|\tau| \leqq \frac{1}{2}\eta^{-1}$. Thus by (11), (13), (14) and (15)

$$\ln \Lambda(i\tau) = \sum_{k=1}^{r-1} \delta_k (i\tau)^k + R_2 \tau^r,$$

where

$$|\delta_k| \leqq c_5 \eta^k \quad (k = 1, \ldots, r-1), \quad |R_2| \leqq 2c_5 \eta^r \tag{16}$$

for $|\tau| \leqq \frac{1}{2}\eta^{-1}$.

For $|\tau| \leqq \frac{1}{2}\eta^{-1}$, $|x| \leqq 1$ consider the functions

$$\Phi_n(i\tau) = \lambda^2 \mu\left(\frac{i\tau}{\lambda}\right) + \frac{\tau^2}{2} + \ln \Lambda\left(\frac{i\tau}{\lambda}\right)$$

$$= \sum_{k=1}^{\infty} \frac{(i\tau)^{k+2}}{(k+2)!\,\lambda^k} + \sum_{k=1}^{r-1} \delta_k \left(\frac{i\tau}{\lambda}\right)^k + R_2 \left(\frac{\tau}{\lambda}\right)^r, \tag{17}$$

$$U(x) = \sum_{k=1}^{\infty} \frac{(i\tau)^{k+2}}{(k+2)!}\left(\frac{x}{\lambda}\right)^k + \sum_{k=1}^{r-1} \delta_k \left(\frac{i\tau x}{\lambda}\right)^k + R_2 \left(\frac{\tau x}{\lambda}\right)^r. \tag{18}$$

By (16) the series for $U(x)$ is dominated by the series

$$(1 + \tau^2) \sum_{k=1}^{\infty} \left(\frac{c_6 |\tau x|}{\lambda}\right)^k, \tag{19}$$

where c_6 is a sufficiently large constant. Hence, in particular, when $|\tau| < (2c_6)^{-1}\lambda$,

$$|U(x)| \leqq 2c_6 \frac{|\tau|}{\lambda}(1 + \tau^2). \tag{20}$$

We raise the series for $U(x)$ and the dominating series (19) to the lth power, where l is a positive integer. For the power of (19) we obtain

$$\left(\frac{c_6|\tau x|}{\lambda}\right)^l (1+\tau^2)^l \sum_{k_1=0}^{\infty} \cdots \sum_{k_l=0}^{\infty} \left(\frac{c_6|\tau x|}{\lambda}\right)^{k_1 + \cdots + k_l}.$$

Since the number of solutions of the equation $k_1 + \cdots + k_l = k$ in non-negative integers $k_1, \ldots, k_l$ does not exceed l^k, the series for $U^l(x)$ is dominated by

$$\left(\frac{c_6|\tau x|}{\lambda}\right)^l (1+\tau^2)^l \sum_{k=0}^{\infty} \left(\frac{c_6 l|\tau x|}{\lambda}\right)^k. \tag{21}$$

Expand $e^{U(x)}$ in a power series in x. We obtain

$$e^{U(x)} = \sum_{k=0}^{r-1} P_k(i\tau)\left(\frac{x}{\lambda}\right)^k + V(x),$$

where $V(x) = O(|x|^r)$, when $x \to 0$, and $P_k(i\tau)$ are polynomials with coefficients not depending on x, τ, n. The degree of $P_k(i\tau)$ is obviously $3k$; $P_0(i\tau) = 1$. On the other hand,

$$e^{U(x)} = \sum_{k=0}^{r-1} \frac{U^k(x)}{k!} + B|U^r(x)|\, e^{|U(x)|}. \tag{22}$$

Thus

$$\sum_{k=0}^{r-1} \frac{U^k(x)}{k!} = \sum_{k=0}^{r-1} P_k(i\tau)\left(\frac{x}{\lambda}\right)^k + V_1(x), \tag{23}$$

where $V_1(x)$ contains the powers of x beginning with the rth. By (21), when $|\tau| \leqq (2c_6 r)^{-1}\lambda$,

$$\begin{aligned} |V_1(x)| &\leqq \sum_{k=1}^{r-1} \left(\frac{c_6|\tau x|}{\lambda}\right)^k (1+\tau^2)^k \sum_{l=r-k}^{\infty} \left(\frac{c_6 k|\tau x|}{\lambda}\right)^l \\ &= B\left(\frac{|\tau|}{\lambda}\right)^r (1+\tau^2)^r. \end{aligned} \tag{24}$$

By (17), (18), (22), (23), (24) and (20) it follows that for $|\tau| \leqq (2c_6 r)^{-1}\lambda$

$$e^{\Phi_n(i\tau)} = \sum_{k=0}^{r-1} \frac{P_k(i\tau)}{\lambda^k} + B\left(\frac{|\tau|}{\lambda}\right)^r (1+\tau^2)^r \exp\left\{c_6 \frac{|\tau|}{\lambda}(1+\tau^2)\right\}.$$

Consequently, for $|\tau| \leqq \varepsilon\lambda$ we have

$$\Lambda\left(\frac{i\tau}{\lambda}\right) e^{\lambda^2 \mu(i\tau/\lambda)} = e^{-(1/2)\tau^2} \sum_{k=0}^{r-1} \frac{P_k(i\tau)}{\lambda^k} + B\left(\frac{|\tau|}{\lambda}\right)^r (1+\tau^2)^r e^{-(1/4)\tau^2}.$$

Substituting this estimate into (10) and taking into account the equality

$$\frac{1}{2\pi}\int_{-\infty}^{\infty} (i\tau)^k e^{-(1/2)\tau^2 - i\tau y}\, d\tau = \varphi^{(k)}(-y) \quad (k = 0, 1, \ldots)$$

and the estimates

$$\int_{\varepsilon\lambda}^{\infty} \tau^k e^{-(1/2)\tau^2}\, d\tau = B(\varepsilon\lambda)^{k-1} e^{-(1/2)\varepsilon^2\lambda^2} \quad (k = 0, 1, \ldots, 3r),$$

$$\int_{-\infty}^{\infty} |\tau|^k e^{-(1/4)\tau^2}\, d\tau = B \quad (k = 0, 1, \ldots, 3r),$$

we obtain that

$$N_n(a) = n\sum_{k=0}^{r-1} \frac{P_k(-\varphi)}{\lambda^{k+1}} + \frac{Bn}{\lambda^{r+1}}.$$

Proof of the theorem 3. We rewrite the formula (3) in the form

$$N_n(a) = \frac{n}{2\pi i}\int_{-\pi i}^{\pi i} \Lambda(z)\, e^{\lambda^2(\mu(z)-\xi z)}\, dz + \frac{Bn\lambda}{\sqrt{\ln n}}, \tag{25}$$

where the integration has to be carried out on the imaginary axis from $-\pi i$ to πi. From the identity

$$\ln h(z) = e^z \sum_p \left[\ln\left(1 - \frac{1}{p}\right) + \frac{1}{p}\right] + \sum_p \left[\ln\left(1 + \frac{e^z}{p}\right) - \frac{e^z}{p}\right]$$
$$+ \sum_p \ln\left(1 + \frac{e^{zf(p)} - e^z}{p + e^z} + \frac{1}{1 + \dfrac{e^z}{p}} \sum_{\alpha=2}^{\infty} \frac{e^{zf(p^\alpha)}}{p^\alpha}\right)$$

it follows that the function $h(z)$ is regular for $|\mathrm{Re}\, z| < c$ and $h(z) = B$ for $|\mathrm{Re}\, z| \leqq \delta < c$, taking into account the conditions (C) and (A) and the uniform convergence of the series for $|\mathrm{Re}\, z| \leqq \delta < c$. Using the regularity of the integrand in (25), we change the path of integration in case $\xi \neq 0$ into the path consisting of the lines:

L_1: $\mathrm{Im}\, z = -\pi, \quad 0 < \mathrm{Re}\, z < \ln(1 + \xi)$

if $\xi > 0$ or $0 > \mathrm{Re}\, z > \ln(1 + \xi)$ if $\xi < 0$;

L_2: $\mathrm{Re}\, z = \ln(1 + \xi), \quad -\pi \leqq \mathrm{Im}\, z \leqq \pi$;

L_3: $\mathrm{Im}\, z = \pi, \quad 0 < \mathrm{Re}\, z < \ln(1 + \xi)$

if $\xi > 0$ or $0 > \mathrm{Re}\, z > \ln(1 + \xi)$ if $\xi < 0$.

On the lines L_1 and L_3 we have by an easy calculation that

$$\mathrm{Re}\,(\mu(z) - \xi z) = -e^{\mathrm{Re}\, z} - 1 - (1 + \xi)\,\mathrm{Re}\, z < -1.$$

Thus in view of the estimate $\Lambda(z) = B$ we obtain

$$\frac{1}{2\pi i}\int_{L_j} \Lambda(z)\, e^{\lambda^2(\mu(z)-\xi z)}\, dz = \frac{B}{\ln n} \quad (j = 1, 3),$$

so that

$$N_n(a) = \frac{n}{2\pi i}\int_{L_2} \Lambda(z)\, e^{\lambda^2(\mu(z)-\xi z)}\, dz + \frac{Bn\lambda}{\sqrt{\ln n}}$$
$$= \frac{n}{2\pi \ln n}\left(\frac{e\lambda^2}{a}\right)^a \int_{-\pi}^{\pi} \Lambda(\ln \zeta)\, e^{\lambda^2(1+\xi)\mu(i\tau)}\, d\tau + \frac{Bn\lambda}{\sqrt{\ln n}}, \tag{26}$$

where $\zeta = (1 + \xi)\, e^{i\tau}$.

Put $c_2 = 7$ and denote by I_1 and I_2 the parts of the integral in (26) along the intervals $|\tau| \leqq \varepsilon$ and $\varepsilon < |\tau| \leqq \pi$ respectively. Applying the estimates $\operatorname{Re} \mu(i\tau) \leqq -\frac{1}{12}\tau^2$ for $|\tau| \leqq \pi$ and $\Lambda(\ln \zeta) = B$ we have

$$I_2 = B \int_{\varepsilon \leqq |\tau| \leqq \pi} e^{-(1/12)\lambda^2(1+\xi)\tau^2} \, d\tau = \frac{B}{\varepsilon\lambda^2} e^{-(1/12)\varepsilon^2\lambda^2(1+\xi)} = \frac{B}{\lambda^3}. \tag{27}$$

For the evaluation of the integral I_1 we need some estimate for the integrand in case $|\tau| \leqq \varepsilon$. At first,

$$e^{\lambda^2(1+\xi)\mu(i\tau)} = \exp\left\{\lambda^2(1+\xi)\left(-\frac{\tau^2}{2} - \frac{i\tau^3}{6} + B\varepsilon^4\right)\right\}$$

$$= e^{-(1/2)\lambda^2(1+\xi)\tau^2}\left\{1 - \frac{i}{6}\lambda^2(1+\xi)\,\tau^3 + \frac{B}{\lambda^2}\ln^3\lambda\right\}. \tag{28}$$

Further,

$$\frac{1}{\Gamma(\zeta)} = 1 + i\gamma\tau + B\left(|\xi| + \frac{\ln\lambda}{\lambda^2}\right). \tag{29}$$

The evaluation of $h(\ln\zeta)$ requires more careful calculations. For brevity's sake we put

$$E_p = \frac{f(p)}{p}, \quad F_p = \sum_{\alpha=2}^{\infty} \frac{f(p^\alpha)}{p^\alpha},$$

$$G_p = \frac{e^{c|f(p)|}}{p}, \quad H_p = \sum_{\alpha=2}^{\infty} \frac{e^{c|f(p^\alpha)|}}{p^\alpha}.$$

We have evidently $E_p = BG_p$, $F_p = BH_p$ and

$$\zeta = e^{i\tau} + B|\xi| = 1 + B(|\xi| + |\tau|),$$

$$\zeta^{f(p^\alpha)} = 1 + i\tau f(p^\alpha) + B(|\xi| + \tau^2)\, e^{c|f(p^\alpha)|}.$$

Hence it follows that

$$\ln \psi_p(\ln\zeta)$$

$$= (\zeta - 1)\ln\left(1 - \frac{1}{p}\right) + \ln\left[1 + i\tau\left(1 - \frac{1}{p}\right)(E_p + F_p) + B(|\xi| + \tau^2)(G_p + H_p)\right]$$

$$= i\tau\left[\ln\left(1 - \frac{1}{p}\right) + \left(1 - \frac{1}{p}\right)(E_p + F_p)\right] + B(|\xi| + \tau^2)\left(\frac{1}{p} + G_p + H_p\right).$$

If $f(p) = 1$, then we have

$\ln \psi_p (\ln \zeta)$

$$= (\zeta - 1) \ln\left(1 - \frac{1}{p}\right) + \ln\left[1 + \left(1 - \frac{1}{p}\right)\left(\frac{\zeta - 1}{p} + F_p\right) + B(|\xi| + \tau^2) H_p\right]$$

$$= i\tau\left[\ln\left(1 - \frac{1}{p}\right) + \left(1 - \frac{1}{p}\right)\left(\frac{1}{p} + F_p\right)\right] + B(|\xi| + \tau^2)\left(\frac{1}{p^2} + H_p\right).$$

From these estimates we obtain

$$\ln h (\ln \zeta) = i\tau \sum_p \left[\ln\left(1 - \frac{1}{p}\right) + \left(1 - \frac{1}{p}\right)(E_p + F_p)\right]$$
$$+ B(|\xi| + \tau^2)\left[1 + \sum_p H_p + \sum_{f(p) \neq 1}\left(\frac{1}{p} + G_p\right)\right],$$

whence, taking into account (C) and (A), we deduce that

$$h (\ln \zeta) = 1 + iM\tau + B(|\xi| + \tau^2), \tag{30}$$

where M is a constant depending only on the function $f(m)$.

Collecting the estimates (28), (29) and (30), we conclude that

$$\Lambda (\ln \zeta) = e^{-(1/2)\lambda^2(1+\xi)\tau^2}\left\{1 + i(M + \gamma)\tau - \frac{i}{6}\lambda^2(1 + \xi)\tau^3 + B\left(|\xi| + \frac{\ln^3 \lambda}{\lambda^2}\right)\right\}.$$

Let us now turn to the estimation of I_1. Taking into account that

$$\int_{-\varepsilon}^{\varepsilon} \tau^k e^{-(1/2)\lambda^2(1+\xi)\tau^2} d\tau = 0$$

for positive odd integers k and that

$$\int_{-\varepsilon}^{\varepsilon} e^{-(1/2)\lambda^2(1+\xi)\tau^2} d\tau = \frac{1}{\lambda\sqrt{1 + \xi}}\left(\int_{-\infty}^{\infty} e^{-(1/2)\tau^2} d\tau + B\int_{\varepsilon\lambda\sqrt{1+\xi}} e^{-(1/2)\tau^2} d\tau\right)$$
$$= \lambda^{-1}(1 + B|\xi|)\left(\sqrt{2\pi} + \frac{B}{\varepsilon\lambda} e^{-(1/2)\varepsilon^2\lambda^2(1+\xi)}\right)$$
$$= \frac{\sqrt{2\pi}}{\lambda} + \frac{B}{\lambda}\left(|\xi| + \frac{1}{\lambda^2}\right),$$

we have

$$I_1 = \frac{\sqrt{2\pi}}{\lambda} + \frac{B}{\lambda}\left(|\xi| + \frac{\ln^3 \lambda}{\lambda^2}\right).$$

Substituting this estimate and (27) into (26) we conclude

$$N_n(a) = \frac{n}{\sqrt{2\pi}\,\lambda \ln n}\left(\frac{e\lambda^2}{a}\right)^a \left\{1 + B\left(|\xi| + \frac{\ln^3 \lambda}{\lambda^2}\right)\right\} + \frac{Bn\lambda}{\sqrt{\ln n}}.$$

The inequality

$$\left(\frac{e\lambda^2}{a}\right)^a > c_7 (\ln n)^{2/3}$$

for $\xi = o(1)$ implies the validity of the theorem 3.

References

[1] H. Delange, Sur un théorème de Rényi, Acta Arithm. **11** (1965), 241—252.

[2] P. Erdös and A. Wintner, Additive arithmetical functions and statistical independence, Amer. J. Math. **61** (1939), 713—721.

[3] S. A. Fainleib, On some asymptotic formulas for sums of multiplicative functions and applications, Lietuvos matematikos rinkinys 7 (1967), 535-546 (in Russian).

[4] M. Kac, A remark on the preceding paper by A. Rényi, Publs. Inst. Math. Acad. serbe sci. **8** (1955), 163—165.

[5] M. Kac, Statistical independence in probability, analysis and number theory, The Carus Mathematical Monographs, No. 12, The Mathematical Association of America, 1959.

[6] I. Kátai, Egy megjegyzés H. Delange „Sur un théorème de Rényi" cimü dolgozatához, Magyar Tudományos Akadémia, Matematikai és fizikai tudományok osztályának közleményei **16** (1966), 269—273 (in Hungarian).

[7] J. Kubilius, Probabilistic methods in the theory of numbers, Translations of Mathematical Monographs, vol. 11, American Mathematical Society, 1964.

[8] B. V. Levin and A. S. Fainleib, On asymptotic behaviour of sums of multiplicative functions, Dokl. Akad. Nauk UzSSR **16** (1965), 5—8 (in Russian).

[9] B. V. Levin and A. S. Fainleib, Generalized problem on numbers with small and large prime divisors and applications. Dokl. Akad. Nauk UzSSR **5** (1966), 3—6 (in Russian).

[10] A. Rényi, On the density of certain sequences of integers. Publs. Inst. Math. Acad. serbe sci. **8** (1955), 157—162.

[11] A. Rényi and P. Turán, On a theorem of Erdös-Kac, Acta Arithm. **4** (1957), 71—84.

[12] A. Rényi, On the distribution of values of additive number-theoretical functions, Publs. Math. **10** (1963), 264—273.

[13] G. J. Rieger, Zum Teilerproblem von Atle Selberg. Math. Nachr. **30** (1965), 181—192.

[14] L. G. Sathe, On a problem of Hardy on the distribution of integers having a given number of prime factors, I, II, III, IV, J. Indian Math. Soc. **17** (1953), 63—82, 83—141: **18** (1954), 27—42, 43—81.

[15] A. Selberg, Note on a paper by L. G. Sathe, J. Indian Math. Soc. **18** (1954), 83—87.

[16] P. Turán, Az egész számok primosztóinak számáról, Mat. és Fiz. Lapok **41** (1934), 103—130 (in Hungarian).

[17] A. Wintner, The distribution of primes, Duke Math. J. **9** (1942), 425—430.

[18] A. Wintner, Eratosthenian averages, Baltimore 1943.

[19] E. Wirsing, Das asymptotische Verhalten von Summen über multiplikative Funktionen, Math. Ann. **143** (1961), 75—102.

J. E. LITTLEWOOD IN CAMBRIDGE

THE "PITS EFFECT" FOR THE INTEGRAL FUNCTION

$$f(z) = \sum \exp\{-\varrho^{-1}(n \log n - n) + \pi i \alpha n^2\} z^n, \quad \alpha = \tfrac{1}{2}(\sqrt{5} - 1)$$

Introduction

If $f_0(z) = \sum c_n z^n$ is any integral function of finite non-zero order ϱ, consider the class $\mathfrak{F}$ of functions

$$f(z) = f(z;t) = \sum r_n(t)\, c_n z^n,$$

where $r_n(t)$ are Rademacher's functions, representing a 'random' factor of the form ± 1. Littlewood and Offord [1] have shown that 'most' $f(z)$ behave with great crudity and violence. If we erect an ordinate $|f(z)|$ at the point z of the z-plane, then the resulting surface is an exponentially rapidly rising bowl, approximately of revolution, with exponentially small 'pits' going down to the bottom. The zeros of f, more generally the w-points where $f = w$, all lie in the pits for $|z| > R(w)$. Finally the pits are very uniformly distributed in direction, and as uniformly distributed in distance as is compatible with the order ϱ.

It will be convenient to separate the existence of pits, which we will call the 'pits effect', from the further property of a maximally uniform distribution.

It has long been known that a factor $e^{i\beta n \log n}$ has very similar effects to a random ± 1 when applied to smooth functions $f = \sum a_n z^n$ with $a_n = O(n^k)$, and that this is true also of a factor $e^{\pi i \alpha n^2}$, where α is a continued fraction with bounded partial quotients, the simplest of which, with which we shall be dealing, being

$$\alpha = \frac{1}{1+}\,\frac{1}{1+}\cdots = \frac{1}{2}(\sqrt{5} - 1).$$

It was natural to examine the effect of a factor $e^{i\beta n \log n}$ on integral functions of order ϱ with sufficiently smooth positive coefficients. Watson [2] gives asymptotic expansions for functions of the form

$$f = \sum \frac{z^n}{\Gamma[1 + (\varrho^{-1} + i\beta)\, n]}$$

and elaborations of them: f is roughly

$$\sum \frac{z^n}{(1+\varrho^{-1}n)} e^{-i\beta n \log n}.$$

It is possible to derive from Watson's results that there exists a pits effect for any ϱ and large enough β. But for large ϱ and small β the results show that $f = O(1)$ in a finite proportion of the z-plane, so that no question of pits effect arises.

The position is complicated by the fact that a sufficiently smooth integral function of order $\varrho < \frac{1}{2}$, including $\varrho = 0$, shows a pits effect without any randomizing factor. Not to extend the present paper unduly I will deal with this, and the result above about pits for large β, on another occasion.

Faced with the part failure of a pits effect for order $\varrho > \frac{1}{2}$ and a factor $e^{i\beta n \log n}$, I suggested in 1949 to Professor M. Nassif, that he should investigate a factor $\exp(i\alpha\pi n^2)$, applied to the standard integral function of order ϱ. Nassif's paper [3] proves, in effect, that there is a pits effect if $\varrho < \varrho_0$, where, fortunately because of the $\varrho < \frac{1}{2}$ phenomenon, $\varrho_0 > 1$. He states by an oversight that the result is true for all $\varrho > 0$, but in fact the method definitely fails for $\varrho > \varrho_0$.

Dr. Y. M. Chen and I have recently made a study (as yet unpublished) of functions

$$f(z) = \sum \exp[-\lambda(n) + i\Lambda(n)]\, z^n,$$

with the aim of discovering what combinations of $\lambda(n)$ and $\Lambda(n)$ yield a pits effect. Our method is a simple one for all combinations; we apply the Poisson Summation Formula to the expression for $f(z)$. The success of this in establishing pits effects depends on a number of conditions being satisfied, and the results, stated below, show curious gaps. One condition for success is $\Lambda(n) > \lambda(n)$, which resolves the former puzzle. Watson's integral functions have $\Lambda(n) \asymp \lambda(n)$; they are a marginal case from this point of view, and since there are pits for any ϱ and large enough β, the results fit in very well.

Dr. Chen's and my results are as follows. We take five typical functions $f_0 = \sum \exp\{-\lambda(n)\}\, z^n$, excluding integral functions with $\varrho < \frac{1}{2}$, and three $\Lambda(n)$'s, sufficiently typical of what we are able to prove:

$$f_0(z) \begin{cases} (1)\ \sum e^{n^a} \quad (0 < a < 1) \text{ (unit circle function of high order)}; \\ (2)\ \sum \exp(-n \log\log n)\, z^n; \\ (3)\ \sum \exp(-n \log^a n)\, z^n \quad (0 < a < 1); \\ (4)\ \sum \exp(-n \log n/\log\log n)\, z^n; \\ (5)\ \sum \exp(-n \log n/\varrho)\, z^n \quad (\varrho > \frac{1}{2}); \end{cases}$$

$$\Lambda(n) \begin{cases} \text{(a)} \ cn \log n & (0 < c < \infty); \\ \text{(b)} \ n \log^{\beta} n & (\beta > 1); \\ \text{(c)} \ n^{1+b} & (0 < b < \tfrac{1}{2}). \end{cases}$$

We find a pits effect in every combination of one of (1) to (5) with one of (a), (b), (c), except that (i) in the combination (c), (1) we have to assume $b < \frac{1}{2}a$, (ii) pits definitively do not exist in the marginal combination (5), (a) unless $c > c_0(\varrho_0)$.

The gap between $\Lambda(n) = n^{1+b}(b < \frac{1}{2})$ and $\Lambda(n) = \pi\alpha n^2$ is very striking. It seems that $\pi\alpha n^2$ is an island in the subject. In respect of results there is the gap. In respect of methods, that appropriate for either of $\Lambda(n) < n^2$ and $\Lambda = \pi\alpha n^2$ seems quite inappropriate for the other.

The present paper is about $\pi\alpha n^2$ and integral functions. I show that there is a pits effect for every finite order ϱ, and that this extends to a limited class of infinite order. The limitation of Nassif's argument (valid only for $\varrho < \varrho_0$) will be explained in its place. In a paper [4] immediately following Nassif's paper [3], Tims proved that the pits of Nassif's function are maximally uniformly distributed (more exactly fall into two such groups superposed, because Nassif's function is one of order 1 in the variable z^2).

In my case (of general ϱ) the question of uniformity runs into a certain difficulty. It is probably possible to overcome this, but this would add further complications to a paper already rather complicated.

§ 1

The problem calls for an intensive study of the elliptic ϑ-function $\vartheta_1(v, \tau)$ and its third derivative $d^3\vartheta/dv^3$, τ being of the form

$$\tau = \alpha + i\varepsilon, \quad \alpha = \tfrac{1}{2}(\sqrt{5} - 1), \quad \varepsilon \text{ small.} \tag{1.1}$$

None of the other ϑ's occur, and we work with the modified form

$$\varphi(v) = \varphi(v, \tau) = e^{-\pi i v}\,\vartheta_1(v, \tau). \tag{1.2}$$

Our notation is that of Tannery and Molk, and the references below are to the tables of Vol. II, subject to trivial translations from ϑ_1 to φ. The following "Dictionary" will be useful.

μ and ν are always integers (positive, negative or zero);

(I) $q = e^{\pi i\tau}$, $\tau = \alpha + i\varepsilon$, $|q| = e^{-\pi\varepsilon}$, $q_0 = \prod_1^\infty (1 - q^{2n})$.

(II) $\varphi(v)$ is an odd function of v with period 1 and zeros at $v = \nu\tau + \mu$ [ϑ_1 has period 2; the factor $e^{-\pi i v}$ reduces this to 1 for φ].

(III) $\varphi(v) = q^{1/4} \sum_{-\infty}^{\infty} (-1)^n \exp\{(n^2 + n)\pi\tau i + 2n\pi v i\}$ [XXXII (1)].

(IV) $\varphi(v) = 2q^{1/4} q_0 e^{-\pi i v} \sin \pi v \prod_1^\infty (1 - 2q^{2n}\cos 2\pi v + q^{4n})$ [XXXII (5)].

(V) $\varphi(v + \nu\tau + \mu) = (-1)^\nu \exp(-\pi\nu\tau i - \pi\nu^2\tau i - 2\nu\pi v i)\,\varphi(v)$ [XXXIV (7)].

For the transformation

$$\tau \to T = \frac{c + d\tau}{a + b\tau}, \quad ad - bc = 1,$$

(VI) $\varphi(v) = \dfrac{\omega}{\sqrt{\Delta}} \exp\left\{-\pi i v + \dfrac{\pi i v}{\Delta} - b v \tau \pi i\right\} \varphi\left(\dfrac{v}{\Delta}, T\right)$,

where $\Delta = a + b\tau$ and ω is an eighth root of unity [XLII (1)].

§ 2

We will use A's for positive absolute constants, B's for positive constants depending only on ϱ. We can also conveniently use the symbols $\succ$, $\prec$, $\asymp$: $X \succ Y$ and $Y \prec X$ mean $Y = o(X)$; $X \asymp Y$ means $B < |X/Y| < B$.

The constants of O's will be of type B.

§ 3

We will take for our standard function, with factor $\exp(\alpha\pi i n^2)$,

$$f(z) = \sum \exp\{-k(n \log n - n) + \alpha\pi i n^2\} z^n$$
$$= \sum c_n z^n, \quad k = 1/\varrho. \tag{3.1}$$

The z-plane (for large z) is $\sum \mathfrak{A}_N$, where $\mathfrak{A}_N$ is the annulus

$$(N - \tfrac{1}{2})^k < |z| \leqq (N + \tfrac{1}{2})^k, \tag{3.2}$$

where N is a large positive integer. Let $\eta = \varepsilon^{\frac{1}{4}}$, $\alpha = +(\sqrt{5} - 1)/2$,

$$\tau = \alpha + i\varepsilon, \quad \varepsilon = k(2\pi N)^{-1}, \tag{3.3}$$

$$z = N^k \exp\{\pi i - (2N - 1)\pi\alpha i + 2\pi v i\}. \tag{3.4}$$

The v-region $\Gamma = \Gamma_N$ corresponding to $\mathfrak{A}_N$ is defined by

$$-\tfrac{1}{2} < \Re v \leqq \tfrac{1}{2}, \quad -\tfrac{1}{2}\varepsilon_1 \leqq \Im v < \tfrac{1}{2}\varepsilon_2 \tag{3.5}$$

where

$$\varepsilon_1 = \varepsilon\left\{2N \log\left(1 + \frac{1}{2N}\right)\right\} = \varepsilon + O(\varepsilon^2), \tag{3.6}$$

$$\varepsilon_2 = \varepsilon\left\{2N \log\left(1 - \frac{1}{2N}\right)^{-1}\right\} = \varepsilon + O(\varepsilon^2).$$

We need to consider formulae valid in more extensive regions than the standard $\Gamma = \Gamma_N$. In one of these $\Im v$ ranges over $|\Im v| < B_1\varepsilon$, where B_1 is a sufficiently large B. To use such a region, however, would involve keeping track of dependencies on B_1, to avoid vicious circles. Fortunately we can work with the larger region (recall $\eta = \varepsilon^{\frac{1}{4}}$)

$$\Gamma_+ = \Gamma_{N+}: \quad |\Im v| \leqq \eta^{3/2}, \quad |\Re v| < \tfrac{1}{2} + \delta. \tag{3.7}$$

The extension of $\Re v$ corresponds merely to extension of ϑ beyond its standard $\vartheta| < \pi$, and is correspondingly trivial.

§ 4

We shall need some crude facts about $M(r)$, the maximum modulus of f.

Lemma 1. (i) *For large z in $\mathfrak{A}_N$*

$$e^{kN(1-\delta)} < M(r) < e^{kN(1+\delta)};$$

(ii) $\quad M(r) \geqq |c_N|\, r^N > M^{1-\delta}(r);$

(iii) $\quad$ *If* $\; r_1 = o(r) \;$ *then* $\; M(r + r_1) < M^{1+\delta}(r);$

(iv) $\quad \dfrac{|f^{(m)}(z)|}{m!} \leqq M(r + 1) < M^{1+\delta}(r).$

In $\mathfrak{A}_N$ the maximum term has rank N or $N \pm 1$, and the moduli of the three terms differ by less than a factor N^B. The contribution of terms with $|n - N| > N^{3/5}$ is small compared with $|c_N z^N|$, and the terms with $|n - N| \leqq N^{3/5}$ contribute at most $2N^{3/5}$ times the maximum term. (i) to (iii) follow easily from these facts.

For $|f^{(m)}(z)|/m!$ we have

$$\frac{|f^{(m)}(z)|}{m!} = \left|\frac{1}{2\pi i}\int_{|\zeta - z|=1} \frac{f(\zeta)\,d\zeta}{(\zeta - z)^{m+1}}\right| \leqq \operatorname{Max}|f(\zeta)| \leqq M(r+1).$$

§ 5

Lemma 2. *For z defined in terms of N and v by* (3.4), *and for $v \in \Gamma_{N+}$, i.e. $|\Re v| < \frac{1}{2} + \delta$, $|\Im v| \leqq \eta^{3/2}$, we have*

$$f_1 = \frac{f(z)}{c_N z^N(-iq^{-1/4})} = \mathfrak{F}(v) + O(\eta),$$

where

$$\mathfrak{F}(v) = \left[1 + ci\varepsilon^2\left(\frac{d}{dv}\right)^3\right]\varphi(v, \tau), \quad \text{and} \quad c = Ak = A/\varrho > 0.$$

Lemma 2 (with $v \in \Gamma_N$) is Nassif's basic formula. What he proves (in effect) is that for $\varrho < \varrho_0$ the modulus $|\varphi|$ dominates $|ci\varepsilon^2\varphi'''|$ in Γ_N except in a small circle round $v = 0$. This requires an elaborate study of φ and φ''' (and is difficult reading). The domination definitely fails for $\varrho > \varrho_0$, and my problem (*inter alia*) is to deal with the combination $\mathfrak{F}(v)$ for v's for which $|ci\varepsilon^2\varphi'''|$ may dominate $|\varphi|$. We do not need the precise value of c; only that it is proportional to $1/\varrho$, which can take any positive value, as also can c.

The proof of lemma 2 (roughly the same as Nassif's) is as follows.

For z given by (3.4), v in Γ_+, we find, by straightforward approximation for $|n| \leqq N_1 = [N^{8/15}]$,

$$\frac{u_{N+n}}{u_N} = \frac{c_{N+n}z^n}{c_N} = (-1)^n \exp\{(n^2 + n)\pi\tau i + 2n\pi v i\}$$
$$\times\left\{1 + \frac{n^3}{6N^2} + O\left(\frac{n^4}{N^3}\right) + O\left(\frac{n^6}{N^4}\right)\right\}\{1 + O(\varepsilon)\}. \qquad (5.1)$$

Now for $r = 4, 6$ we have

$$\sum_{n=-N_1}^{N_1} |(-1)^n n^r \exp\{(n^2 + n)\pi\tau i + 2n\pi v i\}|$$
$$\leqq B\sum_{n=0}^{\infty} n^r \exp\{-Bn^2N^{-1} + BN_1N^{-3/4}\}$$
$$\leqq B\sum_{n=0}^{\infty} n^r \exp(-Bn^2N^{-1}) < BN^{(r+1)/2},$$

since $N_1 N^{-3/4} = O(1)$. Using this in (5.1) for the O-terms $O(n^4 N^{-3})$, and $O(n^6 N^{-4})$ we get

$$\frac{1}{u_N} \sum_{-N_1}^{N_1} u_{N+n} = \sum_{-N_1}^{N_1} (-1)^n \exp\{(n^2+n)\pi\tau i + 2n\pi v i\}\left(1 + \frac{n^3}{6N^2}\right) + O(\eta). \tag{5.2}$$

Further,

$$\left(\sum_{N_1+1}^{\infty} + \sum_{-\infty}^{-N_1-1}\right)\left|(-1)^n \exp\{(n^2+n)\pi\tau i + 2n\pi v i\}\left(1 + \frac{n^3}{6N^2}\right)\right|$$

$$= O\left(\sum + \sum\right) \exp(-Bn^2N^{-1} + BnN^{-3/4})\left(1 + \frac{n^3}{6N^2}\right). \tag{5.3}$$

Since $nN^{-3/4} = o(n^2N^{-1})$ in $\sum + \sum$ the last exponential is $< \exp(-Bn^2N^{-1})$, and (5.3) reduces to $O(N^{-1/2}) = O(\eta)$. Thus, by (5.2),

$$\frac{1}{u_N} \sum_{-N_1}^{N_1} u_{N+n} = \sum_{-\infty}^{\infty} (-1)^n \exp\{(n^2+n)\pi\tau i + 2n\pi v i\}\left(1 + \frac{n^3}{6N^2}\right) + O(\eta).$$

We may finally replace the left hand side here by $\sum_{-N}^{\infty} u_{N+n}/u_N$, the error involved being exponentially small. This last result is equivalent to that of lemma 2.

§ 6

It is natural to estimate φ and φ''' for $v \in \Gamma_+$ by using the most effective transformation

$$\tau \to T = \frac{c + d\tau}{a + b\tau}, \quad ad - bc = 1,$$

available; that is, making $\Im T$ as large as possible and so $|Q| = |e^{\pi i T}|$ as small as possible. Now by calculation, using $ad - bc = 1$, we have

$$\Im T = \frac{\varepsilon}{(a + b\alpha)^2 + b^2\varepsilon^2}. \tag{6.1}$$

We clearly want $a + b\alpha$ to be small, and naturally choose

$$a = p_n, \quad b = -q_n \tag{6.2}$$

for *some* convergent p_n/q_n to α, and then

$$c = -\delta_n p_{n-1}, \quad d = \delta_n q_{n-1}, \quad \delta_n = \operatorname{sgn}(p_n q_{n-1} - p_{n-1} q_n) = \pm 1, \tag{6.3}$$

to make $ad - bc = 1$. This gives us

$$\Im T = \frac{\varepsilon}{q_n^2\left\{\left(\alpha - \frac{p_n}{q_n}\right)^2 + \varepsilon^2\right\}}. \tag{6.4}$$

Now

$$\alpha - \frac{p_n}{q_n} = \frac{(1+\alpha)\,p_n + p_{n-1}}{(1+\alpha)\,q_n + q_{n-1}} - \frac{p_n}{q_n};$$

$$\alpha - \frac{p_n}{q_n} = \frac{(-1)^n}{q_n q_{n+1}} = \frac{(-1)^n}{\beta_n q_n^2}, \tag{6.5}$$

where

$$\beta_n = \frac{q'_{n+1}}{q_n} = (1+\alpha) + \frac{q_{n-1}}{q_n}.$$

Since p_n, q_n are of the form $A\alpha^{-n} + B\alpha^n$ we have $q_{n-1}/q_n = \alpha + O(\alpha^{2n})$,

$$\beta_n = 1 + 2\alpha + O(\alpha^{2n}) = \beta + O(\alpha^{2n}), \quad \beta = \sqrt{5}. \tag{6.6}$$

So

$$\Im T = \frac{\varepsilon}{q_n^2(\beta_n q_n^2)^{-2} + q_n^2\varepsilon^2} = \frac{\varepsilon}{\beta_n^{-2} q_n^{-2} + q_n^2\varepsilon^2} = \frac{\beta_n^2(q_n^2\varepsilon)}{1 + \beta_n^2(q_n^2\varepsilon)^2}. \tag{6.7}$$

We now determine q_n in terms of N, or ε, by

$$q_n^2 \leqq \varepsilon^{-1} < q_{n+1}^2, \quad \text{or, if} \quad \alpha_n = q_n/q_{n+1}, \quad q_n^2 = t^2\varepsilon^{-1},$$
$$1 \geqq t > \alpha_n = \alpha + O(\alpha^{2n}). \tag{6.8}$$

§ 7

The particular choice of T has the following consequences.

Lemma 3. *Let* $a + b\tau = \Delta$, $\Im T = \lambda$.

(i) $$\frac{-ib\varepsilon}{\Delta} = \frac{-\beta_n^2 t^3 \mp i\beta_n t^2}{1 + \beta_n^2 t^4};$$

(ii) $$\frac{\eta}{\Delta} = \frac{\beta_n t(\mp 1 + i\beta_n t^2)}{1 + \beta_n^2 t^4};$$

(iii) $$\frac{\varepsilon}{\Delta^2} = \frac{\beta_n^2 t^2(1 - \beta_n^2 t^4 \mp 2i\beta_n t^2)}{(1 + \beta_n^2 t^4)^2};$$

(iv) $$\lambda = \frac{\beta_n^2 t^2}{1 + \beta_n^2 t^4};$$

(v) $\frac{5}{6} + o(1) < \lambda < \frac{1}{2}\beta + o(1) = \frac{1}{2}\sqrt{5} + o(1)$;

(vi) $|Q| = |e^{\pi i r}| < 0.006, \quad |Q| > A.$

In (i) *to* (iii) *all upper signs or all lower signs are to be taken.*

We have

$$\frac{\eta}{\Delta} = -\frac{q'_{n+1}}{q'_{n+1}} \cdot \frac{\eta}{q_n(\alpha - p_n/q_n) + iq_n\varepsilon} = -\frac{-\beta_n t}{\pm 1 + i\beta_n t^2},$$

by (6.5) and (6.8),

$$= \frac{-\beta_n t(\pm 1 - i\beta_n t^2)}{1 + \beta_n^2 t^4}$$

equivalent to (ii). Squaring this gives (iii). (i) is (ii) multiplied by $-ib\eta = iq_n\eta = it$.

For (iv) we have, by (6.7),

$$\lambda = \Im T = \frac{\beta_n^2 q_n^2 \varepsilon}{1 + \beta_n^2 (q_n^2 \varepsilon)^2} = \frac{\beta_n^2 t}{1 + \beta_n^2 t^4},$$

as desired.

Next, $\Im(t) = \beta_n^2 t^2/(1 + \beta_n^2 t^4)$ has a single maximum at $\beta_n^2 t^4 = 1$ or $t = \beta_g^{-1/2}$. Now we find numerically

$$\alpha < \beta_n^{-1/2} < 1.$$

Hence the minimum of $\Im$ in $\alpha \leqq t \leqq 1$ is, to error $o(1)$, $\min\{\Im(\alpha), \Im(1)\}$, which calculation shows to be $\Im(1) = \frac{5}{6} + o(1)$. This proves (v).

The first part of (vi) follows by calculation; for the second we have $\lambda < \beta_n^2 t^2 < A$.

§ 8

We need an estimate of $\varphi'(0)$.

Lemma 4. $\varphi'(0) = \vartheta'(0) \asymp \eta^{-3/2}$, *i.e.* $\varphi'(0) = B\eta^{-3/2}$.

Differentiating formula (VI) and putting $v = 0$ we have

$$\varphi'(0) = \omega\, \Delta^{-3/2} \varphi'(0, T). \tag{8.1}$$

In this, by formula (III),

$$\varphi'(0, T) = 2\pi i Q^{1/4} \sum_{-\infty}^{\infty} (-1)^n n \exp\{(n^2 + n) Ti\}. \tag{8.2}$$

Now $n^2 + n = 0$ for $n = -1$ and $n = 0$, and otherwise $n^2 + n \geqq 2$. The $\sum$ in (8.2) is $1 + \sum_{n \neq 0, -1}$. Hence, from (8.2),

$$|\varphi'(0, T)| = 2\pi |Q|^{1/4} \left\{1 + O(1) \sum_{n \neq 0, -1} |n| \exp[-(n^2 + n)\pi\lambda]\right\},$$

from which it follows easily that $\varphi'(0, T) \asymp 1$, and so, from (8.1),

$$|\varphi'(0)| \asymp |\Delta|^{-3/2} \asymp \eta^{-3/2},$$

as desired.

§ 9

Let $\Gamma_1 = \Gamma_{N,1}$ be the v-region

$$|\Re v| < \tfrac{1}{2} + \delta, \quad |\Im v| \leqq \eta, \tag{9.1}$$

and let $\Gamma_2 = \Gamma_{N,2}$ be the doubled Γ_1 with

$$|\Re v| < \tfrac{1}{2} + \delta, \quad |\Im v| \leqq 2\eta \tag{9.2}$$

(Γ_1 consists of about $2\eta^{-1}$ concentric Γ_N's).

We have now to introduce a rather sophisticated conception: "the class of w and v associated with $v \in \Gamma_2$".

In the first place let

$$v = x + iy, \quad |y| \leqq 2\eta, \quad x = (p + \xi)/q_n, \tag{9.3}$$

where $|\xi| \leqq \tfrac{1}{2}$ and p is an integer. Since $(p_n, q_n) = 1$ there exists a ν_1 with $|\nu_1| \leqq \tfrac{1}{2} q_n$, and such that $-\nu_1 p_n \equiv p (\text{mod } q_n)$, or

$$-\nu_1 p_n = p + \mu_1 q_n. \tag{9.4}$$

Let

$$w_1 = v + \nu_1 \tau + \mu_1, \tag{9.5}$$

where, since $q_n = \eta^{-1} t$,

$$|\nu_1| \leqq \tfrac{1}{2} t\eta^{-1}. \tag{9.6}$$

Now

$$\begin{aligned} w_1 &= x + iy + \mu_1 + \nu_1 \varepsilon i + \nu_1 \alpha \\ &= \frac{p + \xi}{q_n} + \mu_1 + i(y + \nu_1 \varepsilon) + \nu_1 \left\{ \frac{p_n}{q_n} \pm \frac{1}{\beta_n q_n^2} \right\} \\ &= \frac{p + \mu_1 q_n + \nu_1 p_n}{q_n} + \frac{\xi}{q_n} + i(y + \nu_1 \varepsilon) \pm \frac{\nu_1}{\beta_n q_n^2}. \end{aligned}$$

The first term here being 0, we have, since $|\nu_1| \leqq \tfrac{1}{2} q_n$ and $\eta^{-1} \geqq q_n \geqq \{\alpha + o(1)\}\, \eta^{-1}$,

$$|w_1| \leqq \eta \left\{ \frac{1}{2\alpha} + 2 + \frac{1}{2} + \frac{1}{2\beta\alpha} + o(1) \right\} \leqq A_1 \eta, \tag{9.7}$$

where A_1 is a certain A, which we fix.

Now it may happen, for some $v \in \Gamma_2$, that ν's other than ν_1, with $|\nu| \leqq \frac{1}{2}\eta^{-1}t$, may lead to a $w = v + \nu\tau + \mu$ satisfying $|w| < A_1\eta$ (or even $w = O(\varepsilon)$). We consider *the class of all pairs* (w, ν) *for which, for given* $v \in \Gamma_2$,

$$w = v + \nu\tau + \mu \tag{9.8}$$

and

$$|w| \leqq A_1\eta, \quad |\nu| \leqq \tfrac{1}{2}\eta^{-1}t, \tag{9.9}$$

and we call the class the class of (w, ν) associated with $v \in \Gamma_2$. It has at least one member, namely (w_1, ϱ_1) above.

Nassif's argument needs only the single w_1, but it appears that mine needs the class.

w, ν (and the trivially associated μ that makes $|\Re v| \leqq \frac{1}{2}$) occur continually in what follows, but only when a $v \in \Gamma_2$ occurs also. We shall understand tacitly that then w, ν belong to the class of (w, ν) associated with v.

By formula (V) we have for any w, ν of the class

$$|\varphi(w)| = |(-1)^{\nu}\varphi(v)\exp\{-i\pi\nu\tau - \pi\nu^2\tau i - 2\nu\pi v i\}|.$$

Since the curly bracket is $= O(\nu\varepsilon) + O(\nu^2\varepsilon) + O(\nu\eta) = O(1)$, we have $\varphi(v) \asymp \varphi(w)$.

Summing up we have:

Lemma 5. *For a* $v \in \Gamma_{N,2}$ *all* w, ν *associated with* v *satisfy*

$$w = v = \nu\tau + \mu, \quad |w| \leqq A_1\eta, \quad |\nu| \leqq \tfrac{1}{2}\eta^{-1}t,$$

and we have

$$\varphi(v) = C_\nu e^{2\nu\pi v i}\varphi(w),$$

where $C_\nu = (-1)^\nu \exp(-\pi\nu\tau i - \pi\nu^2\tau i)$ *is a constant satisfying* $A < |C_\nu| < A$.

[The story that follows is unfortunately complicated. I have tried to formulate such a chain of lemmas that the reader can keep in touch by referring to members of the chain alone.]

§ 10

We have next

Lemma 6.

(i) $|\varphi(v)| < A\eta^{-1/2} (v \in \Gamma_{N,2})$;

(ii) $\left|\dfrac{\varphi^{(r)}(v)}{r!}\right| < A\eta^{-1/2-r} \quad (v \in \Gamma_{N,1})$;

(iii) $|\varphi(v)| > A\eta^{-3/2}|v|$ *if* $|v| < \eta$.

These results are true a fortiori for $v \in \Gamma_+$.

(ii) is an easy consequence of (i). For if $v \in \Gamma_1$ the circle C_u, $|u - v| = \eta$, lies in $\Gamma_{N,2}$ with 2δ in place of δ, and we have

$$\left|\frac{\varphi^{(r)}(v)}{r!}\right| = \left|\frac{1}{2\pi i}\int_C \frac{\varphi(u)\, du}{(u - v)^{r+1}}\right| \leqq \eta^{-r} \max_{(C)} |\varphi(u)| \leqq \eta^{-r} A\eta^{-1/2}.$$

For (i) we recall $\varphi(v) \asymp \varphi(w)$ of lemma 5, and we have, by formula (VI),

$$|\varphi(w)| = \frac{1}{|\Delta|^{1/2}}\left|\exp\left\{-\pi i w + \pi i \frac{w}{\Delta} - bw\tau\pi i\right\}\right| \left|\varphi\left(\frac{w}{\Delta}, T\right)\right|. \tag{10.1}$$

The $|\exp|$ is $O(1)$ since $w = O(\eta)$. Also, by formula (IV),

$$\varphi\left(\frac{w}{\Delta}, T\right) = 2Q^{1/4}Q_0 e^{-\pi i w/\Delta} \sin\frac{\pi w}{\Delta}\prod_1^\infty\left(1 - 2Q^{2n}\cos\frac{2\pi w}{\Delta} + Q^{4n}\right). \tag{10.2}$$

Since $\pi w/\Delta = O(1)$, and so $\sin(\pi w/\Delta) = O(1)$, and since $|Q| < \frac{1}{10}$, we get $\varphi(w) = O(\eta^{-1/2})$ from (10.1) and (10.2).

This completes the proof of (i) and (ii).

For (iii) we use (10.1) and (10.2) with v in place of w, and observe that since $v < \eta$, $\pi v/\Delta$ is small and so $|\sin(\pi v/\Delta)| \geqq A|v/\Delta| \asymp v\eta^{-1}$.

§ 11

The regions $\Gamma_{1,2}$ were introduced to prove lemma 6 and are no longer needed: we are now concerned only with $v \in \Gamma_{N+}$, $|v| \leqq \eta^{3/2}$.

Lemma 7. *If $v \in \Gamma_{N+}$, then for any w, ν associated with v:*

(i) $\varphi(v) = C_\nu e^{2\pi\nu v i}\varphi(w)$, $A < |C_\nu| < A$,

(ii) $\varphi(v) \asymp \varphi(w) \asymp \eta^{-3/2}w$,

(iii) $\nu = O(w\varepsilon^{-1}) + O(\eta^{-1/2})$.

(i) comes from lemma 5. (iii) is immediate since $\nu\varepsilon = F(v - w) = O(\eta^{3/2}) + O(w)$ For (ii) we have in the first place

$$\sin\frac{\pi w}{\Delta} = O\left(\frac{w}{\Delta}\right)$$

in (10.2), since $w/\Delta = O(1)$; so

$$\varphi(w) = O(\eta^{-3/2}w)$$

from (10.1) and (10.2). Also $\varphi(v) \asymp \varphi(w)$ by lemma 5.

The opposite inequality, namely $|\varphi(w)| \geqq A\eta^{-3/2}|w|$, is much more delicate. By (10.1) it is enough to prove

$$\left|\sin\frac{\pi w}{\Delta}\right| > A\left|\frac{\pi w}{\Delta}\right|. \tag{11.1}$$

Let

$$w/\eta = r + ij, \tag{11.2}$$

where

$$j = \Im(v - v\tau)/\eta = -v\eta + O(\eta^{1/2}) \tag{11.3}$$

(we do not need the value of r). By lemma 3 (ii) we have, to errors $o(1)$ arising from replacing β_n by $\beta = \sqrt{5}$,

$$\frac{\pi\eta}{\Delta} = -\frac{\pi\beta t}{D}(\pm 1 - i\beta t^2), \quad D = 1 + \beta^2 t^4.$$

Hence, to errors $o(1)$ (now arising also from the error-term in (11.3)),

$$\frac{\pi w}{\Delta} = R + i\mathscr{T},$$

where

$$\left.\begin{aligned} R &= \mp\frac{\pi\beta t}{D}r + v\eta\frac{\pi\beta^2 t^3}{D}, \\ \mathscr{T} &= \frac{\pi\beta^2 t^3}{D}r \pm \frac{\pi\beta t v\eta}{D}. \end{aligned}\right\} \tag{11.4}$$

Now

$$\left|\sin^2\frac{\pi w}{\Delta}\right| = \sin^2 R + \sinh^2\mathscr{T} \geqq \sin^2 R + \mathscr{T}^2.$$

If now $|\mathscr{T}| > a$, where a is some A, then

$$\left|\sin^2\frac{\pi w}{\Delta}\right| \geqq \mathscr{T}^2 > a^2 = a^2\left|\frac{\pi w}{\Delta}\right|^2 \Big/ \left|\frac{\pi w}{\Delta}\right|^2 \geqq Aa^2\left|\frac{\pi w}{\Delta}\right|^2,$$

since the previous denominator $< A$.

We may therefore suppose $|\mathscr{T}| \leqq a$, *and the* a *is at our disposal*. It then follows from (11.4) that

$$r = \mp\frac{v\eta}{\beta t^2} + \vartheta A a + o(1), \quad |\vartheta| < 1,$$

and so, since $|\eta t| \leqq \frac{1}{2} t$ (for any w, ν),

$$|r| < \frac{1}{2\beta t} + Aa. \tag{11.5}$$

From this, (11.4), and $|\nu\eta| \leqq \frac{1}{2} t$, we have

$$|R| \leqq \frac{\pi}{2D} + Aa + \frac{\pi\beta^2 t^4}{2D} \leqq \frac{1}{2}\pi + Aa.$$

By choice of a,

$$|R| < \pi(1 - A), \quad |\sin R| > A|R|,$$

$$\left|\sin^2 \frac{\pi w}{\Delta}\right| \geqq \sin^2 R + \mathcal{T}^2 \geqq A(R^2 + \mathcal{T}^2) = A\left|\frac{\pi w}{\Delta}\right|^2.$$

By (11.1) this completes the proof of lemma 7.

§ 12

Recall

$$F(v) = \varphi(v)\left\{1 + ic\varepsilon^2 \frac{\varphi'''}{\varphi}(v)\right\}, \tag{12.1}$$

$$f_1 = \frac{f(z)}{-iq^{-1/4} c_N z^N} = F(v) + O(\eta) \quad (v \in \Gamma_+) \tag{12.2}$$

(the second from lemma 2). We prove next

Lemma 8. *If $v \in \Gamma_+$, then for any w, ν associated with v:*

(i) $\varphi(w) = O(\eta^{-1/2})$, $\varphi'(w) = O(\eta^{-3/2})$, $\varphi''(w) = O(\eta^{-5/2})$,
$\varphi'''(w) = O(\eta^{-7/2})$,

(ii) $$F(v) = C_\nu e^{-2\pi\nu v i}[\varphi(w) + ic\varepsilon^2 \{\varphi'''(w) + 6\nu\pi i\varphi''(w) + 12\nu^2\pi^2\varphi'(w) + 8\nu^3\pi^3 i\}].$$

(iii) *If further $w = O(\varepsilon)$, then*

$$F(v) = C_\nu e^{-2\pi\nu v i}[\varphi(w) + ic\varepsilon^2\varphi'''(w) + O(\eta)].$$

The constant C_ν in (ii), (iii) *is independent of v, and satisfies $A < |C_\nu| < A$.*

(i) does not follow from lemma 6, for we do not in general have $w \in \Gamma^+$. We have, however, from lemma 7 (i),

$$\varphi(w) = O\{\varphi(v)\}, \quad \varphi'(w) = O\{\varphi'(v)\} + O\{\nu\varphi(v)\},$$

$$\varphi''(w) = O\{\nu^2\varphi(v)\} + O\{\nu\varphi'(v)\} + O\{\varphi''(v)\},$$

$$\varphi'''(w) = O\{\nu^3\varphi(v)\} + O\{\nu^2\varphi(v)\} + O\{\nu\varphi''(v)\} + O\{\varphi'''(v)\},$$

and the results of (i) follow from lemma 6 and $\nu = O(\eta^{-1})$.

For (ii) we have, by lemma 5 (or formula (v))

$$\varphi(v) = (-1)^\nu \exp\left[-\pi\nu\tau i - \pi\nu^2\tau i - 2\nu\pi v i\right] \varphi(w).$$

Differentiating this logarithmically we have, since $d/dv = d/dw$,

$$\frac{\varphi'}{\varphi}(v) = \frac{\varphi'}{\varphi}(w) - 3\nu\pi i.$$

Further differentiation and reduction, rather heavy but straightforward, gives (ii).

For (iii) we have $\nu = O(\eta^{-1/2})$, and so

$$\varepsilon^2\nu\varphi''(w) = O(\eta^{4-1/2-5/2}) = O(\eta),$$

$$\varepsilon^2\nu^2\varphi'(w) = O(\eta^{4-1-3/2}) = O(\eta),$$

$$\varepsilon^2\nu^3 = O(\eta).$$

Substituting these in (ii) gives (iii).

§ 13

Lemma 9. *If* $|f(z)| < M^{1-\delta}(r)$ *for a* $v \in \Gamma_+$, *and if* w, ν *are associated with* v, *then*:

(i) $F(v) = O(\eta)$;

(ii) $w = O(\varepsilon)$;

(iii) $\nu = O(\eta^{-1/2})$.

In lemma 2 we have here $f_1 = O\{M^{-\delta/2}(r)\}$, which is very small; hence $F(v) = O(\eta)$.

Next, by lemmas 7 and 8,

$$\begin{aligned}\eta^{-3/2}|w| &\leqq A|\varphi(v)| \\ &\leqq A|F(v)| + B\varepsilon^2\left\{|\varphi'''(w)| + |\nu\varphi''(w) + \nu^2\varphi'(w)| + |\nu|^3\right\} \\ &\leqq B[\eta + \eta^{4-7/2} + \eta^{4-1-5/2} + \eta^{4-2-3/2} + \eta^{4-3}] \leqq B\eta^{1/2},\end{aligned}$$

and so $w = O(\varepsilon)$, as desired.

(iii) is immediate from this and lemma 7 (ii).

§ 14

Lemma 10. *If $v \in \Gamma_+$, and some w associated with v satisfies $w = O(\varepsilon)$, then for that w*

$$F(v) = C\left(1 + \frac{1}{50}\vartheta\right)\eta^{-3/2}\{w - w_N + O(\varepsilon\eta^{1/2})\},$$

$$f_1 = C\left(1 + \frac{1}{50}\vartheta\right)e^{-2\pi\nu v i}\eta^{-3/2}\{w - w_N + O(\varepsilon\eta^{1/2})\},$$

where C is a constant depending on c and N, $A < |C| < A$, $|\vartheta| < 1$, w_N is given by

$$w_N = W_N\varepsilon, \quad W_N = \left(\frac{ci\varepsilon\pi^2}{\Delta^2}\right)\{-1 + 24Q_1^2 - 3(1 - \alpha b\Delta)^2\},$$

and $|W_N| < B$,

$$Q_1^2 = \sum_1^\infty \frac{Q^{2n}}{(1 - Q^{2n})^2} \quad [Q_1^2 \text{ is a "modified" } Q^2].$$

w_N and W_N depend only on c and N and all $v \in \Gamma_+$ have the same w_N.

In particular all this holds if $|f| < M^{1-\delta}(r)$, in which case we must have also $|w - w_N| < B\varepsilon\eta^{1/2}$.

The result for f_1 follows at once from that for $F(w)$, since the $O(\eta)$ of the difference can be absorbed into the $O(\varepsilon\eta^{1/2})$ of the bracket.

By lemma 8 (ii),

$$F(v) = C_N e^{-2\pi\nu v i}[\varphi(w) + ic\varepsilon^2\varphi'''(w) + O(\eta)]. \tag{14.1}$$

Combining formula (VI) and (10.2) we have

$$\varphi(w) = \frac{C}{\Delta^{1/2}}\exp(-\pi i w - bw\tau\pi i)\sin\frac{\pi w}{\Delta}\Pi, \tag{14.2}$$

where

$$\Pi = \prod_1^\infty\left(1 - 2\cos\frac{2\pi}{\Delta}Q^n + Q^{2n}\right), \quad \text{and} \quad A < |C| < A.$$

Since $w = O(\varepsilon)$, the exp $= 1 + O(\eta)$, and $\sin\frac{\pi w}{\Delta} = \frac{\pi w}{\Delta}\{1 + O(\eta)\}$, so that, from (14.2), with new C,

$$\varphi(w) = C\{1 + O(\eta)\}\Delta^{-3/2}w\Pi, \quad A < |C| < A. \tag{14.3}$$

Next,

$$|\log \Pi| \leqq \sum \log(1 - |Q|^{2n})^{-2} + o(1) < \frac{1}{100}$$

since

$$|Q|^2 < \exp\left\{-\frac{5}{3}\pi + O(1)\right\} < 0{,}0055, \tag{14.4}$$

and from (14.3), since $A\eta^{-3} < |\Delta^{-3}| < A\eta^{-3}$, we get, with new C,

$$\varphi(w) = C\left\{1 + \frac{\vartheta}{50}\right\}\eta^{-3/2}w, \quad |\vartheta| < 1. \tag{14.5}$$

We must now evaluate $\varphi'''(w)/\varphi(w)$. This is found as the sum of a number of separate expressions, which contain terms in γ^3 and γ^2 as well as γ and γ^0, where $\gamma = \cot(\pi w/\Delta)$. The terms in γ^3 cancel. But this is not quite true of those in γ^2; a term $\gamma^2 \sin(2\pi w/\Delta)$ occurs, which $= 2\gamma\{1 + O(\eta)\}$ since $w/\Delta = O(\eta)$. An important point is that we must retain *all* terms until the third differentiation.

Let

$$\gamma = \cot\frac{\pi w}{\Delta}, \quad \sigma = \sin\frac{2\pi w}{\Delta}, \quad \varkappa = \cos\frac{2\pi w}{\Delta}. \tag{14.6}$$

Since $w/\Delta = O(\eta)$ we have

$$\gamma = \frac{\Delta}{\pi w} + O(\eta), \quad \sigma = O(\eta), \quad \gamma\sigma = 2\{1 + O(\eta)\}, \tag{14.7}$$

$$\varkappa = 1 + O(\eta^2).$$

As a preliminary we have

Lemma 11. *Let*

$$\sum\nolimits_1 = \sum_1^\infty \frac{Q^{2n}}{1 - 2Q^{2n} + Q^{4n}}, \quad Q = e^{\pi i \tau}, \quad \varkappa = \cos\frac{2\pi w}{\Delta}.$$

Then:

(i) $\sum_1 = Q_1^2 + O(\eta)$, $Q_1^2 = \sum_1^\infty \frac{Q^{2n}}{(1 - Q^{2n})^2}$;

(ii) $\sum_1' = \frac{d\sum_1}{dw} = O(1)$;

(iii) $\sum_1'' = O\left(\frac{1}{\Delta^2}\right)$.

Since $|Q|^2 < \exp\left(-\frac{5}{3}\pi + o(1)\right)$ these are all straightforward.

§ 15

Differentiating (14.2) logarithmically we have

$$\frac{\varphi'}{\varphi}(w) = -\pi i h + \frac{\pi}{\Delta}\gamma + \frac{4\pi\sigma}{\Delta}\sum\nolimits_1, \tag{15.1}$$

where

$$h = \frac{1}{\Delta} - b\tau = \frac{1}{\Delta}(1 - \alpha b\,\Delta) + o(1). \tag{15.2}$$

Differentiating (15.1) gives, with argument w always,

$$\frac{\varphi''}{\varphi} - \left(\frac{\varphi'}{\varphi}\right)^2 = -\frac{\pi^2}{\Delta^2}(1+\gamma^2) + \frac{8\pi^2\varkappa}{\Delta^2}\sum\nolimits_1 + \frac{4\pi\sigma}{\Delta}\sum\nolimits_1', \tag{15.3}$$

and differentiating again gives

$$\frac{\varphi'''}{\varphi} - 3\left\{\frac{\varphi''}{\varphi} - \left(\frac{\varphi'}{\varphi}\right)^2\right\}\frac{\varphi'}{\varphi} - \left(\frac{\varphi'}{\varphi}\right)^3$$

$$= \frac{2\pi^3\gamma(1+\gamma^2)}{\Delta^3} - \frac{16\pi^3\sigma}{\Delta^3}\sum\nolimits_1 + \frac{16\pi^2\varkappa}{\Delta^2}\sum\nolimits_1' + \frac{4\pi\sigma}{\Delta}\sum\nolimits_1''$$

$$= \frac{2\pi^3\gamma(1+\gamma^2)}{\Delta^3} + O\left(\frac{1}{\Delta^3}\right), \tag{15.4}$$

by lemma 11 and $\sigma = O(\eta)$. We now substitute for the second and third terms two lines above (15.4) from (15.1) and (15.3), rejecting (fortunately!) $O(\Delta^{-3})$, which means that we retain only terms containing γ in the final expression. We find (substituting $\gamma\sigma = 2 + O(\eta)$ in one place)

$$\frac{\varphi'''}{\varphi}(w) = \gamma\left[-\frac{\pi^3}{\Delta^3}(1 - 24Q_1^2 + 3h^2\Delta^2) + O(\varkappa - 1)\frac{\sum_1}{\Delta^3}\right].$$

Since $\gamma = \Delta/\pi w + O(\eta)$ and $h\Delta = 1 - \alpha b\,\Delta$ we find, by combination with lemma 8 and (14.5), the formula of lemma 10 for $F(v)$.

That $|W_N| < B$ is obvious, and this completes the proof of lemma 10.

§ 16

By the definition (3.5) consecutive Γ_N fit together; the open upper horizontal boundary falls on the lower closed one of Γ_{N+1}.

w_N is generally not in Γ_N: let ν_N, μ_N be the integers making

$$v_N = w_N - (\nu_N\tau + \mu_N) \in \Gamma_N.$$

The ν_N, μ_N are unique and depend only on c and N, so that v_N also depends only on c and N. Since $w_N = O(\varepsilon)$,

$$|\nu_N|,\ |\mu_N| < B. \tag{16.1}$$

v_N belongs strictly to Γ_N, and there is just one v_N to each Γ_N [without the half-openness of Γ_N we might have had $v_{N+1} = v_N$].

Now

$$w_N,\ \nu_N \textit{ are a pair } w,\ \nu \textit{ associated with } v_N \tag{16.2}$$

for $|w_N| < A_1\eta$ and $|\nu_N| < B \leqq \frac{1}{2}\eta^{-1}t$.

Consider now the circle

$$S_N\colon |v - v_N| = \varepsilon\eta^{1/4} = \varrho\ (> \varepsilon\eta^{1/2}). \tag{16.3}$$

$w = w_N + v - v_N = v + \nu_N\tau + \mu_N$ is a w associated with ν, since $w = O(\varepsilon)$ and $\nu_N = O(1)$. By lemma 10 we have for v on S_N

$$\begin{aligned} f_1 &= C\left(1 + \frac{\vartheta}{50}\right) e^{-2\nu_N\pi i v}\{w - w_N + O(\varepsilon\eta^{1/2})\} \\ &= C\left(1 + \frac{\vartheta}{50}\right) e^{-2\nu_N\pi i(v - v_N)}\{v - v_N + O(\varepsilon\eta^{1/2})\}, \end{aligned} \tag{16.4}$$

with a new C incorporating a factor $\exp(-2\nu_N\pi i v_N)$, and with $A < |C| < A$.

Now by lemma 10, for all v of Γ_+ (all of which have the same w_N), $|f| > M^{1-\delta}(r)$ unless

$$|w - w_N| = O(\varepsilon\eta^{1/2}). \tag{16.5}$$

Since $\varrho > \varepsilon\eta^{1/2}$ it follows that $|f| > M^{1-\delta}(r)$ for all v of Γ_N not inside S_N, so that

$$|f| > M^{1-\delta}(r) \quad \textit{for all } z \textit{ whose } v \textit{ is not inside some } S_N. \tag{16.6}$$

Further

$$|f| > M^{1-\delta}(r) \quad \textit{on the boundary } S_N. \tag{16.7}$$

As v describes S_N, z describes a simple closed curve S'_N. This is of the form

$$z - z_N = -2\pi i N^k \exp[-(2N-1)\alpha\pi i + 2\pi i v_N](v - v_N)\{1 + O(\varrho)\}. \tag{16.8}$$

Now we have

$$\Delta_{S_N'} \arg f = \Delta_{S_N} \arg f.$$

Abbreviating Δ_{S_N} to Δ we have, by (16.4),

$$\Delta \arg f = \Delta \arg f_1 = \Delta \arg [e^{-2\nu_N\pi i(v - v_N)}\{v - v_N + O(\varepsilon\eta^{1/2})\}].$$

Since $\nu_N = O(1)$, so that $\nu_N(v - v_N)$ is small, and since $\varrho > \varepsilon\eta^{1/2}$, we have $\Delta \arg f = 2\pi + o(1)$, and so $\Delta \arg f = 2\pi$. Thus $f(z)$ has just one, and a simple, zero $\mathfrak{z}_N$ in each S_N, and, by (16.7),

$$|f| > M^{1-\delta}(r) \quad \text{on the boundary } S_N'. \tag{16.9}$$

§ 17

It remains to establish the pits effect, which follows by an argument of very general application; it is used by Nassif.

Let $r_N = |\mathfrak{z}_N|$, $r_N' = |z_N|$, where r_N' corresponds to v_N. After lemma 1, and since $|z - \mathfrak{z}_N|$, $|r - r_N|$ for r of the boundary S_N' are both

$$O(N^k)(v - v_N) = O(r_N^B),$$

$M(r_N)$, $M(r_N')$, and $M(r)$ for a z of S_N' differ by at most an index $1 \pm \delta$. Let

$$\psi(z) = \frac{f(z)}{z - \mathfrak{z}_N}.$$

Since $|z - \mathfrak{z}_N| = O(r_N^B)$, $|\psi(z)| > M^{1-2\delta}(r_N)$ on the boundary S_N', since $|f| > M^{1-\delta}(r_N)$. Since $\psi(z)$ has no zeros in S_N',

$$|f'(\mathfrak{z}_N)| = |\psi(\mathfrak{z}_N)| > M^{1-2\delta}(r_N). \tag{17.1}$$

We now prove:

Theorem. *Let* $0 < p < 1$, $0 < q < 1$, $p + q < 1$. *Then* $|f| > M^q(r)$ *except inside circles* γ_N *round the zeros* $\mathfrak{z}_N$ *of radius* $M^{-p}(r_N)$.

Let $f(z) = (z - \mathfrak{z}_N) f'(\mathfrak{z}_N) + \sum_2^\infty a_m(z - \mathfrak{z}_N)^m$, and let $R = M^{-p}(r_N)$. By lemma 1, $|a_m| \leqq M^{1+\delta}(r_N)$. Hence for z on the boundary γ_N we have, by (17.1),

$$\begin{aligned} |f(z)| &\geqq RM^{1-2\delta}(r_N) - \sum_2^\infty R^m M^{1+\delta}(r_N) \\ &\geqq RM^{1-2\delta}(r_N)\{1 - 2RM^{3\delta}(r_N)\} > M^q(r). \end{aligned} \tag{17.2}$$

Now γ_N has its center in the middle half of S_N'; for we might have used a circle of $\frac{1}{3}$ of the radius of S_N. Hence γ_N is interior to S_N'.

Now a z exterior to all γ_N is either exterior to all S_N', or else interior to some $S_N' - \gamma_N$. In the first case $|f| > M^{1-\delta}(r)$. In the second, since there is no zero of f in $S_N' - \gamma_N$, the minimum there is attained on the boundary, and is either $> M^{1-\delta}(r)$ or $> M^q(r)$, by (17.2).

This completes the proof of the theorem.

§ 18

It did not occur to me until the present paper was completed, that it might be possible to extend the main theorem, *mutatis mutandis*, to *all* sufficiently smooth integral functions, specifically to all functions of infinite order. But with the hindsight provided by the argument of the paper this now seems highly plausible, though a date-line prevents me from taking up the matter here. However, consider the key-lemma 10. The formula for f_1 has $w_N = O(\varepsilon)$ and an error-term $O(\varepsilon\eta^{1/2})$, of lower order than w_N by a power $N^{-1/4}$. Now the typical smooth function of infinite order, with factor $\exp(i\alpha\pi n^2)$, is

$$\sum \frac{z^n}{\Gamma(1 + nk_n)} \exp(i\alpha\pi n^2), \tag{18.1}$$

where the $k = 1/\varrho$ is replaced by a smoothly decreasing k_n. But now, for (18.1) to be an integral function (infinite radius of convergence) it is necessary that $k_n > 1/\log n$. The entire range of infinite order corresponds to the logarithmic range

$$1 > k_n > 1/\log n,$$

and we may reasonably expect the $N^{1/4}$ to spare in the vital formulae to be affected only logarithmically.

References

[1] J. E. Littlewood and A. C. Offord, On the distribution of the zeros and α-values of a random integral funktion I, Journal London Math. Soc. **20** (1965), 130—136; II, Annals of Math. **49** (1948), 885—952.

[2] G. N. Watson, A class of integral functions defined by Taylor Series, Trans. Cambridge Phil. Soc. XXII (II) (1912), 9—37.

[3] M. Nassif, On the behavior of the function $f(z) = \sum e^{\nu 2\pi i n^2} \frac{z^{2n}}{n!}$, Proc. London Math. Soc. (2) **54** (1950), 201—214.

[4] S. R. Tims, Note on a paper by M. Nassif, Proc. London Math. Soc. **54** (1950), 215.

[Added at proof. I have now found a proof of the above conjecture, and it will appear in due course.]

L. J. Mordell in Cambridge

ON NUMBERS WHICH CAN BE EXPRESSED AS A SUM OF POWERS

R. P. Bambah and S. Chowla [1] have proved that there exists in the interval r to $r + 2(2 + \varepsilon)^{1/2} r^{1/4}$ an integer which can be expressed as a sum of two integer squares if $\varepsilon > 0$ and $r > R(\varepsilon)$. It is rather surprising that the order of magnitude of the estimate has never been improved, and it is not the object of this paper to do so.[1])

The idea in their paper can be presented a little more simply and permits of an easy generalization given by the

Theorem. *Let*

$$f_n(x) = a_1 x_1^{l_1} + a_2 x_2^{l_2} + \cdots + a_n x_n^{l_n}, \tag{1}$$

where the a and l are positive numbers, the a are $\geqq 1$, *and the l are* >1, *and* $l_1 \leqq l_2 \leqq \cdots \leqq l_n$. *Then there exists a positive integer set* (x) *such that*

$$r \leqq f_n(x) < r + c_n r^{e_n} + O(r^{(l_n - 2)e_n/(l_n - 1)}), \tag{2}$$

where

$$e_n = \prod_{i=1}^{n} \left(\frac{l_i - 1}{l_i} \right), \quad e_0 = 1, \tag{3}$$

$$c_n = (l_n a_n^{1/l_n})^{e_{n-1}} c_{n-1}, \quad c_0 = 1, \tag{4}$$

or more crudely

$$c_n = \prod_{i=1}^{n} (l_i a_i^{1/l_i}). \tag{5}$$

The constant implied in O depends only on the a and l.

We prove the result by induction starting from $n = 1$, though the result was proved originally for $n = 2$.

Let y, d be positive numbers such that

$$a_1 y^{l_1} = r, \quad a_1 (y + 1)^{l_1} = r + d. \tag{6}$$

---

[1]) I should like to thank Mr. M. Montgomery for comments on my manuscript.

Then there exists an integer $x_1 > 0$ such that $y \leqq x_1 < y + 1$, and then $r \leqq a_1 x_1^{l_1} < r + d$. Hence

$$d = a_1(y + 1)^{l_1} - a_1 y^{l_1} = a_1 l_1 y^{l_1 - 1} + O(y^{l_1 - 2}) \tag{7}$$

if $y > 1$, i.e., $r > a_1$. Then from (6) and (7)

$$d = a_1 l_1 \left(\frac{r}{a_1}\right)^{(l_1 - 1)/l_1} + O(r^{(l_1 - 2)/l_1}) \tag{8}$$

$$= l_1 a_1{}^{1/l_1} r^{(l_1 - 1)/l_1} + O(r^{(l_1 - 2)/l_1}). \tag{9}$$

This is the theorem for the case $n = 1$. We now assume that the theorem holds for n and prove it for $n + 1$. Write

$$f_{n+1}(x) = a_1 x_1^{l_1} + \cdots + a_n x_n^{l_n} + a_{n+1} x_{n+1}^{l_{n+1}}$$

$$= f_n(x) + a_{n+1} x_{n+1}^{l_{n+1}}. \tag{10}$$

In (2), write $r - a_{n+1} x_{n+1}^{l_{n+1}}$ for r. It becomes

$$r \leqq f_{n+1}(x) < r + c_n(r - a_{n+1} x_{n+1}^{l_{n+1}})^{e_n} + O(r - a_{n+1} x_{n+1}^{l_{n+1}})^{e_n(l_n - 2)/(l_n - 1)}$$

$$= r + c_n P^{e_n} + O(Q), \tag{11}$$

say. Take now $x_{n+1} = \left[\left(\frac{r}{a_{n+1}}\right)^{1/l_{n+1}}\right] = \left(\frac{r}{a_{n+1}}\right)^{1/l_{n+1}} - h,\ 0 \leqq h < 1$. Then

$$P = r - a_{n+1}\left(\left(\frac{r}{a_{n+1}}\right)^{1/l_{n+1}} - h\right)^{l_{n+1}}$$

is an increasing function of h if $r > a_{n+1}$. Then we have

$$P < l_{n+1} a_{n+1}^{1/l_{n+1}} r^{(l_{n+1} - 1)/l_{n+1}} + O(r^{(l_{n+1} - 2)/l_{n+1}}).$$

Then

$$P^{e_n} < (l_{n+1} a_{n+1}^{1/l_{n+1}})^{e_n} r^{e_{n+1}}(1 + O(r^{-1/l_{n+1}})). \tag{12}$$

The order of the error term given by this is in accordance with (2) if n is replaced by $n + 1$. For

$$e_{n+1} - 1/l_{n+1} < e_{n+1}(l_{n+1} - 2)/(l_{n+1} - 1),$$

or

$$e_{n+1}/(l_{n+1} - 1) < 1/l_{n+1}$$

is obvious since $e_{n+1} = e_n(l_{n+1} - 1)/l_{n+1}$ and $e_n < 1$.

A similar result holds for the error term given in (11) by (2). For

$$e_n \frac{(l_n - 2)}{(l_n - 1)} \frac{(l_{n+1} - 1)}{l_{n+1}} \leqq e_{n+1} \frac{(l_{n+1} - 2)}{(l_{n+1} - 1)},$$

or

$$\frac{(l_n - 2)}{(l_n - 1)} \leqq \frac{(l_{n+1} - 2)}{(l_{n+1} - 1)},$$

follows since $l_n \leqq l_{n+1}$. The theorem now follows since from (11) and (12),

$$c_{n+1} = c_n(l_{n+1}\, a_{n+1}^{1/l_{n+1}})^{e_n}$$

Reference

[1] R. P. Bambah and S. Chowla, On numbers which can be represented as a sum of two squares, Proceedings of the National Institute of Science of India **13**, Nr. 2 (1947), 101—103.

Note added to the proof sheets.

For improvements on the Theorem and references to recent work, see P. H. Diananda, On integers expressable as a sum of two powers, Proc. Japan Acad. **43** (1967), 417—419, and a further paper to be submitted to the same journal.

L. J. MORDELL IN CAMBRIDGE

ON SOME DIOPHANTINE EQUATIONS $y^2 = x^3 + k$ WITH NO RATIONAL SOLUTIONS (II)

The simplest Diophantine equation of degree greater than 2 is the equation

$$y^2 = x^3 + k, \tag{1}$$

where k is an integer. Two problems arise according as rational solutions or integral solutions are required. It is the problem of rational solutions which will be discussed here. No finite algorithm is known for finding solutions if they exist, except for special values of k.

There are procedures which are useful either in proving that there are no solutions, or in finding the fundamental solutions which generate all the rational solutions by the classic tangent and chord process. Many of these results have been given by Billing [1], Cassels [2], Birch and Swinnerton-Dyer [3], and depend upon the properties of the cubic field $Q(\sqrt[3]{k})$. Some years ago, comparatively simple conditions for insolvability were given for special equations with $k < 0$ by Fueter [4] and Brunner [5], and for $k > 0$ by myself [6] and K.-L. Chang [7]. These depend on the properties of the quadratic fields $Q(\sqrt{-k})$, $Q(\sqrt{-3k})$.

I notice that other results can be found more simply by my method, as is shown by the following theorem.

Theorem. *The equation $y^2 = x^3 + k$ has no rational solutions if all the following conditions are satisfied*:

(A) *k is positive and square free and*

$$k \equiv 2 \text{ or } 3 \pmod 4) \text{ and } k \equiv -2 \text{ or } -4 \pmod 9.$$

(B) *The number h of classes of ideals in the quadratic field $Q(\sqrt{-3k})$ is not divisible by 3.*

(C) *The fundamental solution $(X, Y) = (U, T)$ of the equation $Y^2 - kX^2 = 1$ satisfies the conditions*

(C 1) *if $k \equiv -2 \pmod 9$ then $U \equiv \pm 3$ and $T \equiv \pm 1 \pmod 9$*,

(C 2) *if $k \equiv -4 \pmod 9$ then either $U \equiv \pm 3$ and $T \equiv \pm 1 \pmod 9$ or $U \equiv \pm 4$ and $T \equiv \pm 3 \pmod 9$.*

Instances are $k = 7$ since $U = 3$, $T = 8$, $h = 4$, and $k = 14$ since $U = 4$, $T = 15$, $h = 4$.

To prove the theorem, it obviously suffices to show that the equation

$$y^2 = x^3 + kz^6, \quad (x, y, z) = 1 \tag{2}$$

has no integer solutions with $z \neq 0$. We can suppose that $z > 0$ and $y > 0$. We can exclude $x \equiv 0 \pmod 2$, for this would imply $z \equiv 1 \pmod 2$, whence $k \equiv y^2 \equiv 0$ or $1 \pmod 4$, contrary to (A). Hence x is odd. Also $(x, k) = 1$ since k is square free.

We write the equation as

$$(y + z^3\sqrt{k})(y - z^3\sqrt{k}) = x^3, \tag{3}$$

and study this in the quadratic field $Q(\sqrt{k})$. The two factors on the left have no common divisor, since $(x, 2k) = 1$ and $(x, y, z) = 1$.

By a theorem of Scholz [8], the class number h_1 of the field $Q(\sqrt{k})$ is not divisible by 3, this being a deduction from the hypothesis (B). Hence

$$y + z^3\sqrt{k} = \eta(A + B\sqrt{k})^3, \tag{4}$$

where η is a unit in $Q(\sqrt{k})$, and $(A, B) = 1$ since $(y, z) = 1$. We show that we can take $\eta = 1$ or ε or ε^{-1}, where $\varepsilon = T + U\sqrt{k}$. This is immediate when $k \equiv 3 \pmod 4$, since then ε is the fundamental unit of $Q(\sqrt{k})$ and therefore $\eta = \varepsilon^{3n}$ or ε^{3n+1} or ε^{3n-1}. When $k \equiv 2 \pmod 4$ it is possible that $\varepsilon = \eta_1^2$, where $\eta_1 = T_1 + U_1\sqrt{k}$, with $T_1^2 - kU_1^2 = -1$, is the fundamental unit of $Q(\sqrt{k})$. Then we can take $\eta = 1$ or η_1 or η_1^2; but now $\eta_1^2 = \varepsilon$ and $\eta_1 = \eta_1^3\varepsilon^{-1}$, and the factor η_1^3 can be absorbed in $(A + B\sqrt{k})^3$.

Case 1. *Suppose* $\eta = \varepsilon$ *or* ε^{-1}. Then $\eta = T \pm U\sqrt{k}$, and

$$y + z^3\sqrt{k} = (T \pm U\sqrt{k})(A + B\sqrt{k})^3,$$

and so

$$z^3 = T(3A^2B + kB^3) \pm U(A^3 + 3kAB^2). \tag{5}$$

We show from condition (C) that this is impossible as a congruence (mod 9). For

$$z \equiv kTB \pm UA \pmod 3,$$

whence

$$z^3 \equiv k^3T^3B^3 \pm 3k^2T^2UB^2A + 3kTU^2BA^2 \pm U^3A^3 \pmod 9.$$

On writing $-A$ for A it suffices to take the $+$ sign. By the last congruence and (5),

$$A^3(U - U^3) + 3A^2B(T - kTU^2) + 3AB^2(kU - k^2T^2U)$$
$$+ B^3(kT - k^3T^3) \equiv 0 \pmod 9,$$

that is,

$$\tfrac{1}{3}(U - U^3)\,A + A^2B(T - kTU^2) + AB^2(kU - k^2T^2U)$$
$$+ \tfrac{1}{3}(kT - k^3T^3)\,B \equiv 0 \pmod 3. \tag{6}$$

We prove that $AB \not\equiv 0 \pmod 3$. In the contrary case, just one of A, B would be divisible by 3, since $(A, B) = 1$. If $A \equiv 0 \pmod 3$ we get $kT \equiv k^3T^3 \pmod 9$, whence $kT \equiv 0$ or 1 or $-1 \pmod 9$, contrary to condition (C); and if $B \equiv 0 \pmod 3$ we get $U \equiv 0$ or 1 or $-1 \pmod 9$, again contrary to (C).

We now have $A^2 \equiv B^2 \equiv 1 \pmod 3$, and (6) can be written as

$$A\,\{\tfrac{1}{3}(U - U^3) + kU - k^2T^2U\} + B\,\{T - kTU^2 + \tfrac{1}{3}(kT - k^3T^3)\}$$
$$\equiv 0 \pmod 3, \tag{7}$$

or say as

$$AP + BQ \equiv 0 \pmod 3.$$

If $U \equiv \pm 3 \pmod 9$ and $T \equiv \pm 1 \pmod 9$, as in (C 1) or (C 2), then $P \equiv \tfrac{1}{3}(U - U^3)$ $\equiv \pm 1 \pmod 3$, and

$$Q \equiv T\{1 + \tfrac{1}{3}(k - k^3)\} \equiv 0 \pmod 3,$$

since $k \equiv -2$ or $-4 \pmod 9$. This gives $A \equiv 0 \pmod 3$, contrary to what was proved above.

If $U \equiv \pm 4 \pmod 9$ and $T \equiv \pm 3 \pmod 9$, as in (C 2), then $k \equiv -4 \pmod 9$ and

$$P \equiv \tfrac{1}{3}(U - U^3) + kU \equiv 0 \pmod 3,$$

and

$$Q \equiv \tfrac{1}{3}(kT - k^3T^3) \equiv \pm 1 \pmod 3,$$

giving $B \equiv 0 \pmod 3$, which is again impossible.

Case 2. *Suppose* $\eta = 1$. Then

$$y + z^3\sqrt{k} = (A + B\sqrt{k})^3, \quad (A, B) = 1, \tag{8}$$

and so

$$z^3 = B(3A^2 + kB^2). \tag{9}$$

We recall that $z > 0$, whence $B > 0$.

Suppose first that $B \not\equiv 0 \pmod 3$. Since the factors on the right of (9) are relatively prime, we have

$$B = B_1^3, \quad z = B_1 z_1, \quad z_1^3 = 3A^2 + kB_1^6. \tag{10}$$

Hence

$$(3A)^2 + 3kB_1^6 = 3z_1^3,$$

$$(3A + B_1^3\sqrt{-3k})(3A - B_1^3\sqrt{-3k}) = 3z_1^3. \tag{11}$$

From (10) and the fact that k is square free we have $(z_1, k) = 1$. Also $z_1 \not\equiv 0 \pmod 2$, since if z_1 is even then B_1 cannot be even, so $B_1^6 \equiv 1 \pmod 4$ and

$$3A^2 + k \not\equiv 0 \pmod 4$$

by condition (A). Hence the only possible common divisor of the factors on the left of (11) is a divisor of 3. In fact $(3) = \pi^2$, where π is an ideal in $Q(\sqrt{-3k})$. Hence the ideal equation

$$(3A + B_1^3\sqrt{-3k}) = \pi\mathfrak{a}_1^3, \quad (3A - B_1^3\sqrt{-3k}) = \pi\mathfrak{a}_2^3.$$

Here $\mathfrak{a}_1^3 \sim \mathfrak{a}_2^3$, and since $h \not\equiv 0 \pmod 3$ this implies that $\mathfrak{a}_1 \sim \mathfrak{a}_2$. Hence

$$\frac{3A + B_1^3\sqrt{-3k}}{3A - B_1^3\sqrt{-3k}} = \left(\frac{\xi + \eta\sqrt{-3k}}{n}\right)^3, \quad (\xi, \eta, n) = 1.$$

Then

$$3An^3 = 3A(\xi^3 - 9k\xi\eta^2) - B_1^3(-9k\xi^2\eta + 9k^2\eta^3),$$

$$B_1^3n^3 = 3A(3\xi^2\eta - 3k\eta^3) - B_1^3(\xi^3 - 9k\xi\eta^2).$$

The second equation gives

$$B_1^3n^3 \equiv -B_1^3\xi^3 \pmod 9, \quad n \equiv -\xi \pmod 3.$$

The first equation gives

$$An^3 \equiv A\xi^3 \pmod 3,$$

and since $n \equiv -\xi \pmod 3$ we must have either $n \equiv \xi \equiv 0 \pmod 3$ or $A \equiv 0 \pmod 3$. In the former case, the first equation, considered to the modulus 81, gives $B_1^3k^2\eta^3 \equiv 0 \pmod 3$, whence $\eta \equiv 0 \pmod 3$, contrary to $(\xi, \eta, n) = 1$. Hence $A \equiv 0 \pmod 3$. But then the last equation in (10) gives $z_1^3 \equiv kB_1^6 \equiv k \pmod 9$, which contradicts the hypothesis $k \equiv -2$ or $-4 \pmod 9$.

Suppose finally that $B \equiv 0 \pmod 3$. Then by (9) B is divisible by 9, and after dividing throughout by 27 the factors are relatively prime. Hence

$$B = 9B_1^3, \quad z = 3B_1z_1, \quad z_1^3 = A^2 + 27kB_1^6,$$

whence

$$(A + 3B_1^3\sqrt{-3k})(A - 3B_1^3\sqrt{-3k}) = z_1^3. \tag{12}$$

We cannot have $z_1 \equiv 0 \pmod 3$, since $(A, B) = 1$. Also $(z_1, k) = 1$ and $z_1 \not\equiv 0 \pmod 2$ for similar reasons to those given in the previous treatment. Hence the two factors on the left of (12) are relatively prime. Since $h \not\equiv 0 \pmod 3$, and since the only units in the quadratic field $Q(\sqrt{-3k})$ are ± 1, we have

$$A + 3B_1^3\sqrt{-3k} = (C + D\sqrt{-3k})^3, \quad (C, D) = 1.$$

Hence

$$B_1^3 = C^2D - kD^3.$$

This implies that $D = D_1^3$, $B_1 = B_2D_1$, and

$$B_2^3 = C^2 - kD_1^6.$$

This is the original equation (2), and $D_1 \neq 0$ since $B_1 > 0$. Also

$$|D_1| \leqq B_1 \leqq \tfrac{1}{3}z,$$

and so the method of descent applies. Hence on continuing with the various alternatives, we are led to a solution x, y, z of (2) with $0 < z < 1$, a contradiction. This finishes the proof of the theorem.

Theorem 2. *The equation $y^2 = x^3 + k$ has no rational solutions if the following conditions are satisfied*:

(A) *k is negative and square free, $k \equiv 2$ or $3 \pmod 4$, and $k \equiv 2, 4 \pmod 9$.*

(B) *The number of classes of ideals in the quadratic field $Q(\sqrt{k})$ is not divisible by 3.*

(C) *The fundamental solution $(X, Y) = (U, T)$ of the equation $Y^2 + 3kX^2 = 1$ satisfies the condition $U \not\equiv 0 \pmod 3$.*

Instances are $k = -5, -14, -34, -41$ with $U = 1, 2, 10, 11$, respectively. As in (3), we consider

$$y^2 = x^3 + kz^6, \quad z \neq 0, \tag{13}$$

and suppose that z has its least positive values. As in (4)

$$y + z^3\sqrt{k} = (A + B\sqrt{k})^3, \quad z^3 = B(3A^2 + kB^2).$$

If $(B, 3) = 1$, $B = B_1^3$, $3A^2 + kB_1^6 = z_1^3$.

This gives as a congruence mod 9, $3A^2 + k \equiv z_1^3 \pmod 9$, and is impossible if $k \equiv 2$, or $4 \pmod 9$.

Suppose next that $B = 9B_1^3$. Then

$$A^2 + 27kB_1^6 = z_1^3.$$

Clearly $z_1 \not\equiv 0 \pmod 2$, and $z_1 \not\equiv 0 \pmod 3$ since $(A, B_1) = 1$ and so $A \not\equiv 0 \pmod 3$.

Hence since the class number for $Q(\sqrt{-3k})$ is prime to 3 from Holzer's result,

$$A + B_1^3\sqrt{-3k} = (T + U\sqrt{-3k})^\alpha (C + D\sqrt{-3k})^3, \quad \alpha = 0, \pm 1.$$

Suppose first that $\alpha = \pm 1$. Then

$$0 \equiv C^3 U\sqrt{-3k} \pmod{3\sqrt{-3k}}.$$

But $C \not\equiv 0 \pmod 3$ and so $U \equiv 0 \pmod 3$ and this has been excluded.

Suppose next that $\alpha = 0$. Then

$$B_1^3 = C^2D - kD^3.$$

Hence

$$D = D_1^3, \quad C^2 - kD_1^6 = B_2^3,$$

and this is the same as (13). Also $D_1 = \sqrt[3]{D} \leqq \sqrt[3]{B_1^3} \leqq \sqrt[3]{B/9} \leqq B/\sqrt[3]{9}$ and this contradicts the definition of z.

Theorem 3. *The equation $y^2 = x^3 + 3k$ has no rational solution if all the following conditions are satisfied:*

(A) *k is negative and $3k$ is square free and $k \equiv 2, 1 \pmod 4$, and $k \equiv 1 \pmod 3$.*

(B) *The number h of classes of ideals in the quadratic field $Q(\sqrt{3k})$ is not divisible by 3.*

(C) *The fundamental solution $(X, Y) = (U, T)$ of the equation $Y^2 + kX^2 = 1$ satisfies the condition, either $T \equiv \pm 3 \pmod 9$ or $U \equiv \pm 3 \pmod 9$.*

Instances are given by $3k = -33, -42, -69, -78, -105, -114$ corresponding to $U = 3$, $T = 5$, $T = 24$, $T = 51$, $T = 6$, $U = 6$, respectively.

Here

$$y^2 = x^3 + 3kz^6,$$

$$y + z^3\sqrt{3k} = (A + B\sqrt{3k})^3, \quad (A, B) = 1,$$

$$z^3 = B(3A^2 + 3kB^2).$$

Suppose first that B is prime to 3. Then

$$B = B_1^3, \quad (B, 3) = 1 \quad \text{and} \quad A^2 + kB_1^6 = 9z_1^3.$$

This is impossible if $k \equiv 1 \pmod 3$.

Suppose next that B is not prime to 3. Then

$$B = 9B_1^3, \quad A^2 + 81kB_1^6 = z_1^3.$$

Then

$$A + 9B_1^3\sqrt{-k} = (C + D\sqrt{-k})^3 (T + U\sqrt{-k})^\alpha, \quad \alpha = 0, \pm 1.$$

If $\alpha = 0$,

$$9B_1^3 = D(3C^2 - D^2k).$$

Then

$$D = 3D_1, \quad B_1^3 = D_1(C^2 - 3kD_1^2),$$

and so

$$D_1 = D_2^3, \quad C^2 - 3kD_2^6 = B_2^3.$$

Hence the method of descent applies since

$$D_2 = \sqrt[3]{D_1} \leqq B_1 \leqq \sqrt[3]{B/9} \leqq \sqrt[3]{z/9}.$$

Suppose next $\alpha = \pm 1$. Then

$$9B_1^3 = T(3C^2D - kD^3) \pm U(C^3 - 3CD^2k). \tag{14}$$

Since $T^2 + U^2 \equiv 1 \pmod 3$, $TU \equiv 0 \pmod 3$.

Suppose $U \equiv \pm 3 \pmod 9$. Then (14) becomes $kTD^3 \equiv 0 \pmod 3$. Hence $D \equiv 0 \pmod 3$, and so (14) gives $UC^3 \equiv 0 \pmod 9$ which is impossible. Similarly if $T \equiv \pm 3 \pmod 9$.

The theorem suggests the

Problem. If $k \equiv 1 \pmod 3$, the fundamental solution $(x, y) = (U, T)$ of $y^2 - kx^2 = 1$ satisfies the condition $U \equiv 0 \pmod 3$. To find conditions under which $U \equiv \pm 3 \pmod 9$.

References

[1] G. Billing, Beiträge zur arithmetischen Theorie ebener kubischer Kurven, Nova Acta Reg. Soc. Scient. Upsaliensis, Ser. IV, XI, Nr. 1 (1938), 1—165.

[2] J. W. S. Cassels, The rational solutions of the Diophantine equation $y^2 = x^3 - d$, Acta Mathematica **82** (1950), 243—273.

[3] J. B. Birch and H. P. F. Swinnerton-Dyer, Notes on elliptic curves, J. reine angew. Math. **212** (1963), 7—25.

[4] R. Fueter, Über kubische diophantische Gleichungen, Comm. Math. Helv. **2** (1930), 69—89.

[5] O. Brunner, Lösungseigenschaften der kubischen diophantischen Gleichung $z^3 - y^2 = D$, Inauguraldissertation, Zürich 1933.

[6] L. J. Mordell, One some Diophantine equations $y^2 = x^3 + k$ with no rational solutions, Arch. for Math. og Naturvidenskap **6** (1947).
[7] Kuo-Lung Chang, One some Diophantine equations $y^2 = x^3 + k$ with no rational solutions, Quarterly J. of Math. **19** (1948), 181—188.
[8] A. Scholz, Über die Beziehung der Klassenzahlen quadratischer Körper zueinander, J. reine angew. Math. **166** (1932), 201—203.

G. Pólya in Stanford (Calif.)

Über das Vorzeichen des Restgliedes im Primzahlsatz

[Vorbemerkung. Abgesehen von einigen Abänderungen und einem Zusatz ist das Nachfolgende der redigierte Wiederabdruck einer Abhandlung, welche von E. Landau vorgelegt wurde in der am 24. Januar 1930 gehaltenen Sitzung der Gesellschaft der Wissenschaften zu Göttingen und in deren Nachrichten erschien (Mathematisch-Physikalische Klasse, 1930, S. 19–27).

Die wichtigste Abänderung ersetzt einen fehlerhaften Schluß der Originaldarstellung (im Abschnitt 4, Teilabschnitt b)). Daß da ein Fehler vorliegt, der sich jedoch leicht berichtigen läßt, habe ich bald nach Erscheinen mehreren Kollegen mitgeteilt; vgl. S. 202, Fußnote [3]) in A. E. Ingham's Abhandlung Acta Arithmetica **1** (1936), 201–211. Auch die zur Berichtigung erforderliche Abänderung habe ich mehreren Korrespondenten mitgeteilt.

Der Zusatz betont eine leicht ersichtliche, aber von einigen Kollegen nicht beachtete Folgerung und zitiert Einiges aus der inzwischen erschienenen Literatur.

Abänderungen und Zusatz sind, wie diese Vorbemerkung, in eckige Klammern gesetzt.]

1. Ich setze, wie üblich,

$$\psi(x) = \sum_{p^m \leqq x} \log p$$

(p Primzahl, m ganz, $m \geqq 1$) und bezeichne mit $W(n)$ die Anzahl der Zeichenwechsel in der n-gliedrigen Folge

$$\psi(1) - 1, \quad \psi(2) - 2, \ldots, \psi(n) - n.$$

Offenbar ist $0 \leqq W(n) \leqq n - 1$ und $W(n)$ nicht abnehmend. Bekanntlich ist[1])

$$\lim_{n \to \infty} W(n) = \infty. \tag{1}$$

[1]) Vgl. E. Phragmén, Öfversigt af K. Vetensk. Förhandlingar (Stockholm) **48** (1891–1892), 599–616 und **58** (1901–1902), 189–202, ferner E. Schmidt, Math. Ann. **57** (1903), 195–204.

Mein Anliegen ist, dies zu

$$\overline{\lim_{n\to\infty}} \frac{W(n)}{\log n} \geqq \frac{\gamma}{\pi} > 0 \tag{2}$$

zu verschärfen. Die positive Größe γ hängt mit der (z. Z. noch unbekannten) Konfiguration der Nullstellen der ζ-Funktion in der oberen Halbebene zusammen, und zwar folgendermaßen: Man bezeichne diese Nullstellen mit $\beta_n + i\gamma_n$ (so daß $0 < \beta_n < 1$, $\gamma_n > 0$) und die obere Grenze der Abszissen β_n mit Θ. (Bekanntlich ist $\frac{1}{2} \leqq \Theta \leqq 1$.) Ist Θ ein Maximum, d. h., gibt es Nullstellen von der Form $\Theta + i\gamma_m$, so sei γ das Minimum dieser γ_m. (Dieser Fall würde insbesondere dann eintreten, wenn $\Theta = \frac{1}{2}$, d. h. die Riemannsche Vermutung wahr wäre.) Ist Θ kein Maximum, d. h. sind alle $\beta_n < \Theta$, so sei $\gamma = \infty$. (Dieser Fall würde also insbesondere dann eintreten, wenn $\Theta = 1$ wäre.)

Den eigentlichen Grund der Tatsache (1) hat Landau aufgedeckt, indem er (1) als unmittelbare Folgerung eines allgemeinen funktionentheoretischen Satzes herleitete[1]). Auf ähnliche Weise werde ich zeigen, daß (2) unmittelbar aus einem allgemeinen funktionentheoretischen Satz folgt, zu dessen Formulierung ich nun übergehe.

2. Ich benutze folgende Bezeichnungen:

$\omega(u)$ ist eine reellwertige Funktion der reellen Variablen u, definiert für $u \geqq 1$ und eigentlich integrabel in jedem endlichen Intervall mit linkem Endpunkt 1.

$u_1, u_2, u_3, \ldots$ sind die Zeichenwechselstellen von $\omega(u)$. Genauer gesagt, führe ich eine neue Voraussetzung über $\omega(u)$ ein: Es sei $\omega(u)$ entweder von konstantem Vorzeichen für $u > 1$, oder es seien Zahlen $u_1, u_2, \ldots$ vorhanden, die keinen Häufungspunkt im Endlichen haben (ihre Anzahl kann endlich oder unendlich sein) und so beschaffen sind, daß

$$1 = u_0 < u_1 < u_2 < \cdots,$$

$$(-1)^n \, \omega(u) \geqq 0 \quad \text{für} \quad u_{n-1} < u < u_n \tag{3}$$

($n = 1, 2, 3, \ldots$) ist und daß $\omega(u)$ in keinem der Intervalle $u_{n-1} < u < u_n$ identisch verschwindet.

$W(x)$ ist die Anzahl der Zeichenwechselstellen von $\omega(u)$ bis zur Grenze x, d. h. $W(x) = n$ für $u_n \leqq x < u_{n+1}$. ($W(x)$ verschwindet identisch, wenn $\omega(u)$ konstantes Vorzeichen bewahrt.)

Den hier in Betracht kommenden Teil des vorher angedeuteten Satzes von Landau will ich nun so formulieren:

---

[1]) Math. Ann. **61** (1905), 527—550, und Sitzungsberichte der Akademie München **36** (1906), 151—218.

I. *Es sei*

$$\int_1^\infty \omega(u)\, u^{-s}\, du = \Phi(s) \tag{4}$$

konvergent in einer gewissen Halbebene, die von einer Parallelen zur imaginären Achse von links begrenzt ist. Es sei die dargestellte Funktion $\Phi(s)$ ausnahmslos regulär in der Halbebene $\Re s > \Theta$ jedoch in keiner Halbebene $\Re s > \Theta - \varepsilon$, wo $\varepsilon > 0$ ist. Wenn Θ kein singulärer Punkt für $\Phi(s)$ ist, dann ist $\lim_{x\to\infty} W(x) = \infty$, d. h., $\omega(u)$ hat unendlich viele Zeichenwechsel.

Hierzu will ich folgenden neuen Satz hinzufügen, der mehr über $\Phi(s)$ voraussetzt und mehr über $W(x)$ aussagt:

II. *Es sei das Integral* (4) *konvergent in einer Halbebene, begrenzt von links durch eine Parallele zur imaginären Achse. Es sei $\Phi(s)$ ausnahmslos regulär in der Halbebene $\Re s > \Theta$, jedoch in keiner Halbebene $\Re s > \Theta - \varepsilon$, wo $\varepsilon > 0$ ist, hingegen meromorph in der Halbebene $\Re s \geqq \Theta - b$, wo $b > 0$ gilt.*

Wenn $\Phi(s)$ Pole auf der Geraden $\Re s = \Theta$ besitzt, so sei $s = \Theta + i\gamma$ derjenige darauf und in der Halbebene $\Im s \geqq 0$ gelegene Pol, der den kleinsten Imaginärteil γ hat. ($\Theta + i\gamma$ existiert, da $\omega(u)$ reellwertig ist. Es ist $\gamma \geqq 0$.) *In diesem Falle ist*

$$\overline{\lim_{x\to\infty}} \frac{W(x)}{\log x} \geqq \frac{\gamma}{\pi}. \tag{5a}$$

Wenn $\Phi(s)$ auf der Geraden $\Re s = \Theta$ keinen Pol besitzt, so gilt

$$\overline{\lim_{x\to\infty}} \frac{W(x)}{\log x} = \infty. \tag{5b}$$

Für die Funktion

$$-\frac{\zeta'(s)}{s\zeta(s)} - \frac{\zeta(s)}{s} = \int_1^\infty (\psi(u) - [u])\, u^{-1-s}\, du \tag{6}$$

ist die in den Sätzen I, II erwähnte Zahl Θ zwar nicht genau bekannt, aber auf alle Fälle ist $\frac{1}{2} \leqq \Theta \leqq 1$, ferner ist Θ kein singulärer Punkt, und es kann $\Theta - b = 0$ gewählt werden (etwa). So folgt (1) aus dem Landauschen Satz I und (2) aus dem neuen Satz II.

Satz II hat eine gewisse Verwandtschaft mit bekannten Sätzen von Fabry über die Singularitäten von Potenzreihen und wird mit Methoden bewiesen, die zum Nachweis der Fabryschen Sätze herausgebildet worden sind. Der wesentliche Punkt ist dabei, meiner Ansicht nach, die Verwendung einer linearen Funktionaloperation, vgl. (15). Ich muß beim Beweis auf meine diesbezüglichen Untersuchungen[1]) zurückgreifen, die mühelos zum Satz II führen.

3. Indem ich Satz I als bekannt voraussetze, muß ich mich nur mit dem Fall befassen, in welchem unendlich viele Zeichenwechselstellen $u_1, u_2, u_3, \ldots$ vorliegen. Ich setze (vorderhand!) voraus, daß

$$\overline{\lim_{n\to\infty}} \frac{n}{\log u_n} = \overline{\lim_{x\to\infty}} \frac{W(x)}{\log x} = d \tag{7}$$

endlich ist. Dann ist, bei gegebenem $\varepsilon > 0$ für genügend großes n,

$$\frac{1}{(\log u_n)^2} < \frac{(d+\varepsilon)^2}{n^2}, \tag{8}$$

also das Produkt

$$\prod_{n=1}^{\infty}\left(1 - \frac{z^2}{(\log u_n)^2}\right) = F(z) \tag{9}$$

konvergent. Der Vergleich von (9) mit dem Produkt

$$\prod_{n=1}^{\infty}\left(1 + \frac{(d+\varepsilon)^2\,|z|^2}{n^2}\right) = \frac{e^{\pi(d+\varepsilon)|z|} - e^{-\pi(d+\varepsilon)|z|}}{2\pi(d+\varepsilon)\,|z|}$$

ergibt wegen (8) für genügend großes $|z|$

$$|F(z)| < e^{\pi(d+\varepsilon)|z|}. \tag{10}$$

Ich setze

$$F(z) = a_0 + \frac{a_2 z^2}{2!} + \frac{a_4 z^4}{4!} + \cdots, \tag{11}$$

$$f(z) = \frac{a_0}{z} + \frac{a_2}{z^3} + \frac{a_4}{z^5} + \cdots. \tag{12}$$

Es folgt aus (10), daß die Reihe (12) außerhalb der Kreisfläche

$$|z| \leqq \pi d \tag{13}$$

konvergiert[2]), ferner daß es einen konvexen, in der abgeschlossenen Kreisfläche (13) enthaltenen Bereich $\mathfrak{J}$ gibt, so beschaffen, daß $f(z)$ *außerhalb von* $\mathfrak{J}$ *regulär ist, jedoch*

[1]) Math. Z. **29** (1929), 549—640, im folgenden als LSP zitiert. Es kommt nur Kapitel II, S. 571—610, in Betracht, und zwar insbesondere die Nummern 20—25 und 36—40.

[2]) LSP, S. 578, Satz I.

in jedem extremen Punkt von $\mathfrak{J}$ singulär wird[1]). Es ist übrigens, da die Koeffizienten $a_0, a_2, a_4, \ldots$ reell sind, und $f(z)$ eine ungerade Funktion ist, $\mathfrak{J}$ symmetrisch sowohl in bezug auf die reelle wie auf die imaginäre Achse und enthält den Nullpunkt. Man betrachte diejenige Stützgerade von $\mathfrak{J}$, deren äußere Normale die positive reelle Achse ist. Auf dieser Stützgeraden liegt entweder ein extremer Punkt von $\mathfrak{J}$ – dieser werde j genannt – oder es liegen darauf zwei extreme Punkte von $\mathfrak{J}$ – dann werde derjenige unterhalb der reellen Achse j genannt. Wird $j = \varkappa + i\lambda$ gesetzt ($\varkappa$, λ reell), so ist

$$0 \leqq \varkappa \leqq \pi d, \quad -\pi d \leqq \lambda \leqq 0. \tag{14}$$

Wird unter $\Psi(s)$ irgendeine analytische Funktion verstanden, die in der Halbebene $\Re s > \Theta$ regulär ist, so sei

$$\Psi^*(s) = \frac{1}{2\pi i}\int \Psi(s - w) f(w)\, dw \tag{15}$$

gesetzt. Die Integration in (15) ist in positivem Sinne entlang einer doppelpunktfreien geschlossenen Kurve erstreckt, die den Bereich $\mathfrak{J}$, also sämtliche Singularitäten von $f(w)$ umfaßt und sämtliche Singularitäten von $\Psi(s - w)$ außerhalb läßt; die Integrationskurve läßt sich also ohne Änderung des Wertes beliebig um $\mathfrak{J}$ zusammenschnüren, und hieraus geht hervor, daß $\Psi^*(s)$ durch (15) in der Halbebene

$$\Re s > \Theta + \varkappa \tag{16}$$

erklärt und regulär ist[2]).

Wird

$$\int_1^g \omega(u)\, u^{-s}\, du = \Phi_g(s)$$

gesetzt, so ist, gemäß der Definition (15),

$$\begin{aligned}\Phi_g^*(s) &= \frac{1}{2\pi i}\int f(w) \int_1^\beta \omega(u)\, u^{-s+w}\, du\, dw \\ &= \frac{1}{2\pi i}\int_1^g \omega(u)\, u^{-s} \int f(w)\, e^{w \log u}\, dw\, du \\ &= \int_1^g \omega(u)\, u^{-s} F(\log u)\, du,\end{aligned}$$

---

[1]) LSP, S. 585, Satz III, von dessen Beweis die Überlegung S. 583, unter 23a) zumeist in Betracht kommt. $\mathfrak{J}$ ist jetzt, da die a_n reell sind, zugleich „Indikatordiagramm“ und „konjugiertes Diagramm“, $\mathfrak{J} = \overline{\mathfrak{J}}$.

[2]) Vgl. LSP, S. 598, Nr. 36.

indem die Reihenfolge der beiden (eigentlichen) Integrationen vertauscht wurde[1]). Aus dem Umstand, daß in jedem abgeschlossenen Bereich im Inneren der Konvergenzhalbebene (sagen wir $\Re s > \alpha$) des Integrals (4)

$$\Phi(s) = \lim_{g \to \infty} \Phi_g(s)$$

gleichmäßig gilt, kann geschlossen werden[2]), daß in einer gewissen Halbebene (nämlich für $\Re s > \alpha + \varkappa$)

$$\Phi^*(s) = \lim_{g \to \infty} \Phi_\beta^*(s) = \lim_{g \to \infty} \int_1^g \omega(u)\, F(\log u)\, u^{-s}\, du,$$

$$\Phi^*(s) = \int_1^\infty \omega(u)\, F(\log u)\, u^{-s}\, du \tag{17}$$

ist. Gemäß (3) und (9) ist nun

$$\omega(u)\, F(\log u) \leqq 0 \quad \text{für} \quad u > 1, \tag{18}$$

und Satz I ergibt, daß $\Phi^*(s)$ entweder eine ganze Funktion ist oder einen singulären Punkt mit maximaler Abszisse auf der reellen Achse besitzt. Es fragt sich nur, wie sich dies damit verträgt, was wir sonst über die Singularitäten von $\Phi^*(s)$ wissen.

4. Nach Voraussetzung sind alle im Streifen $\Theta - b < \Re s \leqq \Theta$ gelegenen singulären Stellen von $\Phi(s)$ Pole; abgesehen von den etwaigen reellen sind sie paarweise symmetrisch zur reellen Achse gelegen und haben keinen Häufungspunkt im Endlichen. Bezeichnen wir diese Pole mit $c_1, c_2, c_3, \ldots$ (Die c_n seien nach wachsendem Absolutwert der Ordinaten geordnet, diejenigen mit gleichem Absolutwert der Ordinate nach wachsenden Abszissen, und von zwei spiegelbildlich zur reellen Achse gelegenen gehe der Pol mit positivem Imaginärteil voraus.) Der Hauptteil von $\Phi(s)$ in dem Pol $c_m = c$ sei

$$\sum_{k=0}^{l} \frac{(-1)^k k!\, b_k}{(s-c)^{k+1}} = X_m(s) = X(s) \tag{19}$$

mit $b_l \neq 0$, so daß c ein Pol $(l+1)$-ter Ordnung $(l \geqq 0)$ und

$$\Psi(s) = \Phi(s) - X(s) \tag{20}$$

im Punkte c regulär ist. Gemäß (19) und der Definition (15) findet man[3])

$$b_0 f(s-c) + b_1 f'(s-c) + \cdots + b_l f^{(l)}(s-c) = X^*(s). \tag{21}$$

---

[1]) Ferner wurde die Formel (27) von LSP, S. 580 benutzt.

[2]) LSP, S. 600, Satz VI.

[3]) Vgl. LSP, S. 604, Beispiel d).

Hilfssatz[1]**).** *Die Funktion $X^*(s)$ ist regulär außerhalb des konvexen Bereiches $c + \mathfrak{J}$, hingegen ist jeder extreme Punkt von $c + \mathfrak{J}$, insbesondere $c + j$, ein singulärer Punkt von $\mathfrak{K}^*(s)$.*

Ich habe nur zu zeigen, daß $f(s - c)$ und $X^*(s)$ dieselben singulären Punkte im Endlichen besitzen. Es ist klar, daß wo $f(s - c)$ regulär dort auch $X^*(s)$ regulär ist. Faßt man (21) als eine lineare Differentialgleichung für $f(s - c)$ mit gegebener rechter Seite $X^*(s)$ auf, so zeigt entweder der Existenzsatz, oder noch besser die geläufige Formel, die $f(s - c)$ durch $X^*(s)$ und durch die Integrale der zugehörigen homogenen Gleichung (welche konstante Koeffizienten hat) ausdrückt, daß wo im Endlichen $X^*(s)$ regulär dort auch $f(s - c)$ regulär ist.

Die Funktion (20) ist in der Halbebene $\Re s > \Theta - b$ regulär, abgesehen von Polen, die in den Punkten $c_1, c_2, \ldots, c_{m-1}, c_{m+1}, \ldots$ liegen. Hieraus kann man schließen[2]):

Werden aus der Halbebene

$$\Re s > \Theta - b + \varkappa$$

die kongruenten und gleichgelegenen Bereiche

$$c_1 + \mathfrak{J}, c_2 + \mathfrak{J}, \ldots c_{m-1} + \mathfrak{J}, c_{m+1} + \mathfrak{J}, \ldots$$

herausgeschnitten, so ist in demjenigen übriggebliebenen Teil, der mit $+\infty$ zusammenhängt, $\Psi^*(s)$ regulär. Wir wollen hieraus auf die singulären Punkte von

$$\Phi^*(s) = \Psi^*(s) + X^*(s) \tag{22}$$

schließen. Alle drei Funktionen in (22) sind sicherlich in der Halbebene (16) regulär. Wir haben zweierlei zu untersuchen: Erstens, ob $\Phi^*(s)$ auch in dem reellen Grenzpunkt $\Theta + \varkappa$ der Halbebene (16) regulär ist (um Satz I und (18) anwenden zu können) und zweitens, ob $\Psi^*(s)$ im Punkte $c + j$ (der für $X^*(s)$, wie wir aus dem Hilfssatz wissen, singulär ist) regulär bleibt?

a) Im Falle $\gamma = 0$ ist die Behauptung (5a) inhaltslos. Wir haben uns nur mit dem Fall $\gamma > 0$ zu befassen, in welchem $\Theta + i\gamma = c$ ein Pol von $\Phi(s)$ ist, und zwar von allen denjenigen Polen von $\Phi(s)$, die auf der oberen Hälfte der Geraden $\Re s = \Theta$ liegen, der der reellen Achse nächstgelegene, und wir haben zu entscheiden, ob

$$\pi d < \gamma \tag{?}$$

möglich ist. (Es ist d durch (7) definiert.)

Wird (?) angenommen, so ist $\Phi^*(s)$ in $\Theta + \varkappa$ regulär. Denn der Punkt $\Theta + \varkappa$ ist von keinem Bereich $c_m + \mathfrak{J}$ mit $\Re c_m < \Theta$ überdeckt, unter der Annahme von (?)

[1]) Will man bloß (2) beweisen, so kommt nur der selbstverständliche Fall $l = 0$ dieses Hilfssatzes in Frage, da die Pole der Funktion (6) sämtlich einfach sind.

[2]) LSP, S. 598, Satz V.

auch vom Bereich $c + \mathfrak{J}$ nicht, also um so weniger von einem $c_n + \mathfrak{J}$, $n \neq m$ mit $\mathfrak{R}c_n = \Theta$, da doch diese c_n der reellen Achse nicht näher liegen als c.

Unter der Annahme (?) hat der Bereich $c + \mathfrak{J}$ mit $\bar{c} + \mathfrak{J}$ keinen Punkt gemeinsam. Wenn aber der Punkt $c + j$ von $\bar{c} + \mathfrak{J}$ nicht bedeckt wird, so wird er offenbar auch von keinem anderen Bereich $c_n + \mathfrak{J}$ mit $n \neq m$ bedeckt, ist also ein regulärer Punkt für $\Psi^*(s)$. Er ist aber singulärer Punkt von $X^*(s)$ und daher wegen (22) auch von $\Phi^*(s)$.

Zusammengefaßt ist, unter der Annahme (?), $\Phi^*(s)$ regulär in der Halbebene (16), regulär in dem reellen Grenzpunkt dieser Halbebene, aber nicht regulär auf der ganzen Grenzgeraden davon (nämlich in $c + j = \Theta + \varkappa + i(\gamma + \lambda)$ nicht). Dies ist, gemäß Satz I, mit (17), (18) unvereinbar, und es bleibt uns nichts anderes übrig, als (?) zu verneinen, also (5a) zuzugeben.

b) Im Falle, daß $\Phi(s)$ auf der Geraden $\mathfrak{R}s = \Theta$ regulär, also $\mathfrak{R}c_n < \Theta$ für $n = 1, 2, 3, \ldots$ ist, steht der Ausgangspunkt unserer ganzen Rechnung in Frage, nämlich ob

$$d \text{ endlich.} \tag{??}$$

In diesem Falle hat offenbar kein Bereich $c_n + \mathfrak{J}$ Punkte mit der Geraden $\mathfrak{R}s = \Theta + \varkappa$ gemeinsam, und so ist insbesondere der reelle Punkt davon, $\Theta + \varkappa$, für $\Phi^*(s)$ regulär.

[Es sei gegeben $\varepsilon, \varepsilon > 0$. Wir haben zu untersuchen, ob $\Phi^*(s)$ in der Halbebene

$$\mathfrak{R}s > \Theta + \varkappa - \varepsilon \tag{23}$$

regulär ist.

Man betrachte die Vereinigungsmenge $\mathfrak{B}$ sämtlicher Bereiche $c_n + \mathfrak{J}$ mit c_n in der oberen Halbebene und sämtlicher Punkte der reellen Achse links vom Punkt $\Theta - b$. Man bezeichne mit $\mathfrak{K}$ die kleinste konvexe Hülle von $\mathfrak{B}$. Die Punktmenge $\mathfrak{K}$ ist unendlich, jedoch abgeschlossen im Endlichen. Der Punkt $\Theta - b$ könnte ein extremer Punkt von $\mathfrak{K}$ sein. Ein von $\Theta - b$ verschiedener extremer Punkt von $\mathfrak{K}$ ist notwendigerweise extremer Punkt von irgendeinem Bereich $c_n + \mathfrak{J}$. Es besitzt $\mathfrak{K}$ eine horizontale Stützgerade. Denken wir uns eine bewegliche Gerade $\mathfrak{g}$, welche — ausgehend von der horizontalen Anfangslage und sich stets im positiven Sinne drehend — die Menge aller Stützgeraden von $\mathfrak{K}$ durchläuft; die Grenzlage von $\mathfrak{g}$ ist die vertikale Gerade $\mathfrak{R}s = \Theta + \varkappa$, welche jedoch keine Stützgerade von $\mathfrak{K}$ ist, da sie keinen Punkt von $\mathfrak{B}$ enthält. Die sich drehende Gerade $\mathfrak{g}$ enthält in jeder Lage (abgesehen vielleicht von einer Umgebung der Anfangslage) einen extremen Punkt von irgendeinem der Bereiche $c_n + \mathfrak{J}$: Ist dieser extreme Punkt ein singulärer Punkt von $\Phi^*(s)$? Vielleicht nicht, wenn er mehreren Bereichen $c_n + \mathfrak{J}$ zugleich angehört.

Das eben Gesagte deutet einen anschaulichen Beweis an, dessen springenden Punkt wir jedoch genauer ausführen müssen. Gemäß der Definition von Θ gibt es im gegen wärtigen Falle einen *ersten* Pol c_k von $\Phi(s)$ mit

$$\Re c_k > \Theta - \varepsilon. \tag{24}$$

Falls ε klein genug ist, ist auch

$$\Im c_k > 2\lambda. \tag{25}$$

Der Punkt $c_k + \varkappa$ gehört sowohl der Halbebene (23) als auch $\mathfrak{K}$ an. Man betrachte die Verbindungsgerade der beiden Punkte $\Theta + \varkappa - \varepsilon$ und $c_k + \varkappa$, die dieser Verbindungsgeraden parallele Stützgerade $\mathfrak{g}_0$ von $\mathfrak{K}$, und je nach dem vorliegenden Fall nenne man s_0 entweder den einzigen auf $\mathfrak{g}_0$ liegenden extremen Punkt von $\mathfrak{K}$ oder denjenigen der beiden daraufliegenden extremen Punkte, dessen Ordinate größer ist. Nun ist s_0 auch extremer Punkt eines Bereiches $c_m + \mathfrak{J}$ und liegt in *nur einem* Bereich $c_n + \mathfrak{J}$, selbst wenn wir c_n alle Pole von $\Phi(s)$ durchlaufen lassen, vgl. (25), und nicht nur diejenigen oberhalb der reellen Achse wie bisher im gegenwärtigen Unterabschnitt *b*). Somit ist s_0 ein in der Halbebene (23) gelegener singulärer Punkt von $\Phi^*(s)$.]

Zusammengefaßt: Es hat $\Phi^*(s)$ singuläre Punkte in beliebiger Nähe der Geraden $\Re s = \Theta + \varkappa$, keine darauf oder rechts davon. Dies ist unverträglich mit Satz I, (17) und (18); daher ist (??) zu verneinen und (5b) zu bejahen.

5. Es folgen noch einige Bemerkungen zum Satz II, die ich aber nicht im einzelnen ausführen will.

a) In der Ungleichung (5a) kann für geeignete Funktionen $\Phi(s)$ das Gleichheitszeichen erreicht werden. Ein einfaches Beispiel ist

$$\Gamma(s - i\gamma)\zeta(s - i\gamma) + \Gamma(s + i\gamma)\zeta(s + i\gamma) = \left(\int_0^1 + \int_1^\infty\right) \frac{2\cos(\gamma \log u)}{u\,(e^{1/u} - 1)}\, u^{-s}\, du$$

mit $\gamma > 0$; das Integral $\int_0^1$ stellt eine ganze Funktion dar. Ebenfalls leicht sind Beispiele zu bilden, für welche das Gleichheitszeichen in (5a) nicht erreicht wird.

b) Man kann aus Satz II außer (2) noch andere Beziehungen zwischen dem Verhalten von zahlentheoretischen Funktionen und der Konfiguration der Nullstellen der ζ-Funktion oder verwandter Funktionen gewinnen, indem man Satz II, ganz wie auf (6), noch auf andere ähnlich gebaute Integrale anwendet. Ein naheliegendes Beispiel ist

$$\frac{1}{s\zeta(s)} = \int_1^\infty M(u)\, u^{-1-s}\, du.$$

Um jedoch die Anzahl der Zeichenwechsel des Restgliedes in anderen Formen des Primzahlsatzes abzuschätzen, scheint eine unmittelbare Anwendung des Satzes II noch nicht zu genügen.

c) Auf mehr oder weniger naheliegende Erweiterungen und Analogien des Satzes II soll an dieser Stelle nicht eingegangen werden, abgesehen von einer Ausnahme: Es sei hier der einfachste analoge Satz, nämlich der über Potenzreihen, mit Hinzufügung einer zweiten Ungleichung, ohne Beweis ausgesprochen werden:

Es soll die Potenzreihe

$$c_0 + c_1 z + c_2 z^2 + \cdots + c_n z^n + \cdots$$

mit reellen Koeffizienten $c_0, c_1, c_2, \ldots$ *im Einheitskreis konvergieren, auf dessen Rand außer Polen keine Singularitäten besitzen, und zwar sollen* $e^{i\gamma_1}, e^{i\gamma_2}, \ldots, e^{i\gamma_l}$ *alle auf der abgeschlossenen oberen Halbperipherie liegenden Pole sein,*

$$0 \leqq \gamma_1 < \gamma_2 < \cdots < \gamma_l \leqq \pi.$$

Bezeichnet V_n *die Anzahl der Zeichenwechsel unter den* $n + 1$ *ersten Koeffizienten* $c_0, c_1, c_2, \ldots, c_n$, *so ist*

$$\frac{\gamma_1}{\pi} \leqq \overline{\lim_{n\to\infty}} \frac{V_n}{n} \leqq \frac{\gamma_l}{\pi}.$$

Der Fall $l = 1$, also $\gamma_l = \gamma_1$, ergibt die eine Hälfte eines von J. König herrührenden Satzes[1]), der mir die Anregung zur vorliegenden Untersuchung gab.

[**Zusatz.** Die Kenntnis wohlbekannter Tatsachen betreffend die Funktion $\zeta(s)$ erlaubt es, aus (2) zu folgern, daß auf alle Fälle, unabhängig von jeglicher Vermutung, folgendes gilt:

$$\overline{\lim_{n\to\infty}} \frac{W(n)}{\log n} \geqq \frac{14{,}134\ldots}{3{,}141\ldots}; \tag{2'}$$

es wurde der numerische Wert der ersten nichttrivialen Nullstelle von $\zeta(s)$ berücksichtigt. Meines Wissens ist eine bessere untere Abschätzung der linken Seite in der inzwischen erschienenen Literatur nicht enthalten, obwohl sie in anderer Hinsicht weit über (2′) hinausgeht.[2])

Ich erwähne zwei spätere Veröffentlichungen, welche an die hier wiederabgedruckte Arbeit anknüpfen.[3])

Satz II kann auf manche zahlentheoretische Funktionen angewendet werden.[4])]

[1]) J. König, Math. Ann. **9** (1876), 530—540, auch G. Pólya und G. Szegő, Aufgaben und Lehrsätze aus der Analysis, Bd. 1, Springer, Berlin 1925, S. 131, Nr. 245.

[2]) [Hier nenne ich neben der oben erwähnten Abhandlung von A. E. Ingham auch die neueren Arbeiten u. a. von S. Skewes, S. Knapowski, P. Turán, R. Sherman Lehman und E. Großwald. Für Zitate auf einige dieser Arbeiten vgl. S. Knapowski und P. Turán, Comparative prime-number theory I, Acta Math. Hung. XIII, Nr. 3—4 (1962), 299—314.]

[3]) [G. Pólya, Proc. London Math. Soc., Ser. 2, **33** (1932), 85—101; G. Pólya und A. Bloch, daselbst 102—114.]

[4]) [Vgl. z. B. John Steinig, Comptes Rendus, Paris, **263** (1966), Ser. A, 905—906.]

J. Popken in Amsterdam

A MEASURE FOR THE DIFFERENTIAL-TRANSCENDENCE OF THE ZETA-FUNCTION OF RIEMANN

1. In recent years "measures of transcendence" for several transcendental numbers have been found. The main idea behind all this work is that important theorems like "e and π are transcendental numbers" contain merely negative statements, which properly should be replaced by positive ones.

In this paper we treat a similar question in a different field of research. It is well-known that certain functions, such as the gamma-function (Hölder) and the zeta-function (Hilbert), are differential-transcendental; i.e. they do not satisfy any algebraic differential equation. Therefore the problem arises to find "measures" for the differential-transcendence of these functions. Moreover such a measure should be effectively computable.

From the start it is clear that such a measure could be defined in quite different ways. Here I treat only the zeta-function and closely related functions and the measure which I introduce is defined in a rather formal way. However this must be seen as a first step and many problems in this field are still open.

In the winter-semester of 1930–1931 my late friend Jurjen Koksma and I myself came to Göttingen. It was then that Landau asked us to determine a measure of transcendence for e^{π}, which led to the joint paper [1]. To my best knowledge it was also Landau who coined the expression "Transzendenzmaß". Therefore I think it appropriate to dedicate this particular note to the great mathematician we honour in this book.

The theory developed here is quite elementary but it has the following interesting aspect. The zeta-function is the deep source for our prime number theory; here however the tables are turned in some sense: the elementary Chebyshev theory of primes gives us some information about the zeta-function as we shall see.

2. Consider the ring R of formal Dirichlet series

$$D = D(s) = \sum_{n=1}^{\infty} a_n n^{-s} \tag{1}$$

with complex coefficients a_n. The operations in R are formal addition and Dirichlet multiplication. R becomes a differential ring if for the derivative of the series in (1) is taken:

$$D' = \sum_{n=1}^{\infty} - a_n \log n \, n^{-s}.$$

Moreover we get a valuation $|D|$ in R if we put $|D| = 0$ if all coefficients a_n in (1) are zero and if we put in all other cases $|D| = N^{-1}$, where a_N is the first coefficient in (1) different from zero. Evidently

$$|D| \geqq 0, \quad |D| = 0 \quad \text{iff} \quad D = 0,$$

$$|D_1 \cdot D_2| = |D_1| \cdot |D_2|, \quad |D_1 \pm D_2| \leqq \max(|D_1|, |D_2|),$$

so that this valuation is non-Archimedean. Important are also the following relations which hold for any D in R:

$$|D| \leqq 1, \quad |D'| \leqq |D|, \quad |D'| = |D| \quad \text{if} \quad |D| < 1.$$

Let $f(x_0, x_1, \ldots, x_r)$ be a polynomial with complex coefficients, let $\zeta(s)$ be the zeta-function. Obviously the function

$$f(\zeta(s), \zeta'(s), \ldots, \zeta^{(r)}(s)) \tag{2}$$

is holomorphic for $\operatorname{Re} s > 1$ and is represented there by an absolutely convergent Dirichlet series

$$\sum_{n=1}^{\infty} F(n) \, n^{-s}.$$

If in R we denote the series $\sum_{n=1}^{\infty} n^{-s}$ by ζ, then the series for (2) can also be written

$$f(\zeta, \zeta', \ldots, \zeta^{(r)}). \tag{3}$$

The assertion that $\zeta(s)$ does not satisfy an algebraic differential equation is equivalent to the statement that $\zeta(s)$ does not satisfy such an equation in which the variable s does not occur explicitly (compare Ostrowski [2], p. 246) and this, in its turn, means that for any non-trivial polynomial $f(x_0, x_1, \ldots, x_r)$ not all coefficients $F(n)$ of the series (3) vanish.

Our next theorem gives a refinement of this assertion.

Theorem 1. *Let $f(x_0, x_1, \ldots, x_r) \not\equiv 0$ be a polynomial with complex coefficients of total degree g. Then for $g = 1$*

$$|f(\zeta, \zeta', \ldots, \zeta^{(r)})| \geqq (r + 2)^{-1} \tag{4}$$

and for $g \geqq 2$

$$|f(\zeta, \zeta', \ldots, \zeta^{(r)})| \geqq (cg \log g \, (r + 2) \log^2(r + 2))^{-g}. \tag{5}$$

Here c is a positive absolute constant, which can be computed effectively.

Remarks

1. Observe that the right-hand side of (5) depends only on g and r and not on the values of the coefficients in the polynomial f.

2. The assertion of theorem 1 is not yet completely positive. For it states only that for any polynomial $f(x_0, x_1, \ldots, x_r)$ in a certain range a coefficient $F(n)$ of the Dirichlet series $f(\zeta, \zeta', \ldots, \zeta^{(r)})$ can be found which does not vanish. But the statement that a certain value is not zero is still a negative one. A full solution of our problem would certainly be more complex and would also involve the coefficients of f.

Before we give a proof of theorem 1 we first derive from it the following corollary.

Theorem 2. *Let a Dirichlet series*

$$D(s) = \sum_{n=1}^{\infty} a_n n^{-s}$$

in its half plane of convergence satisfy an algebraic differential equation with constant coefficients of order r and of total degree g. Then for g $= 1$

$$|\zeta - D| \geqq (r + 2)^{-1}$$

and for g $\geqq 2$

$$|\zeta - D| \geqq (cg \log g (r + 2) \log^2 (r + 2))^{-g}.$$

Here c is the constant of theorem 1.

Proof. Since $D(s)$ has also a half plane of absolute convergence the formal series D satisfies an equation over R

$$f(D, D', \ldots, D^{(r)}) = 0,$$

where $f(x_0, x_1, \ldots, x_r)$ is a polynomial of total degree g. By Taylor's formula for polynomials we have therefore

$$\begin{aligned} f(\zeta, \zeta', \ldots, \zeta^{(r)}) = & \sum_{\varrho=0}^{r} (\zeta^{(\varrho)} - D^{(\varrho)}) f_\varrho(D, D', \ldots, D^{(r)}) \\ & + \sum_{0 \leqq \varrho \leqq \sigma \leqq r} (\zeta^{(\varrho)} - D^{(\varrho)})(\zeta^{(\sigma)} - D^{(\sigma)}) f_{\varrho,\sigma}(D, D', \ldots, D^{(r)}) \\ & + \cdots \text{(finitely many terms)}, \end{aligned} \tag{6}$$

where $f, f_{\varrho,\sigma}$, a.s.o. denote the partial derivatives of f, apart from factorial coefficients. Now, for $\varrho, \sigma = 0, 1, \ldots, r$,

$$0 \leqq |\zeta^{(\varrho)} - D^{(\varrho)}| \leqq |\zeta - D| \leqq 1,$$

$$|f_\varrho(D, D', \ldots, D^{(r)})| \leqq 1, \quad |f_{\varrho,\sigma}(D, D', \ldots, D^{(r)})| \leqq 1, \quad \text{a.s.o.}$$

Hence it follows from (6) by the non-Archimedean character of the valuation $|D|$

$$|f(\zeta,\zeta',\ldots,\zeta^{(r)})| \leqq |\zeta - D|$$

and thus theorem 2 follows immediately from theorem 1.

The method we use in the following proof of theorem 1 can be extended to a wide class of Dirichlet series. In this way we are led to the next theorem.

Theorem 3. *Let the function $D(s)$ in some half plane be represented by a Dirichlet series*

$$D(s) = \sum_{n=1}^{\infty} a_n n^{-s},$$

such that $a_p \neq 0$ for every prime p. Let $f(x_0, x_1, \ldots, x_r)$ be a non-trivial polynomial with complex coefficients of total degree $g \geqq 1$. Then there exists a positive absolute constant c_1 such that

$$|f(D, D', \ldots, D^{(r)})| \geqq (c_1 g \log g\, (r+2) \log^2(r+2))^{-g}.$$

(For $g = 1$ the factor $\log g$ on the right-hand side should be omitted.)

Since the proof of theorem 3 differs only in minor points from that of theorem 1 we give only a proof of the latter theorem.

3. For the proof of theorem 1 we need two lemmas. The first one is a special case of a theorem of the author in an earlier paper ([3], Theorem 1, p. 159–160). For the convenience of the reader we give here a proof using Dirichlet series, very much akin to Ostrowski's proof of the differential transcendence of the zeta-function (see [2]).

Lemma 1. *Let $f(x_0, x_1, \ldots, x_r)$ be a polynomial with complex coefficients and let $f_\varrho = \dfrac{\partial f}{\partial x_\varrho}$ denote its first order partial derivatives ($\varrho = 0, 1, \ldots, r$). Put*

$$f(\zeta,\zeta',\ldots,\zeta^{(r)}) = \sum_{n=1}^{\infty} F(n)\, n^{-s},$$

$$f_\varrho(\zeta,\zeta',\ldots,\zeta^{(r)}) = \sum_{n=1}^{\infty} F_\varrho(n)\, n^{-s}.$$

Then for any positive integer n and any prime p, such that $p \nmid n$, we have

$$F(pn) = \sum_{\varrho=0}^{r} \sum_{d|n} (-\log pd)^\varrho\, F_\varrho(n/d).$$

Proof. Let $n = p_1^{\nu_1} p_2^{\nu_2} \cdots p_t^{\nu_t}$ be the decomposition of n into prime factors and let K denote the multiplicative semigroup of all positive integers generated by $p_1, p_2, \ldots, p_t$.

Let $\bar{K}$ be the complementary set of positive integers, so that $p \in \bar{K}$. Now put

$$D(s) = \sum_{m \in K} m^{-s}, \quad R(s) = \sum_{m \in \bar{K}} m^{-s},$$

so that

$$D = \zeta - R, \quad D^{(\varrho)} = \zeta^{(\varrho)} - R^{(\varrho)} \quad (\varrho = 0, 1, \ldots, r).$$

It follows by Taylor's formula

$$f(D, D', \ldots, D^{(r)}) = f(\zeta, \zeta', \ldots, \zeta^{(r)}) - \sum_{\varrho=0}^{r} R^{(\varrho)} f_\varrho(\zeta, \zeta', \ldots, \zeta^{(r)})$$
$$+ \sum_{0 \leqq \varrho \leqq \sigma \leqq r} R^{(\varrho)} R^{(\sigma)} f_{\varrho,\sigma}(\zeta, \zeta', \ldots, \zeta^{(r)}) - \cdots,$$

where, just as in the case of (6), the $f_{\varrho,\sigma}$ denote the second order derivatives apart from factorial coefficients.

Consider the coefficients of $(pn)^{-s}$ on both sides of this identity. At the left such coefficients cannot occur, since the series for $D, D', \ldots, D^{(r)}$ have only terms with m^{-s} such that $m \in K$, while $pn \in \bar{K}$. Therefore the coefficients of $(pn)^{-s}$ on the right must cancel. However terms with $(pn)^{-s}$ occur neither in the sums with partial derivatives of order $\geqq 2$, because for example

$$R^{(\varrho)} R^{(\sigma)} f_{\varrho,\sigma}(\zeta, \zeta', \ldots, \zeta^{(r)})$$

gives only rise to terms with m^{-s} if m contains at least *two* prime factors in $\bar{K}$, while pn contains only one.

Therefore terms with $(pn)^{-s}$ occur only in

a) $f(\zeta, \zeta', \ldots, \zeta^{(r)})$,

b) $-\sum_{\varrho=0}^{r} R^{(\varrho)} f_\varrho(\zeta, \zeta', \ldots, \zeta^{(r)})$.

The coefficient of $(pn)^{-s}$ in a) is $F(pn)$. Since

$$R^{(\varrho)} f_\varrho(\zeta, \zeta', \ldots, \zeta^{(r)}) = \sum_{m \in \bar{K}} (-\log m)^\varrho m^{-s} \cdot \sum_{m=1}^{\infty} F_\varrho(m) m^{-s}$$

the coefficient of $(pn)^{-s}$ in b) is easily seen to be

$$-\sum_{\varrho=0}^{r} \sum_{d \mid n} (-\log pd)^\varrho F_\varrho(n/d).$$

Since these coefficients in a) and b) must cancel the lemma follows.

Lemma 2. *Let $r \geqq 0$ be integral and let $k \geqq 2$. For any given positive integer $n \leqq k$ there exist at least $r + 1$ primes p, such that*

$$p \nmid n, \quad p \leqq \lambda \log k,$$

where

$$\lambda = c_2(r + 2)\log(r + 2)$$

and where $c_2 > e$ is a positive absolute constant.

Remark. In the following c, c_1, c_2, a.s.o. will always denote positive absolute constants ("Welt-Konstanten"), which are moreover effectively computable.

Proof. Let $p_1 = 2$, $p_2 = 3$, $p_3 = 5$, ... denote the natural sequence of primes, $\pi(x)$ the number of primes $\leqq x$.

1. Let $n = 1$ or let n be a power of a prime. Since the Chebyshev inequality

$$p_{r+2} < c_3(r + 2)\log(r + 2)$$

holds, the assertion of our lemma in this case is true if we take c_2 so large that

$$c_2 \log k \geqq c_2 \log 2 \geqq c_3.$$

2. Let us therefore suppose that n contains at least two different prime factors, so that in the decomposition of n

$$n = q_1^{\nu_1} q_2^{\nu_2} \cdots q_t^{\nu_t}$$

we have $t \geqq 2$.

Evidently we have only to show that the coefficient c_2 in $\lambda = c_2(r + 2)\log(r + 2)$ can be taken so large that

$$\pi(\lambda \log k) \geqq t + r + 1. \tag{7}$$

Define

$$n' = p_1 p_2 \cdots p_t, \quad \text{so that} \quad k \geqq n \geqq n' \geqq 6, \quad p_t \geqq 3,$$

then

$$t = \pi(p_t) < c_4 p_t / \log p_t$$

and moreover

$$\log n' = \sum_{p \leqq p_t} \log p = \vartheta(p_t),$$

so that

$$c_5 p_t < \log n' < c_6 p_t.$$

It follows for $\log n' > c_6$

$$t < c_4 p_t / \log p_t < c_4 c_5^{-1} \log n' / \log(c_6^{-1} \log n');$$

hence for sufficiently large c_7 and c_8

$$t + r < c_7 \frac{\log n'}{\log_2 n'} + r < c_8(2 + r)\frac{\log k}{\log_2 k}. \tag{8}$$

We shall take for the c_2 of our lemma a constant $\geqq e^2$ to be specified more precisely below. Now

$$\begin{aligned}\pi(\lambda \log k) &= \pi(c_2(r+2)\log(r+2)\log k)\\ &> c_9 \frac{c_2(r+2)\log(r+2)\log k}{\log c_2 + 2\log(r+2) + \log_2 k}\\ &> c_9 \frac{c_2(r+2)\log(r+2)\log k}{\log c_2 \cdot 3\log(r+2)\cdot 4\log_2 k} \quad \text{(observe } k \geqq 6\text{)},\end{aligned}$$

so that for a sufficiently small c_{10}, independent of the value to be taken for c_2,

$$\pi(\lambda \log k) > c_{10} \frac{c_2}{\log c_2} (r+2) \frac{\log k}{\log_2 k}. \tag{9}$$

If we take c_2 so large that

$$c_{10}c_2/\log c_2 > c_8, \quad c_2 > e^2,$$

then it follows from (8) and (9)

$$\pi(\lambda \log k) > t + r,$$

so that (7) is true.

4. Proof of theorem 1.

a) If a polynomial f is of degree zero, then

$$|f(\zeta, \zeta', \ldots, \zeta^{(r)})| = 1.$$

b) Let f be of degree 1, so that we can write

$$f(x_0, x_1, \ldots, x_r) = a + \sum_{\varrho=0}^{r} a_\varrho x_\varrho,$$

where the coefficients a_ϱ do not vanish simultaneously. Then for $n \geqq 2$ the coefficients $F(n)$ in the Dirichlet series $f(\zeta, \zeta', \ldots, \zeta^{(r)})$ have the form

$$F(n) = \sum_{\varrho=0}^{r} a_\varrho(-\log n)^\varrho.$$

Since the polynomial $a_0 + a_1 t + \cdots + a_r t^r$ has at most r zeros, at least one of the $r+1$ coefficients $F(2), F(3), \ldots, F(r+2)$ differs from zero; i.e.

$$|f(\zeta, \zeta', \ldots, \zeta^{(r)})| \geqq (r+2)^{-1},$$

which proves the first part (4) of our theorem.

c) The second part is shown by mathematical induction. To this end we replace the inequality (5), we have to prove, by the stronger inequality

$$|f(\zeta, \zeta', \ldots, \zeta^{(r)})| \geqq (g \log g \cdot \lambda \log \lambda)^{-g}, \tag{10}$$

where $\lambda = c_2(r+2)\log(r+2)$ is given in lemma 2. Moreover we can take c_2 in lemma 2 so large that

$$\frac{8}{27}\left(\frac{2}{\log c_2} + \frac{2}{\log 2}\right) < 1, \quad \frac{1}{e}\left(\frac{2}{\log c_2} + \frac{2}{\log 3}\right) < 1. \tag{11}$$

Let f be a polynomial of degree $g + 1$ ($g \geqq 1$) and suppose that the assertion already has been proved for polynomials of degree $\leqq g$. Then, unless $f_\varrho = \dfrac{\partial f}{\partial x_\varrho}$ vanishes identically, we have

$$|f_\varrho(\zeta, \zeta', \ldots, \zeta^{(r)})| \geqq k^{-1},$$

where

$$k = \begin{cases} r + 2 & \text{if} \quad g = 1, \\ (g \log g \cdot \lambda \log \lambda)^g & \text{if} \quad g \geqq 2. \end{cases}$$

This implies that not all the following $[k]$ vectors can be null-vectors:

$$\{F_\varrho(1)\}_{\varrho=0}^r, \{F_\varrho(2)\}_{\varrho=0}^r, \ldots, \{F_\varrho([k])\}_{\varrho=0}^r.$$

Let $\{F_\varrho(n)\}_{\varrho=0}^r$ be the first non-vanishing vector in this sequence ($1 \leqq n \leqq k$), so that $F_\varrho(d) = 0$ for $d = 1, 2, \ldots, n-1$; $\varrho = 0, 1, \ldots, r$. This fixes n. Let p denote a prime such that $p \nmid n$, to be specified more precisely below. Then by lemma 1

$$F(pn) = \sum_{\varrho=0}^{r} \sum_{d|n} (-\log pd)^\varrho F_\varrho(n/d)$$

and because of the minimal condition for n

$$F(pn) = \sum_{\varrho=0}^{r} (-1)^\varrho F_\varrho(n) (\log p)^\varrho = \sum_{\varrho=0}^{r} b_\varrho (\log p)^\varrho, \tag{12}$$

with a least one b_ϱ different from zero.

By lemma 2 there exist $r + 1$ different primes p, such that

$$p \nmid n, \quad p \leqq \lambda \log k.$$

It follows then from (12) that the corresponding $r + 1$ values $F(pn)$ cannot all be zero.

Therefore there exists a positive integer N of the form $N = pn$, such that

$$F(N) \neq 0, \quad N \leqq \lambda k \log k$$

so that by the definition of our valuation

$$|f(\zeta, \zeta', \ldots, \zeta^{(r)})| \geqq (\lambda k \log k)^{-1}.$$

Our proof is finished as soon as we can show that

$$\lambda k \log k \leqq ((g+1)\log(g+1) \cdot \lambda \log \lambda)^{g+1} \tag{13}$$

(compare (10)).

This inequality is trivial for $g = 1$, since then $\lambda k \log k = \lambda(r+2)\log(r+2) < \lambda^2$ and $\lambda \geqq c_2 \log 2 > e$, so $\log \lambda > 1$.

Therefore let $g \geqq 2$, so that

$$k = (g \log g \cdot \lambda \log \lambda)^g,$$

hence

$$\lambda k \log k$$
$$= g^{g+1}(\log g)^{g+1}\lambda^{g+1}(\log \lambda)^{g+1}\left(\frac{1}{\log \lambda} + \frac{\log_2 g}{\log g \cdot \log \lambda} + \frac{1}{\log g} + \frac{\log_2 \lambda}{\log g \cdot \log \lambda}\right)$$
$$< ((g+1)\log(g+1) \cdot \lambda \log \lambda)^{g+1}\left(\frac{g}{g+1}\right)^{g+1}\left(\frac{2}{\log c_2} + \frac{2}{\log g}\right).$$

Further, by (11),

$$\left(\frac{g}{g+1}\right)^{g+1}\left(\frac{2}{\log c_2} + \frac{2}{\log g}\right) < \begin{cases} \frac{8}{27}\left(\frac{2}{\log c_2} + \frac{2}{\log 2}\right) < 1 & \text{if } g = 2, \\ \frac{1}{e}\left(\frac{2}{\log c_2} + \frac{2}{\log 3}\right) < 1 & \text{if } g \geqq 3. \end{cases}$$

It follows

$$\lambda k \log k < ((g+1)\log(g+1) \cdot \lambda \log \lambda)^{g+1},$$

so that (13) is true.

References

[1] J. F. Koksma und J. Popken, Zur Transzendenz von e^π, J. reine angew. Math. **168** (1932), 211–230.

[2] A. Ostrowski, Über Dirichletsche Reihen und algebraische Differentialgleichungen, Math. Z. **8**, (1920), 241–298.

[3] J. Popken, Algebraic dependence of arithmetic functions, Proc. Kon. Ned. Akad. Wet. **A65** (= Indag. Math. **24**), (1962), 155–168.

H. Rademacher in New York

COMMENTS ON EULER'S "DE MIRABILIBUS PROPRIETATIBUS NUMERORUM PENTAGONALIUM"

1. In the above mentioned article [1] Euler discusses consequences of his famous identity

$$\prod_{m=1}^{\infty} (1 - x^m) = \sum_{n=-\infty}^{+\infty} (-1)^n x^{\omega_n} = f(x), \tag{1.1}$$

where

$$\omega_n = \frac{n(3n - 1)}{2}$$

are the so-called pentagonal numbers. Some of these consequences, in particular the recursion formula for $p(n)$, the number of unrestricted partitions of n, and the similar recursion formula for $\sigma(n)$, the sum of divisors of n, he had treated extensively in earlier papers [2], [3].

The main part of his memoir, however, is devoted to certain divergent series, connected with the pentagonal numbers. Euler makes the following statements

$$-1^{\lambda} - 2^{\lambda} + 5^{\lambda} + 7^{\lambda} - 12^{\lambda} - 15^{\lambda} + \cdots = 0 \tag{1.2}$$

or in our notation (1.2)

$$\sum_{n=-\infty}^{+\infty} (-1)^n \omega_n^{\lambda} = 0, \tag{1.3}$$

where λ is a positive integer. He states also the generalization

$$\sum_{n} (-1)^n \alpha^{\omega_n} \omega_n^{\lambda} = 0 \tag{1.4}$$

for α a root of unity.

In another set of equations the two "halves" $n > 0$ and $n < 0$ of the sequences of pentagonal numbers are involved. Euler states the equations

$$\begin{aligned} s_1 &= \sum_{n=1}^{\infty} (-1)^n \omega_n = \frac{1}{8}, \\ t_1 &= \sum_{n=1}^{\infty} (-1)^n \omega_{-n} = -\frac{1}{8}, \end{aligned} \tag{1.5}$$

$$s_2 = \sum_{n=1}^{\infty} (-1)^n \omega_n^2 = -\frac{3}{16},$$
$$t_2 = \sum_{n=1}^{\infty} (-1)^n \omega_{-n}^2 = \frac{3}{16} \tag{1.6}$$

and he refers to the results

$$s_1 + t_1 = s_2 + t_2 = 0$$

as corroboration of (1.3). The formulae (1.5) and (1.6) are obtained by Euler through the process of summation named after him,

For (1.3) and (1.4) Euler gives a heuristic argument that treats the function (1.1) as a polynomial of a very high degree, which has multiple zeros at the roots of unity and must therefore vanish together with its derivatives at these roots of unity. The inadequacy of any argument of this sort is obvious, since the power series in (1.1) does not converge on the unit circle and has there a natural boundary, according to Fabry's theorem.

2. We have therefore, in order to make the statements (1.3) and (1.4) acceptable to look for a summation method and compare such a method with the Euler summation applied in (1.5) and (1.6).

Now the product (1.1) appears as the 24th root of the "discriminant" in Dedekind's function

$$\eta(\tau) = e^{\pi i\tau/12} \prod_{n=1}^{\infty} (1 - e^{2\pi i n\tau}), \quad \operatorname{Im}\tau > 0. \tag{2.1}$$

For this function the transformation formula

$$\eta\left(\frac{h+iz}{k}\right) = \varrho_{hk} \frac{1}{\sqrt{z}} \eta\left(\frac{h' + \frac{i}{z}}{k}\right) \tag{2.2}$$

is known [4], where $\frac{h}{k}$ is a reduced fraction, h' is chosen so that

$$hh' \equiv -1 \pmod{k},$$

and where ϱ_{hk} is a certain much discussed 24th root of unity, where finally

$$\operatorname{Re} z > 0, \quad |\arg \sqrt{z}| < \frac{\pi}{4}. \tag{2.3}$$

We have thus, after (1.1) and (2.1)

$$\eta(\tau) = e^{\pi i\tau/12} \sum_{-\infty}^{\infty} (-1)^n e^{2\pi i \omega_n \tau}$$

and therefore in view of (2.2)

$$\sum_n (-1)^n \alpha^{\omega_n} e^{-2\pi\omega_n z/k} \tag{2.4}$$

$$= \varrho_{hk} \left(\frac{\alpha_1}{\alpha}\right)^{1/24} \frac{1}{\sqrt{z}} e^{\pi z/12k} \sum_n (-1)^n \alpha_1^{\omega_n} e^{-2\pi(\omega_n + 1/24)/kz}$$

where we have set

$$\alpha = e^{z\pi ih/k}, \quad \alpha_1 = e^{2\pi ih'/k}.$$

We define now by the Abel method of summation

$$S_0 = \sum_n (-1)^n \alpha^{\omega_n} = \lim_{z\to+0} \sum (-1)^n \alpha^{\omega_n} e^{-2\pi\omega_n z/k}$$

and have

$$S_0 = \varrho_{hk} \left(\frac{\alpha_1}{\alpha}\right)^{1/24} \lim_{z\to+0} \frac{1}{\sqrt{z}} c^{\pi z/12k} \sum_n (-1)^n \alpha_1^{\omega_n} e^{-2\pi(\omega_n + 1/24)/kz}$$

$$= \lim_{z\to+0} O\left(\frac{1}{\sqrt{z}} e^{-(2\pi)/(kz)(1/24)}\right) = 0,$$

which agrees with (1.3) and (1.4) for the case $\lambda = 0$.

Similarly we define

$$S_\lambda = \sum_n (-1)^n \alpha^{\omega_n} \omega_n^\lambda = \lim_{z\to+0} \sum_n (-1)^n \alpha^{\omega_n} \omega_n^\lambda e^{-2\pi\omega_n z/k}$$

$$= \frac{(-k)^\lambda}{(2\pi)^\lambda} \lim_{z\to+0} \left(\frac{d}{dz}\right)^\lambda \sum_n (-1)^n \alpha^{\omega_n} e^{-2\pi\omega_n z/k}.$$

This gives us because of (2.4)

$$S_\lambda = \varrho_{hk} \left(\frac{\alpha_1}{\alpha}\right)^{1/24} \frac{(-k)^\lambda}{(2\pi)^\lambda} \lim_{z\to+0} \left(\frac{d}{dz}\right)^\lambda \left\{\frac{e^{\pi z/12k}}{z} \sum_n (-1)^n \alpha_1^{\omega_n} e^{-2\pi(\omega_n + 1/24)/kz}\right\}$$

$$= \lim_{z\to+0} O\left(\frac{1}{z^{2\lambda+1/2}} e^{-2\pi/24kz}\right) = 0,$$

which gives (1.3) and (1.4) for $\lambda > 0$.

We have applied here the Abel summation by means of a lacunary power series in which only the powers x^{ω_n} appears. Such a method would be designated in Hardy's notation as $A(\omega_n)$.

3. On the other hand, for the functions

$$f_1(x) = \sum_{n=1}^{\infty} (-1)^n x^{\omega_n}, \tag{3.1}$$

$$f_2(x) = \sum_{n=1}^{\infty} (-1)^n x^{\omega_{-n}} \tag{3.2}$$

(the two "halves" of $f(x)$) we do not know anything about their behavior near the unit circle, which is again the natural boundary for them.

Euler, however, remarks that the sequences

$$\omega_n^\lambda \quad \text{and} \quad \omega_{-n}^\lambda, \quad n = 1, 2, 3, \ldots$$

are arithmetical progressions of the order 2λ and as such very suitable to summation by his transformation (see e.g. [5], [6])

$$\begin{aligned}\sum_{n=0}^{\infty}(-1)^n a_n &= \sum_{0}^{\infty}\left(\frac{1}{2}\right)^{l+1} \Delta^l a_0 \\ &= \sum_{l=0}^{\infty}\left(\frac{1}{2}\right)^{l+1} \sum_{n=0}^{l} (-1)^n \binom{l}{n} a_n. \end{aligned} \tag{3.3}$$

In the cases (1.5) he has, for $n \geqq 0$,

ω_n:	0		1		5		12		22		35
$\Delta\omega_n$:		-1		-4		-7		-10		-13	
$\Delta^2\omega_n$:			3		3		3		3		
$\Delta^3\omega_n$:				0		0		0			

and obtains therefore

$$s_1 = \sum_{n=0}^{\infty} (-1)^n \omega_n = \sum_{l=0}^{2} \left(\frac{1}{2}\right)^{l+1} \Delta^l \omega_0 = -\frac{1}{4} + \frac{3}{8} = \frac{1}{8}.$$

For $-n \geqq 0$ the computation appears as

ω_{-n}:	0		2		7		15		26		40
$\Delta\omega_{-n}$:		-2		-5		-8		-11		-14	
$\Delta^2\omega_{-n}$:			3		3		3		3		
$\Delta^3\omega_{-n}$:				0		0		0			

and thus

$$t_1 = \sum_{n=0}^{\infty} (-1)^n \omega_{-n} = \sum_{l=0}^{2} \left(\frac{1}{2}\right)^{l+1} \Delta^l \omega_0 = -\frac{2}{4} + \frac{3}{8} = -\frac{1}{8},$$

both in agreement with (1.5). In the same way the values (1.6) are found. I may add to this list

$$s_3 = \sum_0^\infty (-1)^n \omega_n^3 = \frac{53}{64}.$$

$$t_3 = \sum_0 (-1)^n \omega_{-n}^3 = -\frac{53}{64},$$

Euler finds in the results

$$s_1 + t_1 = s_2 + t_2 [= s_3 + t_3] = 0 \tag{3.4}$$

a corroboration of his statement (1.3), in which the summation is taken over the whole set of pentagonal numbers. It may be observed here that the Euler summation of s_λ and t_λ leads evidently always to rational numbers as limits.

Actually the statements (3.4) and (1.3) have nothing directly to do with each other. The Euler summation applied to the power series

$$g(z) = \sum_0^\infty (-1)^n a_n z^n \tag{3.5}$$

leads to

$$g(z) = \sum_{m=0}^\infty \left(\frac{1}{2}\right)^{m+1} \sum_{n=0}^m (-1)^n \binom{m}{n} a_n z^n, \tag{3.6}$$

and the domain of convergence of (3.5) induces that of (3.6) as is well-known. If in our case we put

$$a_n = \omega_n^\lambda \quad \text{or} \quad a_n = \omega_{-n}^\lambda$$

the function $g(z)$ becomes a rational function which has as its only singularity a pole of order $2\lambda + 1$ at $z = -1$. This follows from the fact that the ω_n^λ and ω_{-n}^λ form arithmetic sequences of order 2λ

It is known [5], [6] that then the series (3.6) for $g(z)$ converges in the circle

$$|z - 1| < 2 \tag{3.7}$$

which has the singularity $z = -1$ on its boundary. Clearly the point $z = +1$ lies in the interior of (3.7). Putting $z = 1$ in (3.5) and (3.6) we have (3.3). We have thus by (3.3) applied Abel's method to the power series (3.5), which is not lacunary and has $z = +1$ as a regular point on its circle of convergence. Euler's method thus has *not* summed the lacunary series (3.1) and (3.2) and their derivatives by Abel's summation near $x = 1$.

4. Nevertheless the results (3.4), generalized immediately to

$$s_\lambda = \sum_0^\infty (-1)^n \omega_n^\lambda, \quad t_\lambda = \sum_0^\infty (-1)^n \omega_{-n}^\lambda, \tag{4.1}$$

$$s_\lambda + t_\lambda = 0 \tag{4.2}$$

cannot be accidental.

Indeed we have

$$s_\lambda + t_\lambda = \sum_{n=0}^\infty (-1)^n (\omega_n^\lambda + \omega_{-n}^\lambda).$$

Now, for $\lambda > 0$,

$$\begin{aligned}\omega_n^\lambda + \omega_{-n}^\lambda &= 2^{-\lambda}\{(3n^2 + n)^\lambda + (3n^2 - n)^\lambda\}\\ &= A_\lambda n^{2\lambda} + B_\lambda n^{2(\lambda-1)} + \cdots + M_\lambda n^2,\end{aligned}$$

a polynomial in n^2 without constant term. Thus $s_\lambda + t_\lambda$ reduces to summation of

$$\sum_{n=1}^\infty (-1)^n n^{2k}.$$

But these sums have been discussed by Euler himself in another connection [7], where he summed them by a correct Abel summation to

$$(2^{2k+1} - 1)\zeta(-2k) = 0.$$

Landau has devoted a paper [8] to this investigation of Euler, in which by the way the Riemann functional equation for special values of the variable s appears for the first time. The statement (4.2) is thus a consequence of properties of $\zeta(s)$, and not of $\eta(\tau)$.

5. We wish, of course, for the sake of consistency, to sum s_λ and t_λ singly by an $A(\omega_n)$ and $A(\omega_{-n})$ method respectively. This can be done by theorems of G. H. Hardy and Miss M. L. Cartwright.

As a preparation we notice

$$24\omega_n + 1 = 36n^2 - 12n + 1 = (6n-1)^2$$

so that

$$\omega_n = \frac{(6n-1)^2 - 1}{24}, \quad \omega_{-n} = \frac{(6n+1)^2 - 1}{24}, \quad n = 1, 2, 3, \ldots \tag{5.1}$$

We may restrict our treatment to ω_n; the treatment of ω_{-n} will be completely analogous.

The sum

$$s_\lambda(x) = \sum_{n=1}^\infty (-1)^n \omega_n^\lambda x^n$$

is thus a linear combination over the rational field of

$$\sum_{1}^{\infty} (-1)^n (6n - 1)^{2q} x^n, \quad q = 0, 1, \ldots, \lambda$$

and we are interested in the limit of these power series for $x \to 1 - 0$. With a new variable we consider

$$\lim_{y \to +0} \sum_{n=1}^{\infty} (-1)^n (6n - 1)^{2q} e^{-(6n-1)y} = l_q.$$

We know that through the Euler summation of

$$\sum_{1}^{\infty} (-1)^n (6n - 1)^{2q},$$

applicable since the sequence $\{(6n - 1)^{2k}\}$ forms an arithmetical progression of order $2k$, the limit l_k is a rational number.

The summation $A(\omega_n)$ would, however, require the limit

$$\begin{aligned} S_\lambda &= \lim_{x \to 1-0} \sum (-1)^n \omega_n^\lambda x^{\omega_n} \\ &= \lim_{y \to +0} \sum_{1}^{\infty} (-1)^n \omega_n^\lambda e^{\frac{-(6n-1)^2 - 1}{24} y} \end{aligned}$$

or with a new variable y instead of $y/24$

$$\lim_{y \to 0} \sum_{1}^{\infty} (-1)^n \omega_n^\lambda e^{-(6n-1)^2 y},$$

which is again a linear combination over the rational field of

$$L_q = \lim_{y \to +0} \sum_{1}^{\infty} (-1)^n (6n - 1)^{2q} e^{-(6n-1)^2 y}, \quad q = 0, 1, \ldots, \lambda.$$

We apply now theorems by Miss M. L. Cartwright [9] and G. H. Hardy [10]:

$$Z_1(s) = \sum_{n=1}^{\infty} (-1)^n (6n - 1)^{-s}$$

is a Dirichlet series (actually the difference of two Hurwitz zetafunctions), which is regular at $s = 1$ and can be continued over the while s-plane as an entire function. If we write it as

$$\sum_{1}^{\infty} a_m m^{-s}$$

with

$$a_m = \begin{cases} 0 & \text{for} \quad m \not\equiv -1 \pmod 6, \\ (-1)^{(m+1)/6} & \text{for} \quad m \equiv -1 \pmod 6 \end{cases}$$

we see that it can be summed for any s by $A(m)$ as well as by $A(m^2)$ and to the same value. The conditions for the growth $Z_1(s)$ in the imaginary direction which Hardy's theorem requires [10] p. 180, are easily fulfilled.

Thus $l_k = L_k$, and by recombination we have proved

$$s_\lambda = \lim_{x \to 1-0} \sum_1^\infty (-1)^n \omega_n^\lambda x^n = \lim_{x \to 1-0} \sum_1^\infty (-1)^n \omega_n^\lambda x^{\omega_n} = S_\lambda.$$

In this way Euler's treatment of (1.3) (our section 2) can be reconciled with that of (1.5), (1.6) in section 3. The fact that

$$S_\lambda + T_\lambda = s_\lambda + t_\lambda = 0$$

is now in a new sense indeed a corroboration of Euler's results (1.3), (1.4).

6. We can add a little to Euler's statement and prove the

Theorem. *If α is a primitive k^{th} root of unity then*

$$\lim_{x \to 1-0} \sum_{n=1}^\infty (-1)^n \alpha^{\omega_n} \omega_n^\lambda x^{\omega_n} = S_\lambda(\alpha)$$

and

$$\lim_{x \to 1-0} \sum_{n=1}^\infty (-1)^n \alpha^{\omega_{-n}} \omega_{-n}^\lambda x^{\omega_{-n}} = T_\lambda(\alpha)$$

exist and are numbers of the field $R(\alpha)$. The limits exist in any Stolz angle $|\Theta| < \frac{\pi}{2}$.

We need the

Lemma.

$$\sum_{n=N+1}^{N+2k} (-1)^n \alpha^{\omega_n} = 0. \tag{6.1}$$

Proof. Two cases have to be distinguished, k odd and k even.

I. Let k be odd. Then

$$2\omega_{n+k} = (n + k)(3n + 3k - 1) = 2\omega_n + k(3k - 1 + 6n) \equiv 2\omega_n \pmod{2k},$$

$$\omega_{n+k} \equiv \omega_n \pmod k$$

and therefore

$$(-1)^n \alpha^{\omega_n} + (-1)^{n+k} \alpha^{\omega_{n+k}} = 0, \tag{6.2}$$

which proves (6.1) for k odd.

II. Let k be even. Then we have

$$2\omega_{n+k} = 2\omega_n + k(3k + 6n) - k \equiv 2\omega_n + k \pmod{2k},$$

$$\omega_{n+k} \equiv \omega_n + \frac{k}{2} \pmod{k}$$

and

$$\alpha^{\omega_{n+k}} = \alpha^{\omega_n}\alpha^{k/2} = -\alpha^{\omega_n},$$

since α is a primitive k^{th} root of unity so that $\alpha^{k/2} \neq 1$. It follows again that

$$(-1)^n \alpha^{\omega_n} + (-1)^{n+k} \alpha^{\omega_{n+k}} = 0, \tag{6.2}$$

which proves (6.1) also in this case.

The lemma shows that

$$Z(s) = \sum_{n=1}^{\infty} (-1)^n \alpha^{\omega_n}(6n - 1)^{-s}$$

is convergent at $s = 1$ and can therefore be continued through the whole s-plane as an entire function.

We conclude again that

$$\lim_{x\to 1-0} \sum_{n=1}^{\infty} (-1)^n \alpha^{\omega_n}(6n - 1)^{2q} x^{6n-1} = l_q(\alpha) \tag{6.3}$$

exists and is equal to

$$\lim_{x\to 1-0} \sum_{n=1}^{\infty} (-1)^n \alpha^{\omega_n}(6n - 1)^{2q} x^{(6n-1)^2} = L_q(\alpha). \tag{6.4}$$

In view of (6.2) the series in (6.3) can be broken up into k series of the sort

$$\alpha^{\omega_l} \sum_{m=1}^{\infty} (-1)^m (6(l + km) - 1)^{2q} x^{6(l+mq)-1}, \tag{6.5}$$

in which the power series is a rational function and converges for $x \to 1$ to a rational number (obtainable, as above e.g. through the Euler transformation). Therefore $l_q(\alpha) = L_q(\alpha) \in R(\alpha)$. This proves the theorem in the beginning of this section.

The approach to the boundary in (6.3) and (6.4) does not have to be made on a radial path. Each component (6.5) converges to a limit for $x \to 1 - 0$ within a Stolz angle $-\frac{\pi}{2} + \varepsilon \leqq \Theta \leqq \frac{\pi}{2} - \varepsilon$. Therefore after Cartwright and Hardy (6.4) converges for $x \to 1 - 0$ within a Stolz angle $-\frac{\pi}{2} + 2\varepsilon \leqq \Theta \leqq \frac{\pi}{2} - 2\varepsilon$.

7. Summarizing we see that Euler was right in all his statements under proper interpretation. In modern notation we have shown that

$$\lim_{z\to 0}\left(\frac{d}{dz}\right)^{\lambda}\eta\left(\frac{h+iz}{k}\right)=0$$

in a Stolz angle $|\Theta|<\frac{\pi}{2}-\varepsilon$, a result which was known from the theory of the modular forms. If we introduce the two "halves" of $e^{-\pi i\tau/12}\eta(\tau)$, viz.

$$H_1(\tau)=\sum_{n=1}^{\infty}(-1)^n e^{2\pi i\omega_n\tau}, \tag{7.1}$$

$$H_2(\tau)=\sum_{n=-1}^{\infty}(-1)^n e^{2\pi i\omega_n\tau}, \tag{7.2}$$

we state the theorem of section 6 in the following way: the limits

$$\lim_{z\to +0}\left(\frac{d}{dz}\right)^{\lambda}H_1\left(\frac{h+iz}{k}\right)=\left(-\frac{2\pi}{k}\right)^{\lambda}S_\lambda, \tag{7.3}$$

$$\lim_{z\to +0}\left(\frac{d}{dz}\right)^{\lambda}H_2\left(\frac{h+iz}{k}\right)=\left(-\frac{2\pi}{k}\right)^{\lambda}T_\lambda \tag{7.4}$$

exist in any Stolz angle $|\Theta|<\frac{\pi}{2}$, where S_λ and T_λ are numbers of the field $R(e^{2\pi ih/k})$. The theory of modular forms would be of no assistance to the proofs of (7.3), (7.4).

References

[1] Leonard Euler, Opera Omnia (1), vol. 2, 480—496.

[2] L. Euler, De partitione numerorum, l.c., 254—294.

[3] L. Euler, Découverte d'une loi tout extraordinaire des nombres par rapport à la somme de leurs diviseurs, l.c. 241—253, and Demonstratio theorematis circa ordinem in summis divisorum observatum, l.c. 390—398.

[4] R. Dedekind, Erläuterungen zu den Fragmenten XXVIII in B. Riemann, Gesammelte Math. Werke, 2nd ed. (1892), 466—478.

[5] K. Knopp, Math. Z. **15** (1922), 226—253.

[6] H. Rademacher, Sitzungsber. Berliner Math. Ges. **21** (1922), 16—24.

[7] L. Euler, Remarques sur un beau rapport entre les séries des puissances tant directes que réciproques, Opera Omnia (1), vol. 15, 70—91.

[8] Edmund Landau, Euler und die Funktionalgleichung der Riemannschen Zetafunktion, Bibliotheca Math. (3) **7** (1906), 69—79.

[9] M. L. Cartwright, On the relation between the different types of Abel summation, Proc. London Math. Soc. (2) **31** (1930), 81—96.

[10] G. H. Hardy, The application of Abel's method of summation to Dirichlet's series, Quarterly J. of Math. **47** (1916), 176—192.

A. Rényi in Budapest

ON THE DISTRIBUTION OF NUMBERS PRIME TO n

§ 0. Introduction

The problem how the numbers prime to n and less than n are distributed asymptotically for large values of n has been raised by P. Erdös ([1], see also [2], [3] and [4]). Recently C. Hooley ([5], [7]) has investigated this problem and obtained very interesting results. The present paper is based entirely on Hooley's work: we shall deduce some further consequences of his results.

§ 1. Notations

Let us put

$$f_n(k) = \begin{cases} 1 & \text{if} \quad (k, n) = 1, \\ 0 & \text{if} \quad (k, n) > 1, \end{cases} \tag{1.1}$$

further

$$F_n(k, h) = \sum_{l=k}^{k+h-1} f_n(l). \tag{1.2}$$

Let us put for $k, h = 1, 2, \ldots; r = 0, 1, \ldots$

$$q_n(k, h, r) = \begin{cases} 1 & \text{if} \quad F_n(k, h) = r, \\ 0 & \text{if} \quad F_n(k, h) \neq r \end{cases} \tag{1.3}$$

and

$$Q_n(h, r) = \frac{1}{n} \sum_{k=1}^{n-h} q_n(k, h, r). \tag{1.4}$$

Let further $1 = a_{n,1} < a_{n,2} < \cdots < a_{n,\varphi(n)} < n$ denote the numbers which are prime to n and less than n; thus $f_n(k) = 1$ for $1 \leqq k \leqq n$ if and only if k is contained in the sequence $\{a_{n,i}\}$. Put

$$G_n(y, d) = \frac{1}{\varphi(n)} \sum_{\substack{a_{n,i+d}-a_{n,i}<y \\ 1 \leqq i \leqq \varphi(n)-d}} 1. \tag{1.5}$$

§ 2. Some results of C. Hooley

In this paragraph we state those results of Hooley which we shall need in what follows.

Theorem 1 (due to Hooley, see [5]). *For $1 \leqq \alpha < 2$ one has*

$$\sum_{i=1}^{\varphi(n)-1} (a_{n,i+1} - a_{n,i})^\alpha \leqq C_\alpha n \left(\frac{n}{\varphi(n)}\right)^{\alpha-1} \tag{2.1}$$

where the constant C_α depends only on α.

Remark to theorem 1. In [6] Hooley has proved that the left hand side of (2.1) is asymptotically equal to $\Gamma(\alpha + 1)\, n \left(\frac{n}{\varphi(n)}\right)^{\alpha-1}$ $(1 \leqq \alpha < 2)$ if n tends to $+\infty$ through such values that $\frac{n}{\varphi(n)} \to +\infty$, but we shall need only (2.1) in what follows. We get immediately from (2.1) and from Hölder's inequality the following

Corollary. For $1 \leqq \alpha < 2$ and for $1 \leqq d < \varphi(n)$ one has

$$\sum_{i=1}^{\varphi(n)-d} (a_{n,i+d} - a_{n,i})^\alpha \leqq C_\alpha d^\alpha n \left(\frac{n}{\varphi(n)}\right)^{\alpha-1}. \tag{2.2}$$

Theorem 2 (due to Hooley, see [6]). *If n tends to infinity through such values that $\frac{n}{\varphi(n)} \to +\infty$, one has for every $x > 0$*

$$\lim_{n\to\infty} G_n\left(x \frac{n}{\varphi(n)}, 1\right) = 1 - e^{-x}. \tag{2.3}$$

Remark to theorem 2. Theorem 2 can be expressed by saying that the differences $a_{n,i+1} - a_{n,i}$ $(i = 1, 2, \ldots, \varphi(n) - 1)$ are asymptotically exponentially distributed if n tends to $+\infty$ through numbers which have "many" prime factors, in the sense that $\frac{n}{\varphi(n)} = \frac{1}{\prod_{p|n}\left(1 - \frac{1}{p}\right)} \to +\infty$. (Here and in what follows p denotes always a prime number.) Let $p_1 < p_2 < \cdots < p_s < \cdots$ be the sequence of all primes. The condition

$\frac{n}{\varphi(n)} \to +\infty$ is evidently fulfilled in the special case when n tends to $+\infty$ through the sequence $N_x = \prod_{j=1}^{x} p_j$ $(x = 1, 2, \ldots)$. Note that in a Poisson process the differences between consecutive points are exponentially distributed, and are independent. In [7] C. Hooley has proved that the consecutive differences $(a_{n,i+1} - a_{n,i})$, $(a_{n,i+2} - a_{n,i+1})$, ..., $(a_{n,i+r} - a_{n,i+r-1})$ are in a certain sense asymptotically independent too. From this result one can deduce – following Hooley[7] – the following theorem which is essentially contained (though not explicitely stated) in his paper [7].

Theorem 3. *If n tends to $+\infty$ through such values that $\frac{n}{\varphi(n)} \to +\infty$, one has for $d = 1, 2, \ldots$ and for every $x > 0$*

$$\lim_{n\to+\infty} G_n\left(x\frac{n}{\varphi(n)}, d\right) = \int_0^x \frac{u^{d-1}e^{-u}}{(d-1)!}\,du. \tag{2.4}$$

Remarks to theorem 3. Clearly for $d = 1$ the statement of theorem 3 reduces to that of theorem 2. Hooley has proved a result very similar to theorem 3, namely that under the same conditions

$$\lim_{n\to+\infty} G_n^*\left(x\frac{n}{\varphi(n)}, d\right) = \int_0^x \frac{u^{d-1}e^{-u}}{(d-1)!}\,du \tag{2.5}$$

where $\varphi(n)\,G_n^*(y, d)$ is equal to the number of those values of $k \leqq \varphi(n) - 2d - 1$ for which $\sum_{l=0}^{d} (a_{n,k+2l+1} - a_{n,k+2l}) < y$. His proof gives even more easily also (2.4).

§ 3. The Poisson distribution of the number of integers prime to n in intervals of length of the order $\frac{n}{\varphi(n)}$

We shall prove now the following

Theorem 4. *One has for every fixed $\lambda > 0$, if n tends to $+\infty$ through values such that $\frac{n}{\varphi(n)} \to +\infty$*

$$\lim_{n/(\varphi(n))\to+\infty} Q_n\left(\lambda\frac{n}{\varphi(n)}, r\right) = \frac{\lambda^r e^{-\lambda}}{r!} \quad (r = 0, 1, \ldots). \tag{3.1}$$

Remark to theorem 4. As $Q_n\left(\lambda \frac{n}{\varphi(n)}, r\right)$ is equal to the relative frequency of those intervals $\left(k, k + \lambda \frac{n}{\varphi(n)}\right)$ $\left(1 \leqq k < n - \lambda \frac{n}{\varphi(n)}\right)$ which contain exactly r integers relatively prime to n, theorem 4 can be expressed by saying that *the distribution of the numbers* $\frac{\varphi(n)\, a_{n,i}}{n}$ $(1 \leqq i \leqq \varphi(n))$ *if* $\frac{n}{\varphi(n)}$ *is large, is in some respect asymptotically like the distribution of points in a typical realization of a (homogeneous) Poisson point process with density* 1.

Recently I have shown (see [8]) that the Poisson process can be characterized among point processes with the same density as the "most random" distribution in the sense that the entropy (uncertainty) of the Poisson process is maximal. Thus the statement of theorem 4 can be expressed somewhat vaguely by saying that *the numbers prime to n and less than n are for values of n for which* $\frac{n}{\varphi(n)}$ *is large, in a certain sense asymptotically as randomly distributed as possible.*

However to show that the distribution of the points $a_{n,i}$ behaves asymptotically *in every respect* like the distribution of the points in a Poisson process, one should prove that these points behave like those in a Poisson process also in another sense, which correponds to the property of the Poisson process that the numbers of points in disjoint intervals are (stochastically) independent. The corresponding statement for the numbers prime to n and less than n does not follow from the mentioned results of Hooley; we want to mention here only that in view of a recent result of the author (see [9]) concerning the characterization of the Poisson process, it would be necessary to prove the following statement. Let $Q_n(b_1, b_1', \ldots, b_\nu, b_\nu'; r)$ denote the relative frequency of those values of k $(1 \leqq k \leqq n - c_\nu)$ for which the number of integers prime to n and lying in the union of the intervals $[k + b_j, k + b_j')$ $(1 \leqq j \leqq \nu)$ is equal to r $(0 \leqq b_1 < b_1' < b_2 < b_2' < \cdots < b_\nu < b_\nu')$; if $\lim_{n\to\infty} \frac{b_j(n)\,\varphi(n)}{n} = B_j$ and $\lim_{n\to\infty} \frac{b_j'(n)\,\varphi(n)}{n} = B_j'$ where $0 \leqq B_1 < B_1' \leqq B_2 < B_2' \leqq \cdots \leqq B_\nu < B_\nu'$ hold and $\sum_{j=1}^{\nu} (B_j' - B_j) = \lambda$ one should have (for $\nu = 1, 2, 3, \ldots$)

$$\lim_{n\to\infty} Q_n(b_1(n), b_1'(n), \ldots, b_\nu(n), b_\nu'(n), r) = \frac{\lambda^r e^{-\lambda}}{r!} \quad (r = 0, 1, 2, \ldots). \tag{3.2}$$

Clearly (3.2) reduces for $\nu = 1$ to (3.1).

We intend to return to this question elsewhere.

Proof of theorem 4. We prove first in detail (3.1) for $r = 0$. The proof of (3.1) for $r \geqq 1$ will follow the same lines. Clearly

$$Q_n(h, 0) = \frac{1}{n} \sum_{\substack{a_{n,i+1} - a_{n,i} > h \\ 1 \leqq i < \varphi(n)}} (a_{n,i+1} - a_{n,i} - h). \tag{3.3}$$

As a matter of fact each interval $[k, k + h)$ $(k \leqq n - h)$ which does not contain any number prime to n is contained in an interval $[a_{n,i}, a_{n,i+1}]$ of length $> h$ and the number of such intervals contained in such an interval is equal to $a_{n,i+1} - a_{n,i} - h$.

Let now $\varepsilon > 0$ be an arbitrary small positive number.

It follows that

$$Q_n\left(\lambda \frac{n}{\varphi(n)}, 0\right) \leqq \sum_{j=0}^{[1/\varepsilon^2]+1} (j + 1)\, \varepsilon \left[G_n\left((\lambda + (j + 1)\,\varepsilon) \frac{n}{\varphi(n)}, 1\right) - G_n\left((\lambda + j\varepsilon) \frac{n}{\varphi(n)}\right)\right]$$
$$+ \frac{1}{n} \sum_{a_{n,i+1} - a_{n,i} > n/\varepsilon\varphi(n)} (a_{n,i+1} - a_{n,i}). \tag{3.4}$$

Now according to theorem 2 it follows that

$$\lim_{n \to \infty} G_n\left((\lambda + j\varepsilon) \frac{n}{\varphi(n)}, 1\right) = 1 - e^{-(\lambda + j\varepsilon)}.$$

On the other hand we get from theorem 1 for $1 < \alpha < 2$

$$\frac{1}{n} \sum_{a_{n,i+1} - a_{n,i} > n/\varepsilon\varphi(n)} (a_{n,i+1} - a_{n,i}) \leqq C_\alpha \frac{1}{n} \left(\frac{\varepsilon\varphi(n)}{n}\right)^{\alpha - 1} n \left(\frac{n}{\varphi(n)}\right)^{\alpha - 1} = C_\alpha \varepsilon^{\alpha - 1}. \tag{3.5}$$

Thus it follows that

$$\lim_{n \to \infty} Q_n\left(\lambda \frac{n}{\varphi(n)}, 0\right) \leqq e^{-\lambda} \sum_{j=0}^{[1/\varepsilon^2]+1} (j + 1)\, \varepsilon (1 - e^{-\varepsilon})\, e^{-j\varepsilon} + C_\alpha \varepsilon^{\alpha - 1}. \tag{3.6}$$

As $\varepsilon > 0$ can be chosen arbitrarily small, and

$$\lim_{\varepsilon \to 0} (1 - e^{-\varepsilon}) \sum_{j=0}^{[1/\varepsilon^2]+1} (j + 1)\, \varepsilon\, e^{-j\varepsilon} = \int_0^\infty x\, e^{-x}\, dx = 1 \tag{3.7}$$

it follows that

$$\lim_{n \to \infty} Q_n\left(\lambda \frac{n}{\varphi(n)}, 0\right) \leqq e^{-\lambda}. \tag{3.8}$$

Similarly we get

$$Q_n\left(\lambda \frac{n}{\varphi(n)}, 0\right) \geqq \sum_{j=0}^{[1/\varepsilon^2]+1} j\varepsilon \left[G_n\left((\lambda + (j + 1)\,\varepsilon) \frac{n}{\varphi(n)}, 1\right) - G_n\left((\lambda + j\varepsilon) \frac{n}{\varphi(n)}, 1\right)\right] \tag{3.9}$$

and thus by theorem 2

$$\lim_{n\to\infty} Q_n\left(\lambda \frac{n}{\varphi(n)}, 0\right) \geqq e^{-\lambda} \sum_{j=0}^{[1/\varepsilon^2]+1} j\varepsilon(1 - e^{-\varepsilon})\, e^{-j\varepsilon} \tag{3.10}$$

and therefore, as $\varepsilon > 0$ is arbitrarily small, and in view of (3.7)

$$\lim_{n\to\infty} Q_n\left(\lambda \frac{n}{\varphi(n)}, 0\right) \geqq e^{-\lambda}. \tag{3.11}$$

From (3.8) and (3.11) we get

$$\lim_{n\to\infty} Q_n\left(\lambda \frac{n}{\varphi(n)}, 0\right) = e^{-\lambda} \tag{3.12}$$

which is the statement of (3.1) for $r = 0$.

Now let us prove (3.1) for $r = 1$. Clearly every interval $[k, k + h)$ which contains *at most* one number relatively prime to n is contained in an interval $[a_{n,i}, a_{n,i+2}]$ of length $> h$; those intervals $[k, k + h)$ which contain exactly one number prime to n are contained in exactly one such interval $[a_{n,i}, a_{n,i+2}]$ while those intervals $[k, k + h)$ which contain no number prime to n are contained in two such intervals. Thus we have, if $h = \lambda \dfrac{n}{\varphi(n)}$

$$Q_n\left(\lambda \frac{n}{\varphi(n)}, 1\right) + 2Q_n\left(\lambda \frac{n}{\varphi(n)}, 0\right) = \frac{1}{n} \sum_{a_{n,i+2} - a_{n,i} > h} (a_{n,i+2} - a_{n,i} - h). \tag{3.13}$$

It follows in the same way as in the proof of (3.12), only using theorem 3 with $d = 2$ instead of theorem 2, and corollary to theorem 1 instead of theorem 1 that

$$\lim_{n\to\infty} Q_n\left(\lambda \frac{n}{\varphi(n)}, 1\right) = \int_0^\infty x(\lambda + x)\, e^{-(\lambda+x)}\, dx - 2e^{-\lambda} = \lambda\, e^{-\lambda}. \tag{3.14}$$

Thus (3.1) is proved for $r = 1$. Now we apply induction. Suppose that (3.1) is proved for $r < s$; we shall show that it holds for $r = s$ too ($s \geqq 2$). Clearly every interval $[k, k + h)$ ($k \leqq n - h$) which contains exactly s numbers prime to n is contained in exactly one interval $[a_{n,i}, a_{n,i+s+1}]$ of length $> h$, while if the interval $[k, k + h)$ contains $r < s$ numbers prime to n it is contained in exactly $s - r + 1$ such intervals. Thus we get, by using theorem 3,

$$\lim_{n\to\infty} \sum_{r=0}^{s} (s - r + 1)\, Q_n\left(\lambda \frac{n}{\varphi(n)}, r\right) = \int_\lambda^\infty (x - \lambda) \frac{x^s e^{-x}}{s!}\, dx. \tag{3.15}$$

Thus by the induction hypothesis

$$\lim_{n\to\infty} Q_n\left(\lambda\frac{n}{\varphi(n)}, s\right) = \int_\lambda^\infty (x-\lambda)\frac{x^s e^{-x}}{s!}\,dx - \sum_{r=0}^{s-1}(s-r+1)\frac{\lambda^r e^{-\lambda}}{r!}. \tag{3.16}$$

Now we use the well-known identity

$$\sum_{k=0}^{m}\frac{\lambda^k e^{-\lambda}}{k!} = \int_\lambda^\infty \frac{x^m e^{-x}}{m!}\,dx \quad (m = 0, 1, 2, \dots). \tag{3.17}$$

It follows that

$$\int_\lambda^\infty (x-\lambda)\frac{x^s e^{-x}}{s!}\,dx = (s+1)\sum_{k=0}^{s+1}\frac{\lambda^k e^{-\lambda}}{k!} - \lambda\sum_{k=0}^{s}\frac{\lambda^k e^{-\lambda}}{k!} = \sum_{k=0}^{s}(s+1-k)\frac{\lambda^k e^{-\lambda}}{k!} \tag{3.18}$$

and thus

$$\lim_{n\to\infty} Q_n\left(\lambda\frac{n}{\varphi(n)}, s\right) = \frac{\lambda^s e^{-\lambda}}{s!} \tag{3.19}$$

what was to be proved.

Thus the proof of theorem 4 is completed.

We add one final remark. It is interesting to compare the statement (3.1) for $r = 0$ with the statement of theorem 2. While (3.1) for $r = 0$ states that the relative frequency of those intervals $\left[k, k + \lambda\frac{n}{\varphi(n)}\right)$ which do not contain any number prime to n, tends to $e^{-\lambda}$, theorem 2 states the same for the relative frequency of those of these intervals for which $(k, n) = 1$. Thus the probability that an interval $\left[k, k + \lambda\frac{n}{\varphi(n)}\right)$ does not contain any number prime to n is asymptotically independent of k being relatively prime to n. This is clearly a consequence of theorem 2 and the characteristic property of the exponential distribution that if the random variable ξ has an exponential distribution then $P(\xi > t + s|\xi > s) = P(\xi > t)$ if $s > 0$ and $t > 0$ where $P(A|B)$ denotes the conditional probability of the event A under condition B.

References

[1] P. Erdös, The difference of consecutive primes, Duke Math. J. **6** (1940), 438—441.

[2] P. Erdös, On the integers relatively prime to n and on a number-theoretic function considered by Jacobsthal, Math. Scand. **10** (1962), 163—170.

[3] P. Erdös, Some unsolved problems, MTA Mat. Kut. Int. Közl. **6** (1961), 221—254.

[4] P. Erdös, Some recent advances and current problems in number theory, Lectures on Modern Mathematics, Vol. 3. p. 196—244, Wiley, 1965.
[5] C. Hooley, On the difference of consecutive numbers prime to *n*, Acta Arithmetica **8** (1963), 343—347.
[6] C. Hooley, On the difference between consecutive numbers prime to *n*, II, Publicationes Mathematicae **12** (1965), 39—49.
[7] C. Hooley, On the difference between consecutive numbers prime to *n*, III, Math. Z. **90** (1965), 355—364.
[8] A. Rényi, On an extremal property of the Poisson process, Annals of the Institute of Stat. Math. **16** (1964), 129—133.
[9] A. Rényi, Remarks on the Poisson process, Studia Sci. Math. Hung. **2** (1967), 119-124.

I. J. SCHOENBERG IN MADISON (WISC.)

SPLINE INTERPOLATION AND THE HIGHER DERIVATIVES

Introduction

Edmund Landau was very much interested in the theory of approximation. It was he who inspired Dunham Jackson to do his fundamental work on the orders of best approximation. His proof of Weierstrass' theorem by means of Landau's singular integral is now classical. A new generalization of Weierstrass' theorem found by a graduate student at the University of Wisconsin, Martin Marsden, plays an essential role in the present paper.

In 1910 F. Riesz published the following theorem.

Theorem I. *Let $I = [a, b]$ be a finite interval and let $f(x)$ be real-valued and defined in I. Let $p > 1$. Finally, let $a = x_0 < x_1 < \cdots < x_n = b$ be a division of the interval. Then $f(x)$ is absolutely continuous in I and*

$$f'(x) \in L_p(a, b)$$

if and only if there is a constant K such that

$$\sum_{i=1}^{n} \frac{|f(x_i) - f(x_{i-1})|^p}{(x_i - x_{i-1})^{p-1}} < K^p$$

for all possible divisions.

See [6]; a proof can also be found in Natanson's book [4, 259–261]. In particular, for $p = 2$, we obtain the following special case.

Theorem II. *The function $f(x)$ is absolutely continuous and $f'(x) \in L_2(a, b)$ if and only if*

$$\sum_{i=1}^{n} \frac{(f(x_i) - f(x_{i-1}))^2}{x_i - x_{i-1}} < K^2 \tag{1}$$

for all divisions.

The main objective of the present paper is to generalize theorem II so as to obtain conditions which are to insure that

$$f^{(m)}(x) \in L_2(a, b).$$

For a precise formulation of our results, which were announced without proofs in [7], see § 3.

Let us slightly alter theorem II by considering divisions

$$\Delta\colon a \leqq x_0 < x_1 < \cdots < x_n \leqq b \tag{2}$$

whose norms we define by

$$\|\Delta\| = \max\{x_0 - a, x_1 - x_0, \ldots, x_n - x_{n-1}, b - x_n\}. \tag{3}$$

Again the boundedness of the sums (1) is necessary and sufficient for $f'(x) \in L_2(a, b)$, the new statement being logically equivalent to the old one.

Let $S_\Delta(x)$ be the continuous function in I which interpolates $f(x)$ at all points x_i and such that $S_\Delta(x)$ is linear in each of the n intervals $[x_{i-1}, x_i]$, while being constant in $[a, x_0]$ and also in $[x_n, b]$. Neither Riesz nor Natanson mention the obvious fact that the sum appearing in (1) is equal to the integral

$$\int_I (S'_\Delta(x))^2\, dx.$$

However, $S_\Delta(x)$ is the so-called natural linear spline function which interpolates $f(x)$ in the points x_i and this remark suggests the correct generalization. Accordingly, spline interpolation of degree $2m - 1$ will be our tool in extending Riesz's theorem to mth derivatives.

An essential part of our discussion deals with the convergence of spline interpolation on indefinite refinement of Δ (theorem 1, § 3). This problem was discussed by several writers since the appearance of our paper [7] in which the present results were announced (see [5] for references). Here we are discussing the convergence of spline interpolation of the fixed degree $2m - 1$ for a function $f(x) \in C^{m-1}$ such that $f^{(m)}(x) \in L_2(a, b)$. The problem becomes much more delicate if $f(x)$ is less smooth and much remains to be done in this direction. In [5] Stig Nord constructs a continuous function for which cubic spline interpolation diverges.

An introductory discussion of spline interpolation (§ 1) is unavoidable also because we wish to add an essential remark (§ 2) to our previous paper [8] on this subject. We must also describe briefly the variation dimishing spline approximation methods (§ 5) for the following reason: M. Marsden's result mentioned in our first paragraph and which is to appear in his paper [3], is derived by and expressed in terms of such approximation methods.

1. Spline interpolation

A function $S(x)$ defined in a finite or infinite interval I is called a *spline function* of degree k provided that $S(x) \in C^{k-1}(I)$, while I can be so partitioned into subintervals that $S(x) \in \pi_k$ in each of the subintervals. Here and below we denote by π_k the class of real polynomials of degree not exceeding k. The common endpoints of pairs of adjacent intervals are called the *knots* of the spline function.

One of the convenient approaches to spline interpolation is the solution of the following minimum problem: We consider the class of functions

$$\mathscr{F}_m = \{f(x); f \in C^{m-1}[a, b], f^{(m-1)} \text{ absolutely continuous}, \ f^{(m)} \in L_2(a, b)\}. \quad (1.1)$$

If $f(x) \in \mathscr{F}_m$ and the division Δ is such that

$$n \geqq m, \quad (1.2)$$

we wish to find within the class $\mathscr{F}_m$ the function $S(x)$ satisfying the interpolatory conditions

$$S(x_i) = f(x_i) \quad (i = 0, \ldots, n) \quad (1.3)$$

which also solves the problem

$$\int_I (S^{(m)}(x))^2 \, dx = \text{minimum}. \quad (1.4)$$

The solution of this problem is always unique and is obtained as the restriction to $[a, b]$ of the function $S(x)$ which is uniquely defined by the equations (1.3) and the following additional conditions:

$$S(x) \in C^{2m-2}(-\infty, \infty), \quad (1.5)$$

$$S(x) \in \pi_{2m-1} \quad \text{in each of the intervals } (x_{i-1}, x_i) \quad (i = 1, \ldots, n), \quad (1.6)$$

$$S(x) \in \pi_{m-1} \quad \text{in} \quad (-\infty, x_0) \quad \text{and in} \quad (x_n, +\infty). \quad (1.7)$$

A function $S(x) = S_\Delta(x)$ satisfying the conditions (1.5), (1.6) and (1.7) is called a *natural spline function* of degree $2m - 1$ for the knots $x_0, \ldots, x_n$, the term "natural" referring to the peculiar drop in degree described by (1.7). We shall denote their class by the symbol S_Δ.

Since $S(x) \equiv f(x)$ evidently satisfies (1.3) we conclude from the minimal property (1.4) of $S_\Delta(x)$ the following:

$$\int_I (S^{(m)}(x))^2 \, dx \leqq \int_I (f^{(m)}(x))^2 \, dx, \quad (1.8)$$

with equality only if $f(x) = S_\Delta(x)$.

Let us now consider

$$s(x) = S_\Delta^{(m)}(x). \tag{1.9}$$

This is evidently a spline function of degree $m - 1$, having the same knots $x_0, \ldots, x_n$, with the additional property, implied by (1.7), that the support of $s(x)$ is contained in (x_0, x_n). We shall refer to $s(x)$ as a *confined spline function* which is *confined to* (x_0, x_n). Let $\mathscr{C}_\Delta$ denote the class of confined spline functions of degree $m - 1$ which are confined to (x_0, x_n).

A minimal property of the mth derivative (1.9) is as follows: *If*

$$\sigma(x) \in \mathscr{C}_\Delta \tag{1.10}$$

and $s(x)$ *is defined by* (1.9) *then*

$$\int_I (s(x) - f^{(m)}(x))^2\, dx \leqq \int_I (\sigma(x) - f^{(m)}(x))^2\, dx. \tag{1.11}$$

In words: The confined spline (1.9) is the unique solution of the problem of best approximation of $f^{(m)}(x)$ in the L_2-norm within the class $\mathscr{C}_\Delta$.

Spline interpolation enjoys further properties of best approximation in the sense of Sard of linear functionals, but these do not concern us here.

2. On the minimal value of the integral (1.4)

Let

$$M(x; y) = \begin{cases} m(y - x)^{m-1} & \text{if } y \geqq x, \\ 0 & \text{if } y < x. \end{cases}$$

Using Steffensen's divided difference notation we consider the functions

$$M_j(x) = M(x; x_j, x_{j+1}, \ldots, x_{j+m}) \quad (j = 0, 1, \ldots, n - m). \tag{2.1}$$

These are the so-called *B*-splines, or basis splines, which are very likely familiar to many readers because they allow to express the divided difference $f(x_j, \ldots, x_{j+m})$ of $f(x)$ in the form

$$f(x_j, x_{j+1}, \ldots, x_{j+m}) = \frac{1}{m!} \int_I M_j(x) f^{(m)}(x)\, dx. \tag{2.2}$$

$M_j(x)$ is easily seen to be a spline function of degree $m - 1$ with the knots $x_j, \ldots, x_{j+m}$ and which is confined to (x_j, x_{j+m}). Thus $M_j(x) \in \mathscr{C}_\Delta$ $(j = 0, \ldots, n - m)$.

The significance of the functions (2.1) is due to the fact that every element of $\mathscr{C}_\Delta$ may be represented uniquely as a linear combination of the $M_j(x)$ (see [1]). In particular we may represent the confined spline (1.9) in the form

$$s(x) = \sum_{0}^{n-m} c_j M_j(x). \tag{2.3}$$

The least square property of $s(x)$ described by (1.11) leads to $n - m + 1$ "normal" equations for the unknowns c_j. Using the property (2.2) these normal equations are readily seen to become

$$\sum_{j=0}^{n-m} L_{ij} c_j = m! f_i \quad (i = 0, \ldots, n - m) \tag{2.4}$$

where

$$L_{ij} = \int_I M_i(x)\, M_j(x)\, dx \tag{2.5}$$

and

$$f_i = f(x_i, x_{i+1}, \ldots, x_{i+m}). \tag{2.6}$$

For a full discussion see [8, 112–114].

Now the determination of

$$J^2 = \int_I (S_\Delta^{(m)}(x))^2\, dx = \int_I (s(x))^2\, dx \tag{2.7}$$

presents no difficulty. From (2.3), (2.5) and (2.4) we obtain

$$J^2 = \sum L_{ij} c_i c_j = \sum_i c_i \sum_j L_{ij} c_j = m! \sum c_i f_i.$$

Eliminating the c_j among the $n - m + 2$ equations

$$J^2 = \sum_j m! f_j c_j,$$

$$m! f_i = \sum_j L_{ij} c_j \quad (i = 0, \ldots, n - m),$$

we obtain

$$\begin{vmatrix} J^2 & m! f_j \\ m! f_i & L_{ij} \end{vmatrix} = 0$$

and solving for J^2 we get

$$\int_I (S_\Delta^{(m)}(x))^2\, dx = -\frac{(m!)^2}{|L_{ij}|} \begin{vmatrix} 0 & f_j \\ f_i & L_{ij} \end{vmatrix}. \tag{2.8}$$

In terms of the positive definite matrix

$$\Lambda = \|L_{ij}\|_{i,j=0,\ldots,n-m} \tag{2.9}$$

and its inverse

$$\Omega = \Lambda^{-1} = \|\omega_{ij}\| \tag{2.10}$$

we can write (2.8) as

$$\int_I \left(\frac{1}{m!} S_\Delta^{(m)}(x)\right)^2 dx = \sum_{i,j=0}^{n-m} \omega_{ij} f_i f_j. \tag{2.11}$$

The right-hand side is again a positive definite quadratic form of the variables f_i. However, since the f_i are by (2.6) divided differences, we see that (2.11) is a positive form of the functional values $f(x_0), \ldots, f(x_n)$, which is of rank $n - m + 1$.

We may state the following

Corollary 1. *If* $f(x) \in \mathscr{F}_m$ *and in terms of the notations introduced by* (2.1), (2.5), (2.9) *and* (2.10) *we have*

$$\int_I \left(\frac{1}{m!} f^{(m)}(x)\right)^2 dx \geqq \sum_0^{n-m} \omega_{ij} f_i f_j \tag{2.12}$$

with equality if and only if $f(x)$ *is a natural spline function of degree* $2m - 1$ *and knots* $x_0, \ldots, x_n$.

Examples. 1. If $m = 1$ then (2.1) shows that

$$M_j(x) = \begin{cases} (x_{j+1} - x_j)^{-1} & \text{if } x \in (x_j, x_{j+1}), \\ 0 & \text{if } x \notin (x_j, x_{j+1}). \end{cases}$$

By (2.9) and (2.5) Λ is seen to be the diagonal matrix $\Lambda = \|(x_{i+1} - x_i)^{-1} \cdot \delta_{ij}\|$. Therefore $\Omega = \Lambda^{-1} = \|(x_{i+1} - x_i)\,\delta_{ij}\|$. Since $f_i = (f(x_{i+1}) - f(x_i))/(x_{i+1} - x_i)$ we obtain

$$\sum \omega_{ij} f_i f_j = \sum_0^{n-1} \frac{(f(x_{i+1}) - f(x_i))^2}{x_{i+1} - x_i}$$

which is identical with Riesz's expression (1).

2. If $n = m$, which by (1.2) is the lowest possible value for n, the matrix inversion (2.10) is again trivially performed and corollary 1 reduces to

Corollary 2. *If* $f(x) \in \mathscr{F}_m$ *then*

$$\int_I \left(\frac{1}{m!} f^{(m)}(x)\right)^2 dx \geqq (f(x_0, \ldots, x_m))^2 \Big/ \int_I (M(x; x_0, \ldots, x_m))^2 \, dx,$$

with equality if and only if $f(x)$ *is a natural spline function of degree* $2m - 1$ *and knots* $x_0, \ldots, x_m$.

3. Statement of main results

We use the notations so far introduced. Let

$$f(x) \in \mathscr{F}_m \tag{3.1}$$

and let $S_\Delta(x)$ be the natural spline function of degree $2m - 1$ which interpolates $f(x)$ in the points x_i of the division Δ of norm $\|\Delta\|$ $(n \geqq m)$.

Theorem 1. (i) *If* (3.1) *holds then*

$$\lim_{\|\Delta\| \to 0} \int_I (S_\Delta^{(m)}(x) - f^{(m)}(x))^2 \, dx = 0. \tag{3.2}$$

(ii) *For each* $\nu = 0, 1, \ldots, m - 1$ *we have*

$$|f^{(\nu)}(x) - S_\Delta^{(\nu)}(x)| \leqq \frac{2}{\sqrt{m}} \frac{m!}{\nu!} \|\Delta\|^{m-\nu-1/2} \cdot \|f^{(m)}\|_2 \quad (a \leqq x \leqq b), \tag{3.3}$$

where

$$\|f^{(m)}\|_2 = \left(\int_I (f^{(m)}(x))^2 \, dx \right)^{1/2}.$$

In particular (3.3) *imply that*

$$\lim_{\|\Delta\| \to 0} S_\Delta^{(\nu)}(x) = f^{(\nu)}(x) \quad \textit{uniformly in } x \in I, \tag{3.4}$$

if $\nu = 0, \ldots, m - 1$.

We now turn to our generalization of Riesz's theorem II. If (3.1) holds then the minimum property (1.8) shows that

$$\int_I (S_\Delta^{(m)}(x))^2 \, dx$$

must be below a fixed bound independent of Δ. The converse is the subject of our next theorem.

Theorem 2. *Let* $f(x)$ *be finite valued and defined in* I. *Let* $S_\Delta(x)$ *be the natural spline function of degree* $2m - 1$ *which interpolates* $f(x)$ *in the points* x_i *of the division* Δ $(n \geqq m)$. *If there is a constant* K, *independent of* Δ, *such that*

$$\int_I (S_\Delta^{(m)}(x))^2 \, dx \leqq K^2 \quad \textit{for all } \Delta, \tag{3.5}$$

then $f(x) \in \mathscr{F}_m$.

We observe that if the condition (3.5) is satisfied then the conclusions of theorem 1 are valid.

We shall also establish the following variant of theorem 2:

Theorem 3. *Let*

$$f(x) \in C[a, b], \tag{3.6}$$

and let $\Delta = \Delta_n$ *be the set of* $n + 1$ *equidistant points* $x_i = a + ih\,(i = 0, \ldots, n)$, $h = (b - a)/n$. *If*

$$\int_I (S_\Delta^{(m)}(x))^2\, dx = O(1) \quad as \quad n \to \infty, \tag{3.7}$$

then $f(x) \in \mathscr{F}_m$.

Finally, we refer to [7, 28] for a sketch of the proof that theorem 3 implies the following

Theorem 4. *Let* (3.6) *hold and let the notations of theorem* 3 *stand. Then* $f(x) \in \mathscr{F}_m$ *if and only if there is a constant K, independent of n, such that*

$$h \sum_{i=0}^{n-m} (\Delta_h^m f(x_i)/h^m)^2 \leqq K^2 \quad for\ all \quad n \geqq m.$$

4. An approximation property of the class $\mathscr{C}_\Delta$ which implies theorem 1 (i)

Evidently

$$\mathscr{C}_\Delta \subset L_2(a, b),$$

in fact $\mathscr{C}_\Delta$ is an $(n - m + 1)$-dimensional subspace of L_2. If $h(x) \in L_2(a, b)$, we define its distance to $\mathscr{C}_\Delta$ as usual by

$$d_2(h, \mathscr{C}_\Delta) = \inf_\sigma \|h(x) - \sigma(x)\|_2 \quad (\sigma \in \mathscr{C}_\Delta).$$

The approximation property of $\mathscr{C}_\Delta$ is as follows.

Theorem 5. *Let* $h(x) \in L_2(a, b)$. *To every* ε *corresponds a* $\delta = \delta_h$ *such that*

$$d_2(h, \mathscr{C}_\Delta) < \varepsilon \quad if \quad \|\Delta\| < \delta. \tag{4.1}$$

We shall establish theorem 5 in § 5. Here we observe that it immediately implies the relation (3.2) or

$$\lim_{\|\Delta\| \to 0} \|S_\Delta^{(m)}(x) - f^{(m)}(x)\|_2 = 0. \tag{4.2}$$

Indeed $f^{(m)}(x) \in L_2(a, b)$ and $S_\Delta^{(m)}(x) = s(x) \in \mathscr{C}_\Delta$. The minimum property (1.11) implies

$$d_2(f^{(m)}, \mathscr{C}_\Delta) = \|s(x) - f^{(m)}(x)\|_2$$

and now (4.1) shows that

$$\|S_\Delta^{(m)}(x) - f^{(m)}(x)\|_2 < \varepsilon \quad \text{if} \quad \|\Delta\| < \delta.$$

5. Variation diminishing spline approximations imply theorem 5

It is well-known that S. Bernstein derived constructively Weierstrass' approximation theorem by means of the polynomials which bear his name. Generalizations of these in terms of spline functions have recently been constructed (see [9], [2] and [3]). They are called *variation diminishing* (or v.d.) because they share this property with the Bernstein polynomials. Using these v.d. spline approximations M. Marsden has recently found a generalization of Weierstrass's theorem which will easily imply theorem 5, actually in a modified form, theorem 5′ below, which we describe now.

Let $C[a, b]$ have the usual meaning and let

$$C_0[a, b] = \{g(x);\ g(x) \in C[a, b],\ g(a) = g(b) = 0\}. \tag{5.1}$$

Evidently

$$\mathscr{C}_\Delta \subset C_0[a, b] \subset C[a, b] \subset L_2(a, b).$$

Within $C[a, b]$ we can use the usual sup-norm $\|g(x)\|$ and the distance

$$d(g, \mathscr{C}_\Delta) = \inf_\sigma \|g(x) - \sigma(x)\| \quad (\sigma \in \mathscr{C}_\Delta).$$

Since $C_0[a, b]$ is dense in $L_2(a, b)$, it suffices to assume in theorem 5 that $h(x) = g(x) \in C_0[a, b]$. Moreover

$$\|g(x) - \sigma(x)\|_2 \leqq \sqrt{b - a}\,\|g(x) - \sigma(x)\|.$$

Therefore theorem 5 will be implied by the following

Theorem 5′. *Let $g(x) \in C_0[a, b]$. To every ε corresponds a $\delta = \delta_g$ such that*

$$d(g, \mathscr{C}_\Delta) < \varepsilon \quad \textit{if} \quad \|\Delta\| < \delta. \tag{5.2}$$

We shall now describe Marsden's approximation theorem and show how it implies theorem 5′. To fix the ideas we shall discuss only the case when

$$\Delta:\ a = x_0 < x_1 < \cdots < x_{n-1} < x_n = b, \tag{5.3}$$

the changes necessary if $a < x_0$, or $x_n < b$, or both, being of a trivial nature.

Let $\mathcal{T}_\Delta(m)$ denote the class of spline functions of degree $m-1$ defined in $[a, b]$ and having the knots $x_1, \ldots, x_{n-1}$. We extend the sequence x_j by setting

$$x_{-m+1} = x_{-m+2} = \cdots = x_{-1} = a, \quad b = x_{n+1} = x_{n+2} = \cdots = x_{n+m-1},$$

so that x_j is non-decreasing and defined for $j = -m+1, -m+2, \ldots, n+m-1$. A linear basis for $\mathcal{T}_\Delta(m)$ is provided by the following sequence of functions

$$M_j(x) = M(x; x_j, x_{j+1}, \ldots, x_{j+m}) \quad (j = -m+1, \ldots, n-1), \tag{5.4}$$

which are now divided differences with coalescent arguments if $j < 0$ or $j > n - m$. Again $M_j(x)$ vanishes everywhere outside the non-vanishing interval (x_j, x_{j+m}) in which they are positive.

Setting

$$N_j(x) = \frac{1}{m}(x_{j+m} - x_j)\, M_j(x), \tag{5.5}$$

$$\xi_j = \frac{1}{m-1}(x_{j+1} + x_{j+2} + \cdots + x_{j+m-1}) \quad (j = -m+1, \ldots, n-1), \tag{5.6}$$

the following inequalities and identities hold:

$$a = \xi_{-m+1} < \xi_{-m+2} < \cdots < \xi_{n-2} < \xi_{n-1} = b, \tag{5.7}$$

$$1 = \sum_j N_j(x) \quad (x \in I), \tag{5.8}$$

$$x = \sum_j \xi_j N_j(x) \quad (x \in I). \tag{5.9}$$

In terms of the "nodes" ξ_j and the "fundamental functions" $N_j(x)$ we may now define the spline approximation to a prescribed function $g(x)$ by the formula

$$t(x) = \sum_{-m+1}^{n-1} g(\xi_j)\, N_j(x) \tag{5.10}$$

which is an element of $\mathcal{T}_\Delta(m)$.

Formulae (5.8) and (5.9) show that $t(x) \equiv g(x)$ whenever $g(x)$ is a linear function. Let us also mention that (5.10) reduces to the $(m-1)$st Bernstein polynomial for $[a, b]$ in the case when $n = 1$ and there are no knots in (a, b) as seen from (5.3).

We may now state

Marsden's Approximation Theorem. *Let* $g(x) \in C[a, b]$. *To every* ε *corresponds a* $\delta = \delta_g$ *such that*

$$d(g, \mathcal{T}_\Delta(m)) < \varepsilon \tag{5.11}$$

provided that

$$\frac{1}{m-1}\|\Delta\| < \delta \quad (j = -m+1, \ldots, n-1). \tag{5.12}$$

If $n = 1$ this statement reduces to Weierstrass' theorem because then $\mathscr{T}_\Delta(m) = \pi_{m-1}$ while the left side of (5.12) reduces to $(b-a)/(m-1)$. Marsden establishes his theorem by showing that the spline approximation (5.10) converges uniformly to $g(x)$ whenever the left side of (5.12) converges to zero. Here m and n are arbitrary natural numbers except that we require that $m > 1$. The convergence condition

$$\frac{1}{m-1}\|\Delta\| \to 0 \tag{5.13}$$

can be realized in a variety of ways. Surely $\|\Delta\| \to 0$ is sufficient whatever m may be. Likewise $m \to \infty$ is sufficient in which case the behavior of Δ becomes irrelevant.

In our present application of Marsden's theorem to the proof of theorem 5′ we assume as here-to-fore that

$$m \text{ is fixed}, \quad m \geqq 2,$$

in which case Marsden's convergence condition (5.13) is equivalent to

$$\|\Delta\| \to 0. \tag{5.14}$$

Proof of theorem 5′. We truncate the sum (5.10) and define

$$t^*(x) = \sum_{j=0}^{n-m} g(\xi_j)\, N_j(x). \tag{5.15}$$

The first thing to notice is that now

$$t^*(x) \in \mathscr{C}_\Delta \tag{5.16}$$

because all functions $M_j(x)$ $(j = 0, \ldots, n-m)$ are elements of $\mathscr{C}_\Delta$. Moreover, (5.8) and $N_j(x) \geqq 0$ surely imply that

$$0 \leqq N_j(x) \leqq 1 \quad \text{for all } j \quad \text{and} \quad x \in I. \tag{5.17}$$

We now observe the following:

(i) The function $g(x)$ of theorem 5′ vanishes if $x = a$ or $x = b$.

(ii) The first $m-1$ terms of (5.10) (which were discarded) correspond to nodes ξ_j which all converge to a as $\|\Delta\| \to 0$. Likewise, the nodes ξ_j of the last $m-1$ terms all converge to b as $\|\Delta\| \to 0$. But then it is clear by (5.17), (i) and (ii) that

$$\|t(x) - t^*(x)\| \to 0 \quad \text{as} \quad \|\Delta\| \to 0. \tag{5.18}$$

Since (5.14) implies $\|g(x) - t(x)\| \to 0$, we conclude from (5.18) that

$$\|g(x) - t^*(x)\| \to 0 \quad \text{as} \quad \|\Delta\| \to 0.$$

A fortiori $d(g, \mathscr{C}_\Delta) \to 0$ and theorem 5′ is established.

6. Proof of theorem 1 completed

The relations (3.3) will easily follow from (3.2). We need

Lemma 1. *If $\varphi(x) \in C^{m-1}[a, b]$ and if $\varphi(x)$ has a zero in every open subinterval of length δ (open relative to $[a, b]$), then $\varphi^{(\nu)}(x)$ has a zero in every subinterval of length $(\nu + 1)\,\delta$ $(\nu = 1, \ldots, m-1)$.*

Proof. Let J be a subinterval of length $(\nu + 1)\,\delta$. We divide it into $\nu + 1$ open subintervals of length δ. By assumption we can find $\nu + 1$ zeros of $\varphi(x)$, say y_0, $y_1, \ldots, y_\nu$ in increasing order, in these subintervals. The mean value theorem

$$\varphi(y_0, y_1, \ldots, y_\nu) = \varphi^{(\nu)}(\eta)/\nu! \quad (y_0 < \eta < y_\nu),$$

shows that $\varphi^{(\nu)}(\eta) = 0$ because the divided difference certainly vanishes. Thus $\varphi^{(\nu)}(x)$ vanishes at the point η of J and the lemma is proved.

Let

$$\delta > \|\Delta\| \tag{6.1}$$

and let us consider the difference

$$\varphi(x) = f(x) - S_\Delta(x).$$

Since $S_\Delta(x)$ interpolates $f(x)$ in the points of Δ, (6.1) implies that $\varphi(x)$ vanishes in every open subinterval of length δ. To every $x \in I$ there corresponds a ξ such that $\varphi(\xi) = 0$ and therefore

$$\varphi^{(m-1)}(x) = \int_\xi^x (f^{(m)}(x) - S_\Delta^{(m)}(x))\,dx \quad \text{and} \quad |x - \xi| < m\delta,$$

whence

$$|\varphi^{(m-1)}(x)| \leq \left|\int_\xi^x f^{(m)}(x)\,dx\right| + \left|\int_\xi^x S_\Delta^{(m)}(x)\,dx\right|$$

$$\leq |x - \xi|^{1/2} \left|\int_\xi^x (f^{(m)}(x))^2\,dx\right|^{1/2} + |x - \xi|^{1/2} \left|\int_\xi^x (S_\Delta^{(m)}(x))^2\,dx\right|^{1/2}.$$

By (1.8) we conclude that

$$|\varphi^{(m-1)}(x)| < 2(m\delta)^{1/2} J \quad \text{where} \quad J = \|f^{(m)}\|_2. \tag{6.2}$$

This reasoning may be repeated: To every x will correspond a ξ such that

$$\varphi^{(m-2)}(x) = \int_\xi^x \varphi^{(m-1)}(u)\,du \quad \text{and} \quad |x - \xi| < (m-1)\,\delta.$$

Now (6.2) implies

$$|\varphi^{(m-2)}(x)| < 2(m\delta)^{1/2}\,(m-1)\,\delta \cdot J.$$

Continuing in like manner we obtain

$$|\varphi^{(\nu)}(x)| < 2(m\delta)^{1/2}\,\frac{(m-1)!}{\nu!}\,\delta^{m-\nu-1} \cdot J.$$

Since δ was arbitrary subject to (6.1), we may let $\delta \to \|\Delta\|$ to obtain

$$|f^{(\nu)}(x) - S_\Delta^{(\nu)}(x)| \leqq \frac{2}{\sqrt{m}}\,\frac{m!}{\nu!}\,\|\Delta\|^{m-\nu-1/2} \cdot \|f^{(m)}\|_2$$

which completes our proof.

7. Proofs of theorems 2 and 3

By assumption $f(x)$ is finite valued in I and such that

$$\|S_\Delta^{(m)}(x)\|_2 \leqq K \quad \text{for all } \Delta. \tag{7.1}$$

Let Δ be a division and Δ^* a refinement of Δ, i.e. $\Delta \subset \Delta^*$. Since (7.1) holds for Δ as well as for Δ^* we may argue as in § 6 as follows. Let

$$\varphi(x) = S_\Delta(x) - S_{\Delta^*}(x)$$

which vanishes in every open interval of length $> \|\Delta\|$ because $S_\Delta(x)$ interpolates $S_{\Delta^*}(x)$ in all points of Δ. Again let $\delta > \|\Delta\|$. As in § 6 we conclude that

$$|S_\Delta^{(\nu)}(x) - S_{\Delta^*}^{(\nu)}(x)| \leqq \frac{2}{\sqrt{m}}\,\frac{m!}{\nu!}\,\|\Delta\|^{m-\nu-1/2} \cdot K \quad (\nu = 0, \ldots, m-1;\, x \in I). \tag{7.2}$$

Let now Δ and Δ' be any two divisions. Let $\Delta^* = \Delta \cup \Delta'$ so that Δ^* is a common refinement of Δ and Δ'. But then (7.2) remains valid if we replace Δ by Δ' while Δ^* has remained the same. We therefore conclude that

$$|S_\Delta^{(\nu)}(x) - S_{\Delta'}^{(\nu)}(x)| \leqq \frac{2}{\sqrt{m}}\,\frac{m!}{\nu!}\,K(\|\Delta\|^{m-\nu-1/2} + \|\Delta'\|^{m-\nu-1/2}) \tag{7.3}$$

$(\nu = 0, 1, \ldots, m-1;\, x \in I)$.

Evidently (7.3) implies that the limits

$$\lim_{\|\Delta\|\to 0} S_\Delta^{(\nu)}(x) = f_\nu(x) \quad (\nu = 0, \ldots, m-1), \tag{7.4}$$

exist uniformly in x for $x \in I$. Therefore $f_\nu(x) \in C[a, b]$. Now

$$\int_a^t f_1(x)\,dx = \lim_{\|\Delta\|\to 0} \int_a^t S'_\Delta(x)\,dx = \lim\,(S_\Delta(t) - S_\Delta(a)) = f_0(t) - f_0(a)$$

showing that $f'_0(t) = f_1(t)$ and similarly we establish that $f'_{\nu-1}(t) = f_\nu(t)$ ($\nu = 2, \ldots, m-1$) whence

$$f_\nu(x) = f_0^{(\nu)}(x). \tag{7.5}$$

Δ being an arbitrary division, (7.4), for $\nu = 0$, implies that $f(x) = f_0(x)$ for all x and now (7.5) implies that $f(x) \in C_\nu[a, b]$ and

$$f_\nu(x) = f^{(\nu)}(x) \quad (\nu = 0, \ldots, m-1). \tag{7.6}$$

We are still to show that

$$f^{(m-1)}(x) \quad \text{is absolutely continuous} \tag{7.7}$$

and that

$$f^{(m)}(x) \in L_2(a, b). \tag{7.8}$$

If $(\alpha, \beta) \subset I$ then

$$(S_\Delta^{(m-1)}(\beta) - S_\Delta^{(m-1)}(\alpha))^2 = \left(\int_\alpha^\beta S_\Delta^{(m)}(x)\,dx\right)^2 \leqq (\beta - \alpha)\int_\alpha^\beta (S_\Delta^{(m)}(x))^2\,dx$$

hence

$$\frac{(S_\Delta^{(m-1)}(\beta) - S_\Delta^{(m-1)}(\alpha))^2}{\beta - \alpha} \leqq \int_\alpha^\beta (S_\Delta^{(m)}(x))^2\,dx.$$

For any division Δ' this implies

$$\sum_1^n \frac{(S_{\Delta'}^{(m-1)}(x_i) - S_{\Delta'}^{(m-1)}(x_{i-1}))^2}{x_i - x_{i-1}} \leqq \sum \int_{x_{i-1}}^{x_i} (S_{\Delta'}^{(m)}(x))^2\,dx \leqq K^2$$

and letting $\|\Delta'\| \to 0$ by (7.4) and (7.6), for $\nu = m - 1$, we obtain

$$\sum_1^n \frac{(f^{(m-1)}(x_i) - f^{(m-1)}(x_{i-1}))^2}{x_i - x_{i-1}} \leqq K^2.$$

This is precisely Riesz's condition which by his theorem II implies the desired conclusions (7.7) and (7.8).

Let us finally establish theorem 3. Its assumption (3.7) holds in particular for the sequence of binary divisions when $n = 2^s$. All reasonings leading to the relations (7.4) and (7.5) remain valid if we use exclusively only these binary divisions. We may therefore consider (7.4) and (7.5) as established for binary divisions. We conclude the proof as follows: Surely

$$f_0(x) = f(x) \quad \text{for all binary fractions} \quad x = r/2^s. \tag{7.9}$$

$f(x)$ being continuous by our assumption (3.6), we conclude that (7.9) holds for all x of I. The remainder of the proof of theorem 2 is unchanged.

Observe that the assumption that $f(x)$ is continuous is essential in theorem 3. Indeed, let us assume that $a = 0$ and $b = 1$, and let $f(x)$ be a discontinuous solution of Cauchy's functional equation

$$f(x + y) = f(x) + f(y).$$

For such $f(x)$ it is known that $f(x) = Cx$ for all *rational* x and therefore

$$S_{\Delta_n}(x) = Cx \quad \text{for all } x.$$

But then our condition (3.7) is trivially verified while the conclusion of theorem 3 does not hold.

References

[1] H. B. Curry and I. J. Schoenberg, On Polya frequency functions IV: The fundamental spline functions and their limits, J. d'Analyse Math. XVII (1966), 71—107.

[2] M. Marsden and I. J. Schoenberg, On variation diminishing spline approximation methods, Mathematica (Cluj) **8** (1966), 61—82.

[3] M. Marsden, An identity for spline functions and its applications to variation diminishing spline approximations (to appear).

[4] I. P. Natanson, Theorie der Funktionen einer reellen Veränderlichen, 2. Aufl., Akademie-Verlag, Berlin 1961 (Übersetzung aus dem Russischen).

[5] S. Nord, Approximation properties of the spline fit, to appear in Nordisk Tidskrift for Informations-Behandling during 1967.

[6] F. Riesz, Systeme integrierbarer Funktionen, Math. Ann. **69** (1910), 449—497; also Collected Works I, 454.

[7] I. J. Schoenberg, Spline interpolation and the higher derivatives. Proc. Nat. Acad. of Sciences **51** (1964), 24—28.

[8] I. J. Schoenberg, On interpolation by spline functions and its minimal properties, I.S.N.M. (Birkhäuser Verlag) **5** (1964), 109—129.

[9] I. J. Schoenberg, On spline functions, with a supplement by T.N.E. Greville, Inequalities, Proceedings of a Symposium held at Wright-Patterson Air Force Base, Ohio, August 19—27, 1965, edited by Oved Shisha, Academic Press, New York and London 1967, 255—291.

C. L. Siegel in Göttingen

ZU DEN BEWEISEN DES VORBEREITUNGSSATZES VON WEIERSTRASS

Ich stamme noch aus dem großen neunzehnten Jahrhundert und hatte dann das weitere Glück, während meiner Studienzeit in Berlin und Göttingen durch hervorragende Persönlichkeiten in der damals lebendigen Tradition aus der Glanzzeit der Mathematik erzogen zu werden. Von diesen akademischen Lehrern hat mich besonders Edmund Landau beeinflußt, dessen Sorgfalt in der Vorbereitung von Vorlesungen und schriftlichen Veröffentlichungen mir dauernd ein unerreichtes Vorbild geblieben ist. „Er war der pflichttreueste von uns allen" sagte Hilbert zu mir, als ich 1938 die Nachricht vom Tode Landaus überbrachte. Jetzt, 30 Jahre später, seien die folgenden Zeilen seinem Andenken gewidmet.

Der Vorbereitungssatz von Weierstraß entstand aus der Frage nach den Nullstellen von analytischen Funktionen mehrerer komplexen Veränderlichen. Es sei $F(y, x_1, \ldots, x_n)$ eine Potenzreihe in den $n + 1$ Variablen $y, x_1, \ldots, x_n$ mit komplexen Koeffizienten, die in einer Umgebung des Nullpunktes $y = 0$, $x_1 = 0, \ldots, x_n = 0$ konvergiert. Aus einem bald ersichtlichen Grunde haben wir dabei eine der Variablen ausgezeichnet und y genannt. Es sollen dann sämtliche Lösungen der Gleichung $F = 0$ betrachtet werden, welche in einer hinreichend klein gewählten Umgebung des Nullpunktes gelegen sind. Wir wollen die trivialen Fälle ausschließen, daß entweder die Reihe F identisch 0 ist oder aber mit einer von 0 verschiedenen Konstanten anfängt. Es beginnt dann also F mit Gliedern einer gewissen positiven Ordnung r, und wir können nach einer geeigneten umkehrbaren homogenen linearen Transformation der $n + 1$ Variablen annehmen, daß in F das Glied y^r auftritt. Wir fassen nun die sämtlichen Glieder r-ter Ordnung zusammen und können also

$$F = F_r + G_r, \quad F_r = y^r + P_1 y^{r-1} + \cdots + P_r$$

setzen, wobei G_r die sämtlichen Glieder höherer Ordnung aufnimmt und $P_1, \ldots, P_r$ homogene Polynome der n Variablen $x_1, \ldots, x_n$ von den Graden $1, \ldots, r$ bedeuten.

Zunächst wollen wir den einfachen Fall betrachten, daß der Rest G_r identisch 0 ist. Bei jeder beliebigen Wahl der n Variablen $x_1, \ldots, x_r$ ist dann $F = 0$ eine algebraische Gleichung r-ten Grades für die Unbekannte y und hat daher nach dem Fundamentalsatz der Algebra genau r Lösungen $y = \eta_1, \eta_2, \ldots, \eta_r$, die aber nicht alle verschieden zu sein brauchen. Da ferner die Koeffizienten $P_1, \ldots, P_r$ für $x_1 = 0, \ldots, x_n = 0$ sämtlich verschwinden, kann man zu beliebig vorgegebenem positiven ϱ stets einen positiven Radius σ derart bestimmen, daß für $|x_1| < \sigma, \ldots, |x_n| < \sigma$ die r Wurzeln $\eta_1, \ldots, \eta_r$ alle im Kreise $|y| < \varrho$ liegen.

Nun gehen wir zum allgemeinen Fall über, wobei wir uns wegen des Auftretens von G_r von vornherein auf genügend kleine absolute Beträge der $n + 1$ Variablen beschränken müssen. Man wird vermuten, daß dann das vorangehende Resultat bestehen bleibt, daß es also bei hinreichend kleinem ε zu beliebigem positiven $\varrho < \varepsilon$ eine positive Zahl $\sigma < \varepsilon$ gibt, für welche bei beliebiger Wahl von $x_1, \ldots, x_n$ im Kreise $|x| < \sigma$ die Gleichung $F = 0$ im Kreise $|y| < \varrho$ genau r Wurzeln $y = \eta_1, \ldots, \eta_r$ besitzt. Dies ist in der Tat richtig und wurde von Cauchy 1831 im Falle $n = 1$ bewiesen, sodann 1879 allgemein von Poincaré [1]. Beim Beweise benutzt Poincaré die von Cauchy eingeführten Methoden und insbesondere den Residuensatz. Ferner bemerkt Poincaré, daß die elementaren symmetrischen Polynome von $\eta_1, \ldots, \eta_r$ bei veränderlichen Werten der n Parameter $x_1, \ldots, x_n$ konvergente Potenzreihen werden, daß also das Produkt

$$\Phi = (y - \eta_1)(y - \eta_2) \cdots (y - \eta_r) = y^r + Q_1 y^{r-1} + \cdots + Q_r$$

ist, worin sich die Koeffizienten $Q_1, \ldots, Q_r$ als Potenzreihen in $x_1, \ldots, x_n$ ergeben.

In jeder genügend kleinen Umgebung des Nullpunktes im Raume der $n + 1$ unabhängigen komplexen Variablen $y, x_1, \ldots, x_n$ stimmen die Lösungen der Gleichung $F = 0$ mit denen der Gleichung $\Phi = 0$ überein. Dieses zusammenfassende Ergebnis von Poincaré ist nun aber eine unmittelbare Folgerung aus dem weitergehenden Satz von Weierstraß, nämlich dem sogenannten Vorbereitungssatz. Der Weierstraßsche Vorbereitungssatz besteht in folgender Identität:

$$F = \Phi E, \quad E = 1 + \cdots;$$

dabei ist E eine mit dem konstanten Gliede 1 beginnende Potenzreihe in den $n + 1$ unabhängigen Variablen $y, x_1, \ldots, x_n$, welche in einer genügend kleinen Umgebung des Nullpunktes $y = 0, x_1 = 0, \ldots, x_n = 0$ konvergiert, während Φ die obige Bedeutung hat und also ein Polynom bezüglich y ist. Aus dieser Identität folgt tatsächlich sofort, daß in hinreichender Nähe des Nullpunktes die Nullstellen von F genau mit denen von Φ übereinstimmen.

Weierstraß [2] hat seinen Vorbereitungssatz zuerst 1886 im Druck veröffentlicht und gibt dabei an, daß er ihn bereits seit 1860 in seinen Vorlesungen mitgeteilt hätte. Bekanntlich hat Weierstraß bei seinen Publikationen nach Möglichkeit vermieden, von der Cauchyschen Integralformel und den daran anschließenden Methoden Gebrauch zu machen, und so ist es von Interesse, seinen Beweis zu betrachten. Das wollen wir jetzt tun und dann zum Vergleich spätere Beweise anderer Mathematiker besprechen.

Bei dem von Weierstraß angegebenen Beweis ist es wesentlich, die Variable y zunächst auf einen kleinen Kreisring $\varrho^* < |y| < \varrho$ zu beschränken, wobei wir der Einfachheit halber noch $\varrho^* = \frac{\varrho}{2}$ wählen können. Setzt man

$$v = \frac{P_1}{y} + \frac{P_2}{y^2} + \cdots + \frac{P_r}{y^r}, \quad F_r = y^r(1 + v),$$

so kann man dann noch bei beliebigem ϱ den Radius σ so klein festlegen, daß für $|x_1| < \sigma, \ldots, |x_n| < \sigma$ die Größe $|v| < \frac{1}{2}$ und folglich $|F_r - y^r| < \frac{1}{2}|y|^r$ wird. Da ferner der Rest G_r nur Glieder von höherer als r-ter Ordnung enthält, wird bei genügend kleinem ϱ und geeignetem σ in dem Gebiete $\frac{\varrho}{2} < |y| < \varrho$, $|x_1| < \sigma, \ldots, |x_n| < \sigma$ auch $|G_r| < \frac{1}{2}|y|^r$ werden. Setzt man dann

$$F = y^r(1 + w),$$

so ist also

$$w = v + G_r y^{-r}, \quad |w| < 1, \quad F \neq 0$$

und $F - y^r = y^r w$ eine konvergente Potenzreihe in den $n + 1$ Variablen $y, x_1, \ldots, x_n$. Zunächst seien nun $x_1, \ldots, x_n$ festgehalten, so daß F eine Potenzreihe in y allein wird. Deutet man durch einen Strich die Differentiation nach y an, so wird im Kreisring die logarithmische Ableitung

$$\frac{F'}{F} = \frac{r}{y} + \frac{w'}{1 + w}, \quad \frac{w'}{1 + w} = w' - w'w + w'w^2 - \cdots = \left(w - \frac{w^2}{2} + \frac{w^3}{3} - \cdots\right)'.$$

Hierbei ist zu beachten, daß w und die Potenzen $w^2, w^3, \ldots$ Laurentsche Reihen in y sind, in denen übrigens jeweils nur endlich viele negative Potenzen von y auftreten. Nach dem Doppelreihensatz ist dann auch $\frac{w'}{1 + w}$ eine konvergente Laurentsche Reihe, und zwar tritt in dieser die Potenz y^{-1} nicht auf. Andererseits hat F als reguläre analytische Funktion von y im abgeschlossenen Kreis $|y| \leqq \frac{\varrho}{2}$ nur endlich viele Nullstellen, die jetzt mit $\eta_1, \ldots \eta_h$ bezeichnet seien. Im Kreisring $\frac{\varrho}{2} < |y| < \varrho$ gilt dann

$$\frac{F'}{F} - \left(\frac{1}{y - \eta_1} + \frac{1}{y - \eta_2} + \cdots + \frac{1}{y - \eta_h}\right) = \frac{r - h}{y} + \frac{w'}{1 + w}$$
$$- \sum_{k=1}^{\infty} \frac{\eta_1^k + \eta_2^k + \cdots + \eta_h^k}{y^{k+1}},$$

worin nun aber links eine im gesamten Kreis $|y| < \varrho$ reguläre Funktion auftritt. Die linke Seite kann man daher in eine Reihe nach nicht-negativen Potenzen von y entwickeln, während rechts eine Laurentsche Reihe steht.

Da für Laurentsche Reihen wie bei gewöhnlichen Potenzreihen der Eindeutigkeitssatz gilt, dürfen also rechts keine negativen Potenzen von y vorkommen, woraus zunächst $h = r$ folgt. Weiter ergibt sich aber jede Potenzsumme $\eta_1^k + \cdots + \eta_r^k$ $(k = 1, 2, \ldots)$ als der Koeffizient von y^{-k-1} in der Reihe für $\dfrac{w'}{1+w}$. Geht man jetzt auf die Abhängigkeit von $x_1, \ldots, x_n$ zurück und macht diese Parameter wieder im Kreis $|x| < \sigma$ variabel, so erweisen sich jene Koeffizienten sämtlich als konvergente Potenzreihen in $x_1, \ldots, x_n$. Also sind auch die elementaren symmetrischen Polynome der r Wurzeln $\eta_1, \ldots, \eta_r$ konvergente Potenzreihen in $x_1, \ldots, x_n$, womit das Ergebnis von Poincaré bewiesen ist. Schließlich ist noch die logarithmische Ableitung von $\dfrac{F}{\Phi}$, nämlich die Differenz $\dfrac{F'}{F} - \dfrac{\Phi'}{\Phi}$, eine Reihe nach nicht-negativen Potenzen aller Variablen, welche dann als Ableitung A' einer Reihe A ohne konstantes Glied angesetzt werden kann. Für $e^A = E$ wird $A' = \dfrac{E'}{E}$ und durch Integration $F = \Phi E$, womit die Aussage des Vorbereitungssatzes gewonnen ist. Allerdings kommt bei der Integration zunächst ein von y unabhängiger Faktor hinzu, über dessen Bestimmung Weierstraß keine genaue Angabe macht, und hierfür wäre eine sorgfältige Überlegung nötig, wenn man mit Weierstraß die logarithmische Funktion ganz umgehen will. Es ist ferner vom prinzipiellen Standpunkt aus einzuwenden, daß beim Beweis des Eindeutigkeitssatzes der Laurentschen Entwicklung die Cauchyschen Ideen nicht zu vermeiden sind.

Im Jahre 1927 hat Wirtinger [3] den von Weierstraß gegebenen Beweis erheblich vereinfacht, indem er die Wurzeln $\eta_1, \ldots, \eta_r$ vollkommen eliminiert und insbesondere von ihrer Existenz keinen Gebrauch macht; allerdings scheute er sich nicht, den Logarithmus explizit einzuführen. In der oben benutzten Formel $F = y^r(1 + w)$ setzt er $1 + w = e^{\log(1+w)}$ und bildet die Laurentsche Reihe

$$\log(1 + w) = w - \frac{w^2}{2} + \frac{w^3}{3} - \cdots = A + B,$$

worin B die sämtlichen Glieder mit negativen Exponenten von y zusammenfaßt, während A eine gewöhnliche Potenzreihe in y wird. Es folgt

$$Fe^{-A} = y^r e^B, \quad F(1 - A + \cdots) = y^r(1 + B + \cdots).$$

Links erhält man offenbar eine gewöhnliche Potenzreihe in y, dagegen rechts eine mit y^r beginnende Reihe nach fallenden Potenzen von y. Wegen der Eindeutigkeit der Laurentschen Entwicklung im Kreisring $\frac{\varrho}{2} < |y| < \varrho$ muß dann aber rechts ein Polynom Φ in y stehen, dessen Koeffizienten sich als konvergente Potenzreihen in $x_1, \ldots, x_n$ erweisen. Setzt man wieder $e^A = E$, so ergibt sich der Vorbereitungssatz.

Wirtinger hat nicht bemerkt, daß Stickelberger [4] schon 1887 im wesentlichen denselben eleganten Beweis veröffentlicht hat. Dieser Beweis war inzwischen auch von Hartogs [5] im Jahre 1909 wiederentdeckt worden.

Im Jahre 1929 gab Späth [6] einen Beweis des Vorbereitungssatzes, der insofern axiomatisch etwas Neues zu bieten schien, als er nur Hilfsmittel aus der lokalen Theorie der Potenzreihen benötigte. Späth behandelt dabei sogleich eine Verallgemeinerung des Vorbereitungssatzes, für welche neuerdings der Name Divisionssatz vorgeschlagen wurde. Dieser Satz besagt, daß für jede konvergente Potenzreihe L in den $n + 1$ Variablen $y, x_1, \ldots, x_r$ mit der obigen Bedeutung von F die Formel

$$L = FQ + R$$

besteht, wobei Q und R wiederum konvergente Potenzreihen sind, von denen aber R die Variable y höchstens bis zur Potenz y^{r-1} enthält. Wählt man speziell $L = y^r$ und setzt $L - R = \Phi$, $Q^{-1} = E$, so bekommt man die Aussage des Vorbereitungssatzes. Umgekehrt läßt sich aber wiederum der Divisionssatz durch einen naheliegenden einfachen Schluß aus dem Vorbereitungssatz ableiten, und zwar findet ein aufmerksamer Leser dies bereits in Stickelbergers Arbeit aus dem Jahre 1887, was anscheinend bisher übersehen worden ist.

Späth beweist den Divisionssatz, indem er Q und R als Potenzreihen in $y, x_1, \ldots, x_n$ mit unbekannten Koeffizienten ansetzt und diese dann in geeigneter Reihenfolge durch Koeffizientenvergleich rekursiv bestimmt, worauf die Konvergenz mit der Cauchyschen Majorantenmethode gezeigt wird. Der von Späth gegebene Beweis, soweit er sich auf den Vorbereitungssatz bezieht, ist nun aber ebenfalls mehrere Jahrzehnte früher bekannt gewesen. In einer kürzlich erschienenen Festschrift zum 150. Geburtstag von Weierstraß bemerkt Henri Cartan [7], daß bereits Brill [8] 1910 denselben Beweis mittels Koeffizientenvergleich und Majorantenmethode ausgeführt hat, und weist sodann darauf hin, daß schon vorher 1905 durch Lasker [9] die gleiche Idee ohne nähere Einzelheiten skizziert worden war, während aber Lasker nicht von Brill erwähnt wird. Tatsächlich war nun aber schon im Jahre 1891 dieselbe Beweisskizze für $n = 1$ durch Brill [10] angegeben worden, was nun wiederum von Lasker nicht zitiert wurde. Damals erklärte Brill außerdem, der Satz sei ihm schon einige Zeit bekannt gewesen und er verzichte aber nach den Publikationen von Weierstraß und Stickelberger auf eine Ausführung seines eigenen Beweises.

Einige gegenwärtige Verfasser bezeichnen übrigens den Divisionssatz als Weierstraßschen Vorbereitungssatz, was natürlich historisch und sachlich nicht berechtigt ist, aber so geschieht es unter anderem in der unlängst herausgekommenen Funktionentheorie von Hörmander [11] und dem Abhyankarschen Buche mit dem irreführenden Titel „Local analytic geometry". Ich sage „irreführend"; denn zum mindesten seit 1797 haben die bisherigen Mathematiker unter analytischer Geometrie etwas ganz Bestimmtes verstanden, was eben dem Genius von Descartes zu verdanken ist, und es ist ein grober Unfug, das jetzt ändern zu wollen. Die schöpferischen Mathematiker früherer Generationen sind in der Wahl von Bezeichnungen meist sehr behutsam vorgegangen, wie etwa Gauß bei der Einführung des Kongruenzbegriffes in die Zahlentheorie oder Dedekind bei der Erklärung eines Körpers in der Algebra. Dagegen vergreifen sich moderne Epigonen an Begriffen, Denkweisen und Formelzeichen, welche durch Jahrzehnte und Jahrhunderte für Lehre und Forschung von Nutzen gewesen sind, und ertränken uns dafür mit einer Sintflut von komischen Merkbildern, sprachwidrigen Abkürzungen und absurden Zauberworten. Der Nachwuchs wird überhaupt nicht mehr imstande sein, etwa in Riemanns oder Hilberts Werken zu lesen, wenn er nur auf exakte Sequenzen und kommutative Diagramme dressiert ist. Der gegenwärtige bedrohliche Zustand in der Mathematik erinnert durchaus an die Zeiten des Nationalsozialismus unter Hitler, als solange marschiert wurde, bis alles in Scherben fiel. In diesem Zusammenhang sei noch erwähnt, daß in ebenfalls irreführender Weise in dem Abhyankarschen Buche [12] als ursprünglich von Weierstraß gegebener Beweis des Vorbereitungssatzes ein Gedankengang vorgeführt wird, welcher in den bekannten Lehrbüchern von Picard, Goursat und Osgood zu finden ist und sehr wenig mit Weierstraß zu tun hat. Diese Beweisführung benutzt nämlich gerade die Cauchyschen Methoden der Funktionentheorie, im Gegensatz zur Einstellung von Weierstraß.

Nach dieser langen Einleitung komme ich endlich zum eigentlichen Gegenstand meiner Veröffentlichung. Es ist behauptet worden, der sogenannte Späthsche Beweis des Vorbereitungssatzes sei den anderen Beweisen dadurch überlegen, daß er nur lokale Eigenschaften von Potenzreihen verwendet und deswegen auch auf den Fall übertragen werden kann, in welchem die Koeffizienten der Potenzreihe F einem nicht-archimedisch bewerteten Körper angehören. Es ist zuzugeben, daß der Stickelbergersche Beweis in der oben angegebenen Form von der Eindeutigkeit der Laurentschen Entwicklung Gebrauch macht, welche nicht durch rein lokale Untersuchung von Potenzreihen bewiesen werden kann. Ich werde aber jetzt auseinandersetzen, wie man durch eine naheliegende und höchst einfache Änderung des Stickelbergerschen Ansatzes erreichen kann, daß man mit lokalen Betrachtungen auskommt und dann auch die Übertragung auf andere Bewertungen ausführen kann, wenn man daran überhaupt interessiert sein sollte.

Wir machen die Substitutionen $x_k = yz_k$ $(k = 1, \ldots, n)$ und führen dadurch anstelle von $y, x_1, \ldots, x_n$ die neuen unabhängigen Variablen $y, z_1, \ldots, z_n$ ein. In der Gleichung

$F = y^r(1 + w)$ erhalten wir für w eine gewöhnliche Potenzreihe in $y, z_1, \ldots, z_n$, welche sicherlich im Bereich $|y| < \varrho$, $|z_k| < \frac{\sigma}{\varrho}$ $(k = 1, \ldots, n)$ konvergiert und kein konstantes Glied hat. Es wird dann auch der Ausdruck

$$\log(1 + w) = w - \frac{w^2}{2} + \cdots$$

eine konvergente Potenzreihe in $y, z_1, \ldots, z_n$ ohne konstantes Glied. Für jeden Summanden $cy^k z_1^{k_1} \cdots z_n^{k_n}$ bilden wir den Wert

$$k - (k_1 + \cdots + k_n) = l,$$

den wir als Gewicht dieses Reihengliedes bezeichnen wollen. Es ist dabei klar, daß das Gewicht des Produktes zweier Reihenglieder gleich der Summe ihrer einzelnen Gewichte ist und daß durch Angabe der Exponenten $k_1, \ldots, k_n$ zusammen mit l auch wiederum der Exponent $k = l + (k_1 + \cdots + k_n)$ festgelegt wird. Wir zerlegen nun

$$\log(1 + w) = A + B,$$

wobei B genau alle Glieder negativen Gewichts $l = -1, -2, \ldots$ umfaßt und A die Glieder mit $l = 0, 1, \ldots$. Für die betreffenden gewöhnlichen Potenzreihen in den Variablen $y, z_1, \ldots, z_n$ gilt dann wieder die Identität

$$F e^{-A} = y^r e^B$$

mit $e^{-A} = 1 - A + \cdots$, $e^B = 1 + B + \cdots$.

Ist nun $cy^l x_1^{k_1} \cdots x_n^{k_n}$ ein Glied von F in den alten Variablen, so geht es durch die Substitution in $cy^k z_1^{k_1} \cdots z_n^{k_n}$ über und hat dann also das nicht-negative Gewicht l. Daher bekommt man auf der linken Seite der Identität kein Glied negativen Gewichts. Auf der rechten Seite sind aber alle auftretenden Gewichte $\leqq r$ und genauer $< r$ mit Ausnahme des einzigen Gliedes y^r selbst. Also können überhaupt nur die Gewichte 0 bis r wirklich auftreten und liefern beim Übergang zu den alten Variablen $y, x_1, \ldots, x_n$ die gesuchte Potenzreihe Φ. Da außerdem absolute Konvergenz für $y = \frac{\varrho}{2}$, $z_k = \frac{\sigma}{\varrho}$, $x_k = \frac{\sigma}{2}$ $(k = 1, \ldots, n)$ vorliegt, so auch für $|y| < \frac{\varrho}{2}$, $|x_k| < \frac{\sigma}{2}$, womit alles bewiesen ist.

Es ist klar, daß in der neuen Fassung des Beweises der Eindeutigkeitssatz nur für gewöhnliche Potenzreihen und nicht mehr für die Laurentsche Entwicklung benötigt wird. Ferner ergeben sich A, B, Φ und auch $e^A = 1 + A + \cdots = E$ durch sehr einfaches Rechnen mit Potenzreihen, während der Brill-Lasker-Späthsche Koeffizientenvergleich zu einer etwas mühsameren Rekursion führt, die auch in der abgeänderten Darstellung bei Abhyankar, Cartan und Kneser [13] noch nicht so ganz befriedigen wird. Da außerdem der Konvergenzbeweis völlig evident ist, scheint mir der 80 Jahre alte Stickelbergersche Beweis auch heute noch der kürzeste, durchsichtigste und zweckmäßigste zu sein.

Literatur

[1] H. Poincaré, Oeuvres **1**, LIII—LVIII.
[2] K. Weierstrass, Mathematische Werke **2**, 135—142.
[3] W. Wirtinger, Über den Weierstrass'schen Vorbereitungssatz, J. reine angew. Math. **158** (1927), 260—267.
[4] L. Stickelberger, Ueber einen Satz des Herrn Noether, Math. Ann. **30** (1887), 401—409.
[5] F. Hartogs, Über die elementare Herleitung des Weierstrass'schen „Vorbereitungssatzes", Bayer. Akad. Wiss. Math.-Naturw. Kl. S.—B. 1909, Nr. 3, 12 S.
[6] H. Späth, Der Weierstraßsche Vorbereitungssatz, J. reine angew. Math. **161** (1929), 95—100.
[7] H. Cartan, Sur le théorème de préparation de Weierstraß, Festschr. zur Gedächtnisfeier für K. Weierstraß (1966), 155—168.
[8] A. Brill, Über den Weierstraßschen Vorbereitungssatz, Math. Ann. **69** (1910), 538—549.
[9] E. Lasker, Zur Theorie der Moduln und Ideale, Math. Ann. **60** (1905), 20—116, insbes. 89.
[10] A. Brill, Ueber das Verhalten einer Funktion von zwei Veränderlichen in der Umgebung einer Nullstelle, Bayer. Akad. Wiss. Math.-Naturw. Kl. S.—B. 1891, 207—220.
[11] L. Hörmander, An introduction to complex analysis in several variables, D. van Nostrand Comp., Inc., Princeton (N. Y.) 1966, 144.
[12] S. S. Abhyankar, Local analytic geometry, Academic Press, New York 1964, 100—105.
[13] H. Kneser, Funktionentheorie, 2. Aufl., Vandenhoeck & Ruprecht, Göttingen 1966, 127 bis 130.

Arnold Walfisz† und Anna Walfisz in Tbilissi

Über Gitterpunkte in mehrdimensionalen Kugeln IV

§ 1. Bezeichnungen. Problemstellung

Im folgenden bezeichnen a, h, m ganze Zahlen; c, d, j, n, q, r positive ganze Zahlen; α nichtnegative ganze Zahlen; k ganze Zahlen $\geqq 3$; u, v, M, H positive ungerade ganze Zahlen; w ungerade ganze Zahlen; y, X, Y reelle Zahlen; s, t positive Zahlen. Ferner sei $x \geqq 3$, $0 < \varepsilon < 1$.

Der Buchstabe B ohne Indizes bezeichnet unterschiedslos Zahlen, die ihrem absoluten Betrag nach unterhalb von Schranken liegen, die nur von k abhängen dürfen.

$d|m$ bedeutet, daß d in m als Teiler aufgeht; dagegen bedeutet d/m einen Bruch mit dem Zähler d und Nenner m.

$\left(\frac{a}{u}\right)$ ist für $u > 1$ das Jacobische Symbol; es ist gleich 1 für $u = 1$. Ferner ist $\left(\frac{-1}{-u}\right) = -\left(\frac{-1}{u}\right)$.

Weiter sei

$$e(y) = e^{2\pi i y}.$$

In der Summe $\sum\limits_{a \bmod q}$ durchläuft a ein vollständiges Restsystem mod q und in der Summe $\sum\limits_{a \bmod q}{}'$ ein reduziertes System. In allen übrigen Summen ist die untere Summationsgrenze, falls sie nicht ausdrücklich angegeben ist, gleich Eins.

Es sei $r_q(y)$ die Anzahl der Darstellungen von y als Summe von q Quadraten ganzer Zahlen;

$$A_q(t) = \sum_{0 \leqq m \leqq t} r_q(m) = 1 + \sum_{n \leqq t} r_q(n)$$

die Anzahl der Gitterpunkte $(a_1, \ldots, a_q)$ in der q-dimensionalen Kugel

$$y_1^2 + \cdots + y_q^2 \leqq t.$$

Ferner bezeichne

$$V_q(t) = \frac{\pi^{q/2}}{\Gamma(q/2+1)} t^{q/2} \tag{1}$$

das Volumen dieser Kugel und

$$P_q(t) = A_q(t) - V_q(t) \tag{1_1}$$

den Gitterrest. Weiter sei

$$D_q = \frac{\pi^{q/2}}{\Gamma(q/2)}. \tag{2}$$

Für $s > 1$ sei

$$\zeta(s) = \sum_{n=1}^{\infty} n^{-s}, \quad L(s) = \sum_{u=1}^{\infty} \left(\frac{-1}{u}\right) u^{-s},$$

$$Z(s) = Z(s; k) = \sum_{u=1}^{\infty} \left(\frac{-1}{u}\right)^k u^{-s},$$

also

$$Z(s) = \begin{cases} L(s) & (k \text{ ungerade}), \\ (1-2^{-s})\,\zeta(s) & (k \text{ gerade}); \end{cases} \tag{3}$$

$$\{Z(s)\}^{-1} = \sum_{u=1}^{\infty} \left(\frac{-1}{u}\right)^k \mu(u)\, u^{-s}. \tag{4}$$

In der vorliegenden Arbeit benutzen wir oft Ergebnisse der Monographie [5] „Gitterpunkte in mehrdimensionalen Kugeln" von Arnold Walfisz. Daher ist, wenn wir z. B. auf Hilfssatz 5.1.5 hinweisen, Hilfssatz 5 von Kapitel V, § 1 in [5] gemeint. Dasselbe gilt natürlich auch für Sätze und Formeln in [5]. Dagegen sind Sätze, Hilfssätze und Formeln in dieser Arbeit durchnumeriert.

Für die Gaußsche Summe

$$S(h, q) = \sum_{a \bmod q} e\left(\frac{ha^2}{q}\right) \tag{5}$$

gelten bekanntlich, sofern $(h, q) = 1$ ist, folgende einfache Tatsachen (vgl. z. B. (1.1.2) und (1.1.9)):

$$|S(h, q)| \leqq (2q)^{1/2}; \tag{6}$$

$$S^2(h, q) = \begin{cases} \left(\frac{-1}{q}\right) q & \text{für} \quad q \equiv 1 (\bmod 2), \\ 0 & \text{für} \quad q \equiv 2 (\bmod 4), \\ 2i^h q & \text{für} \quad q \equiv 0 (\bmod 4). \end{cases} \tag{7}$$

Für beliebige h, q gilt außerdem noch

$$S(dh, dq) = dS(h, q). \tag{8}$$

Ferner werden wir im folgenden auch die sogenannte Ramanujansche Summe

$$C(h, q) = \sum_{a \bmod q}' e\left(\frac{ha}{q}\right) = \sum_{a \bmod q}' e\left(\frac{-ha}{q}\right) \tag{9}$$

verwenden. Aus ihren Eigenschaften brauchen wir nur die Beziehung

$$C(h, q) = \sum_{d \mid (h,q)} d\mu\left(\frac{q}{d}\right) \tag{10}$$

(vgl. z. B. Hilfssatz 5.1.7).

Es sei $B_n(y)$ das n-te Bernoullische Polynom in der üblichen Bezeichnungsweise. Ferner werde

$$\overline{B}_n(y) = B_n(y - [y])$$

gesetzt, also insbesondere

$$\overline{B}_1(y) = \psi(y) = y - [y] - \frac{1}{2}. \tag{11}$$

Bekanntlich gilt

$$\frac{d}{dy}\overline{B}_{n+1}(y) = (n+1)\,\overline{B}_n(y). \tag{12}$$

Es sei

$$F_r(y) = \overline{B}_r\left(\frac{y}{2}\right) - 2^r\overline{B}_r\left(\frac{y}{4}\right), \tag{13}$$

$$G_r(y) = \overline{B}_r(y) - 2^{r-1}\overline{B}_r\left(\frac{y}{2}\right) - 2^{2r-1}\overline{B}_r\left(\frac{y-1}{4}\right). \tag{14}$$

Also haben die Funktionen $\overline{B}_r(y)$ und $\psi(y)$ die Periode 1, die Funktionen $F_r(y)$ und $G_r(y)$ die Periode 4.

Weiter sei

$$\Psi_{k,r}(y) = \sum_{u=1}^{\infty} u^{r-k}\overline{B}_r\left(\frac{y}{u}\right) + (-1)^{k/2+1}\,2^{r-k}\sum_{d=1}^{\infty} d^{r-k}F_r\left(\frac{y}{d}\right)$$

$$(k \geqq 4 \text{ gerade}, r \leqq k-2),$$

$$\Phi_{k,r}(y) = \sum_{u=1}^{\infty}\left(\frac{-1}{u}\right)u^{r-k}\overline{B}_r\left(\frac{y}{u}\right) + (-1)^{(k+1)/2}\,2^{1-k}\sum_{d=1}^{\infty} d^{r-k}G_r\left(\frac{y}{d}\right) \tag{15}$$

$$(k \geqq 3 \text{ ungerade}, r \leqq k-2),$$

$$\Omega_{k,r}(y) = \begin{cases} \Psi_{k,r}(y) & \text{für gerade } k, \\ \Phi_{k,r}(y) & \text{für ungerade } k; \end{cases} \tag{16}$$

$$\Psi^*_{k,r}(y) = \sum_{u=1}^{\infty} u^{r-k-1} \sum_{a \bmod u} \overline{B}_r\left(\frac{y-a^2}{u}\right)$$
$$+ (-1)^{k/2+1}\, 2^{r-k-2} \sum_{d=1}^{\infty} d^{r-k-1} \sum_{a \bmod 4d} F_r\left(\frac{y-a^2}{d}\right) \tag{17}$$
$$(k \geqq 4 \text{ gerade}, \ r \leqq k-2),$$

$$\Phi^*_{k,r}(y) = \sum_{u=1}^{\infty} \left(\frac{-1}{u}\right) u^{r-k-1} \sum_{a \bmod u} \overline{B}_r\left(\frac{y-a^2}{u}\right)$$
$$+ (-1)^{(k+1)/2}\, 2^{-k-1} \sum_{d=1}^{\infty} d^{r-k-1} \sum_{a \bmod 4d} G_r\left(\frac{y-a^2}{d}\right) \tag{18}$$
$$(k \geqq 3 \text{ ungerade}, \ r \leqq k-2)$$

$$\Omega^*_{k,r}(y) = \begin{cases} \Psi^*_{k,r}(y) & \text{für gerade } k, \\ \Phi^*_{k,r}(y) & \text{für ungerade } k. \end{cases} \tag{19}$$

Lursmanaschwili [2] hat zwei Näherungssätze für $P_{2k}(x)$ ($k \geqq 4$) gefunden, sie sind in [5], §§ 5.1 – 5.3, wiedergegeben. Arnold Walfisz hat sie auf folgende Weise in einem Satz zusammengefaßt (siehe [6], Satz 5):

Satz 1. *Es sei $k \geqq 4$. Dann ist*

$$P_{2k}(x) - D_{2k}\{kZ(k)\}^{-1} \sum_{r \leqq k/2-1} (-1)^r \binom{k}{r} \Omega_{k,r}(x)\, x^{k-r}$$
$$= \begin{cases} Bx^{k/2} \log x & \textit{für gerade } k, \\ Bx^{(k+1)/2} \log x & \textit{für ungerade } k. \end{cases}$$

In [3] hat Lursmanaschwili einen Näherungssatz für $P_{2k+1}(x)$ ($k \geqq 4$) bewiesen, er ist in [5], § 5.4, und [7], Satz 4, in leicht veränderter Form wiedergegeben. Wir bringen die Formulierung aus [7]:

Satz 2. *Es sei $k \geqq 4$. Dann ist*

$$P_{2k+1}(x) = \pi^{k+1/2}\{Z(k)\}^{-1} \sum_{r \leqq k/2-1} \frac{(-1)^r}{r!} \frac{x^{k-r+1/2}}{\Gamma\left(k-r+\frac{3}{2}\right)} \Omega^*_{k,r}(x)$$
$$= \begin{cases} Bx^{(k+1)/2} \log x & \textit{für gerade } k, \\ Bx^{k/2+1} \log x & \textit{für ungerade } k. \end{cases}$$

Es sei weiter

$$\mathfrak{R}_k(x, q, r) = \sum_{h \bmod q}{}' \left(\frac{S(h, q)}{q}\right)^k \sum_{\substack{m=-\infty \\ m \neq -h/q}}^{\infty} \frac{e\{-(h/q + m)\, x\}}{(h/q + m)^r},$$

$$\mathfrak{R}_k(x, r) = \sum_{q=1}^{\infty} \mathfrak{R}_k(x, q, r) \quad \left(k > 4,\ r < \frac{k}{2} - 1\right). \tag{20}$$

Die Reihe (20) konvergiert absolut und gleichmäßig in x; es ist nämlich (vgl. [5], Hilfssatz 4.1.3)

$$\mathfrak{R}_k(x, q, r) = Bq^{r-k/2} \log 3q.$$

Petersson in [4] hat folgenden Satz für $k > 4$ bewiesen (siehe auch [5], Satz 4.1.1):

Satz 3.

$$\frac{1}{2}\{P_k(x+0) + P_k(x-0)\}$$

$$= -\sum_{r \leqq (k-1)/4} \frac{1}{(2\pi i)^r} \frac{\pi^{k/2}}{\Gamma(k/2 - r + 1)} x^{k/2 - r} \mathfrak{R}_k(x, r) + Bx^{k/4} \log x.$$

Arnold Walfisz hat einen Zusammenhang zwischen den Peterssonschen Funktionen (20) im Falle eines geraden k bzw. eines ungeraden k und den Lursmanaschwilischen Funktionen $\Omega_{k,r}(x)$ bzw. $\Omega^*_{k,r}(x)$ hergestellt (siehe [6], Satz 6, und [7], Satz 5). Durch Verknüpfung dieser beiden Sätze mit dem Peterssonschen Satz 3 erhielt er als Folgerungen:

Satz 4 ([6], Satz 7). *Für $k \geqq 3$ ist*

$$P_{2k}(x) = D_{2k}\{kZ(k)\}^{-1} \sum_{r \leqq (k-1)/2} (-1)^r \binom{k}{r} \Omega_{k,r}(x)\, x^{k-r} + Bx^{k/2} \log x. \tag{21}$$

und

Satz 5 ([7], Satz 6). *Für $k \geqq 3$ ist*

$$P_{2k+1}(x) = \pi^{k+1/2}\{Z(k)\}^{-1} \sum_{r \leqq k/2} \frac{(-1)^r}{r!} \frac{x^{k-r+1/2}}{\Gamma(k - r + 3/2)} \Omega^*_{k,r}(x)$$

$$+ Bx^{k/2+1/4} \log x.$$

Satz 4 fällt für gerade k mit Satz 1 zusammen. Für ungerade k ist er in zweifacher Hinsicht günstiger: Einmal wird der Wert $k = 3$ zugelassen und dann enthält die r-Summe ein Glied mehr, und das Restglied ist um $x^{1/2}$ besser. Satz 5 ist auch in zweifacher Hinsicht günstiger als Satz 2: Einmal wird der Wert $k = 3$ zugelassen und sodann enthält die r-Summe ein Glied mehr, und das Restglied ist für gerade k um $x^{1/4}$, für ungerade k um $x^{3/4}$ besser.

Wie schon bemerkt wurde, mußte Arnold Walfisz, um die Sätze 4 und 5 in [6] bzw. [7] beweisen zu können, die Gültigkeit des Peterssonschen Satzes 3 annehmen und dann den Zusammenhang zwischen den oben erwähnten Peterssonschen und Lursmanaschwilischen Funktionen herstellen. Der Beweisgang dieses Zusammenhanges ist aber ziemlich schwierig und lang.

Deshalb ist das Ziel der vorliegenden Arbeit, die Sätze 4 und 5 unmittelbar zu beweisen, und zwar durch Verbesserung der Methode, die zu den Sätzen 1 und 2 führt (siehe [5], Kap. V).

Im folgenden werden Kugeln gerader und ungerader Dimensionen im einzelnen behandelt. Zuvor stellen wir einige Hilfsbetrachtungen an, die für beide Fälle gelten.

§ 2. Hilfsbetrachtungen

Es sei stets $j \leqq k/2$.

Hilfssatz 1. *Für $k \geqq 4$ ist*

$$P_k(x) = D_k \sum_{n \leqq x} n^{k/2-1} - \frac{2}{k} D_k x^{k/2} + D_k \sum_{2 \leqq q \leqq x^{1/2}} \sum_{h \bmod q}{}' \left(\frac{S(h,q)}{q}\right)^k \sum_{n \leqq x} n^{k/2-1} e\left(\frac{-nh}{q}\right) + Bx^{k/4} \log x. \tag{22}$$

Beweis. Da nach einem Resultat von Landau [1] (siehe auch [5], Satz 1.4.1)

$$A_k(x) = D_k \sum_{q \leqq x^{1/2}} \sum_{h \bmod q}{}' \left(\frac{S(h,q)}{q}\right)^k \sum_{n \leqq x} n^{k/2-1} e\left(\frac{-nh}{q}\right) + Bx^{k/4} \log x$$

ist, folgt die Behauptung (22) aus (1), (1_1) und (2).

Der folgende Hilfssatz ist eine Variante der bekannten Euler-Soninschen Summenformel (siehe [2], Hilfssatz 21, sowie [5], Hilfssatz 1.3.2).

Hilfssatz 2. *Die (nicht notwendig reelle) Funktion $f(y)$ habe eine stetige j-te Ableitung im Intervall $X \leqq y \leqq Y$, $X < Y$.*

Dann ist

$$\sum_{X < m \leqq Y} f(m) = \int_X^Y f(y)\,dy + \sum_{r=1}^{j} \frac{(-1)^r}{r!} \{\bar{B}_r(Y) f^{(r-1)}(Y) - \bar{B}_r(X) f^{(r-1)}(X)\} + \frac{(-1)^{j+1}}{j!} \int_X^Y \bar{B}_j(y) f^{(j)}(y)\,dy.$$

Hilfssatz 3. *Es sei* $l = k$ *oder* $l = k + \frac{1}{2}$. *Dann ist*

$$\sum_{\substack{n \leqq s \\ n \equiv h (\bmod c)}} n^{l-1} = \frac{1}{l}\frac{s^l}{c} + \frac{1}{l}\sum_{r=1}^{j}(-1)^r\binom{l}{r}s^{l-r}c^{-l+r}\overline{B}_r\left(\frac{s-h}{c}\right) + Bs^{l-1-j}c^j.$$

Beweis. Es ist

$$\sum_{\substack{n \leqq s \\ n \equiv h (\bmod c)}} n^{l-1} = \sum_{\substack{m \\ 0 < h + mc \leqq s}} (h + mc)^{l-1} = c^{l-1} \sum_{-h/c < m \leqq (s-h)/c} (h/c + m)^{l-1}. \tag{22$_1$}$$

Wendet man Hilfssatz 2 an mit $X = -\frac{h}{c}$, $Y = \frac{s-h}{c}$, $f(y) = \left(\frac{h}{c} + y\right)^{l-1}$, so ergibt sich

$$\sum_{-h/c < m \leqq (s-h)/c} \left(\frac{h}{c} + m\right)^{l-1} = \frac{1}{l}\left(\frac{s}{c}\right)^l + \frac{1}{l}\sum_{r=1}^{j}(-1)^r\binom{l}{r}s^{l-r}c^{-l+r}\overline{B}_r\left(\frac{s-h}{c}\right)$$
$$+ B\int_{-h/c}^{(s-h)/c} \overline{B}_j(y)\,(h/c + y)^{l-1-j}\,dy. \tag{22$_2$}$$

Hierin ist nach (12) und dem zweiten Mittelwertsatz das Integral auf der rechten Seite gleich $B\left(\frac{s}{c}\right)^{l-1-j}$. Der Hilfssatz 3 folgt also aus (22_1) und (22_2).

Hilfssatz 4. *Es sei*

$$\sigma_q(n) = \sum_{d \mid n} d^q \tag{23}$$

die Summe der q-ten Potenzen der Teiler von n. Dann ist

$$\sigma_q(n) = \begin{cases} Bn \log 2n & (q = 1), \\ Bn^q & (q > 1). \end{cases} \tag{23$_1$}$$

Beweis. Die Behauptung folgt aus (23), da n/d zugleich mit d alle Teiler von n durchläuft.

Hilfssatz 5. *Es sei*

$$S_q(x) = \sum_{n \leqq x} n^{k-1} \sum_{d \mid (n,q)} d\mu\left(\frac{q}{d}\right). \tag{24}$$

Für $q > 1$ *ist*

$$S_q(x) = \frac{1}{k}\sum_{r=1}^{j}(-1)^r\binom{k}{r}x^{k-r}\sum_{d \mid q}\mu\left(\frac{q}{d}\right)d^r\overline{B}_r\left(\frac{x}{d}\right) + Bx^{k-j-1}q^{j+1}. \tag{25}$$

Beweis. Dieser Hilfssatz stellt den Hilfssatz 5.1.5 dar mit etwas besserem Restglied. Er wird aber ohne die Hilfssätze 5.1.3 und 5.1.4 bewiesen. Wendet man nämlich den

Hilfssatz 3 mit $l = k$, $s = x/d$, $h = 0$, $c = 1$ an, so ergibt sich aus (24)

$$S_q(x) = \sum_{d|q} d\mu\left(\frac{q}{d}\right) \sum_{\substack{n \leqq x \\ d|n}} n^{k-1} = \sum_{d|q} d^k \mu\left(\frac{q}{d}\right) \sum_{n \leqq x/d} n^{k-1}$$

$$= \sum_{d|q} d^k \mu\left(\frac{q}{d}\right) \left\{\frac{1}{k}\left(\frac{x}{d}\right)^k + \frac{1}{k}\sum_{r=1}^{j} (-1)^r \binom{k}{r} \left(\frac{x}{d}\right)^{k-r} \bar{B}_r\left(\frac{x}{d}\right) + B\left(\frac{x}{d}\right)^{k-1-j}\right\}. \tag{26}$$

Nach einer bekannten Eigenschaft der Möbiusschen Funktion fällt das erste Glied auf der rechten Seite weg, das dritte Glied ist auf Grund von Hilfssatz 4 gleich $Bx^{k-j-1}q^{j+1}$. Die Behauptung (25) ergibt sich daher aus (26).

Hilfssatz 6. *Es sei*

$$T_u(x) = \sum_{v \leqq x} \left(\frac{-1}{v}\right) v^{k-1} \sum_{d|(u,v)} d\mu\left(\frac{u}{d}\right). \tag{27}$$

Dann ist für $x \geqq 3u$

$$T_u(x) = -\frac{1}{k}\sum_{r=1}^{j} (-1)^r \binom{k}{r} x^{k-r} \sum_{n|u} \left(\frac{-1}{n}\right) \mu\left(\frac{u}{n}\right) n^r G_r\left(\frac{x}{n}\right) + Bx^{k-j-1}u^{j+1},$$

wobei $G_r(y)$ *durch Formel* (14) *definiert ist.*

Beweis. Man kann hier Ähnliches bemerken wie beim Hilfssatz 5. Hilfssatz 6 hat nämlich ein etwas besseres Restglied als Hilfssatz 5.3.4, außerdem werden die Hilfssätze 5.3.1 und 5.3.2 beim Beweis nicht benutzt. Es ist

$$T_u(x) = \sum_{d|u} d\mu\left(\frac{u}{d}\right) \sum_{\substack{v \leqq x \\ d|v}} \left(\frac{-1}{v}\right) v^{k-1} = \sum_{d|u} \left(\frac{-1}{d}\right) d^k \mu\left(\frac{u}{d}\right) \sum_{v \leqq x/d} \left(\frac{-1}{v}\right) v^{k-1}. \tag{28}$$

Da Hilfssatz 5.3.3 auch für unsere Werte von j gilt, kann man auf die v-Summe in (28) die Formel (5.3.13) mit x/d statt x anwenden. Die dort vorkommende Funktion $G_r(y, q)$ ist aber unsere Funktion $G_r\left(\frac{y}{q}\right)$ aus (14). Es ergibt sich daher

$$T_u(x) = \sum_{d|u} \left(\frac{-1}{d}\right) \mu\left(\frac{u}{d}\right) d^k \left\{-\frac{1}{k}\sum_{r=1}^{j} (-1)^r \binom{k}{r} \left(\frac{x}{d}\right)^{k-r} G_r\left(\frac{x}{d}\right) + B\left(\frac{x}{d}\right)^{k-j-1}\right\}.$$

Hierin ist das Restglied auf Grund von Hilfssatz 4 gleich

$$Bx^{k-j-1}\sum_{d|u} d^{j+1} = Bx^{k-j-1}u^{j+1},$$

womit der Hilfssatz 6 bewiesen ist.

Es sei weiter

$$T(m, q) = \sum_{\substack{u=1 \\ (u,q)=1}}^{4q-1} \left(\frac{-1}{u}\right) e\left(\frac{mu}{4q}\right). \tag{29}$$

Hilfssatz 7.

$$T(m, 2^\alpha H) = 0 \quad \textit{für} \quad m \neq 2^\alpha w, \tag{30}$$

$$T(2^\alpha w, 2^\alpha H) = 2^{\alpha+1}\left(\frac{-1}{wH}\right) i \sum_{u|(w,H)} u\mu\left(\frac{H}{u}\right). \tag{31}$$

Beweis. Die Behauptung (30) für $m = 0$ folgt aus der Formel

$$T(0, q) = \sum_{\substack{u=1 \\ (u,q)=1}}^{4q-1} \left(\frac{-1}{u}\right),$$

die Glieder u und $4q - u$ heben sich nämlich gegenseitig auf. Da ferner $T(-n, q)$ die zu $T(n, q)$ konjugiert komplexe Zahl ist, folgen die restlichen Behauptungen aus Hilfssatz 5.3.1.

Hilfssatz 8. *Es sei*

$$\sum_{a \bmod q} \psi\left(\frac{y - a^2}{q}\right) = E(y, q), \tag{32}$$

wobei $\psi(y)$ *die Funktion* (11) *ist. Dann gilt die Abschätzung*

$$E(y, q) = B\sigma_{1/2}(q) \log 2q. \tag{32_1}$$

Beweis. Da $E(y, q)$ für y die Periode q hat, darf man $y \geqq 1$ annehmen. Außerdem kann man nur ganzzahlige $y = n$ betrachten. Es genügt also, statt (32_1) die Abschätzung

$$E(n, q) = B\sigma_{1/2}(q) \log 2q \tag{33}$$

zu beweisen.

Ist andererseits y keine ganze Zahl, so gelten bekanntlich die Beziehungen

$$\psi(y) = -\frac{1}{\pi}\sum_{m=1}^{\infty} \frac{\sin 2\pi my}{m} \tag{34}$$

und

$$s(y, r) = \sum_{m=r+1}^{\infty} \frac{\sin 2\pi m y}{m} = \frac{B}{r\{y\}}, \tag{35}$$

wobei hier $\{y\}$ den Abstand zwischen y und der nächsten ganzen Zahl bedeutet.

Aus (32) folgt

$$E(n, q) = \sum_{a \bmod q} \psi\left(\frac{n-a^2}{q}\right) = E_1(n, q) + E_2(n, q) + E_3(n, q). \tag{36}$$

Hierin ist

$$E_1(n, q) = \sum_{\substack{a \bmod q \\ a^2 \equiv n (\bmod q)}} \psi\left(\frac{n-a^2}{q}\right) = B \sum_{\substack{a \bmod q \\ a^2 \equiv n (\bmod q)}} 1 = Bq^{1/4}, \tag{37}$$

da die Summe rechts der Lösungszahl der Kongruenz

$$a^2 \equiv n \ (\bmod q)$$

gleich ist. Es könnte statt $\frac{1}{4}$ auch eine beliebige Zahl $\varepsilon < \frac{1}{2}$ genommen werden.

Ferner ist wegen (34), (5), (8) und (6)

$$E_2(n, q) = -\frac{1}{\pi} \sum_{m=1}^{q^2} \frac{1}{m} \sum_{\substack{a \bmod q \\ a^2 \not\equiv n (\bmod q)}} \sin \frac{2\pi m(n-a^2)}{q} = B \sum_{d \mid q} \sum_{\substack{m=1 \\ (m,q)=d}}^{q^2} \frac{1}{m} |S(m, q)|$$

$$= B \sum_{d \mid q} \sum_{\substack{m=1 \\ (m, q/d)=1}}^{q^2} \frac{1}{m} \left| S\left(m, \frac{q}{d}\right) \right|$$

$$= B \sum_{d \mid q} \left(\frac{q}{d}\right)^{1/2} \sum_{m=1}^{q^2} \frac{1}{m} = B \log 2q \sum_{d \mid q} d^{1/2}.$$

Berücksichtigt man die Definition (23) von $\sigma_a(n)$, so erhält man hieraus für $E_2(n, q)$ die Abschätzung

$$E_2(n, q) = B\sigma_{1/2}(q) \log 2q. \tag{38}$$

Weiter ist nach (35)

$$E_3(n, q) = -\frac{1}{\pi} \sum_{\substack{a \bmod q \\ a^2 \not\equiv n (\bmod q)}} s\left(\frac{n-a^2}{q}, q^2\right)$$

$$= -\frac{1}{\pi} \sum_{\substack{a \bmod q \\ a^2 \not\equiv n (\bmod q)}} \frac{1}{q^2} \frac{1}{\left\{\frac{n-a^2}{q}\right\}} = B \tag{39}$$

wegen $\left\{\frac{n-a^2}{q}\right\} \geqq \frac{1}{q}$. Setzt man jetzt (37), (38) und (39) in (36) ein, so ergibt sich die zu beweisende Abschätzung (33).

Auf Grund von (11), (13) und (14) erhält man für $r > 1$ folgende Abschätzungen:

$$\sum_{a \bmod d} \bar{B}_r\left(\frac{y-a^2}{d}\right) = Bd, \tag{40}$$

$$\sum_{a \bmod 4d} F_r\left(\frac{y-a^2}{d}\right) = Bd, \tag{41}$$

$$\sum_{a \bmod 4d} G_r\left(\frac{y-a^2}{d}\right) = Bd. \tag{42}$$

Für $r = 1$ ergeben sich dagegen auf Grund von Hilfssatz 8 bessere Abschätzungen:

$$\sum_{a \bmod d} \bar{B}_1\left(\frac{y-a^2}{d}\right) = \sum_{a \bmod d} \psi\left(\frac{y-a^2}{d}\right) = E(y, d) = B\sigma_{1/2}(d)\log 2d, \tag{43}$$

$$\sum_{a \bmod 4d} F_1\left(\frac{y-a^2}{d}\right) = 2E(y, 2d) - 2E(y, 4d) = B\sigma_{1/2}(d)\log 2d, \tag{44}$$

$$\begin{aligned}\sum_{a \bmod 4d} G_1\left(\frac{y-a^2}{d}\right) &= 4E(y, d) - 2E(y, 2d) - 2E(y-d, 4d) \\ &= B_{1/2}(d)\log 2d.\end{aligned} \tag{45}$$

Schließlich beweisen wir noch den

Hilfssatz 9. *Es ist*

$$\sum_{q > x^{1/2}} q^{-k} \sum_{d \mid q} \sigma_{1/2}(d)\log 2d = Bx^{3/4-k/2}\log x. \tag{46}$$

Beweis.

$$\begin{aligned}\sum_{q > x^{1/2}} q^{-k} \sum_{d \mid q} \sigma_{1/2}(d)\log 2d &= \sum_{d=1}^{\infty} \sigma_{1/2}(d)\log 2d \sum_{\substack{q > x^{1/2} \\ d \mid q}} q^{-k} \\ &= \sum_{d \leqq x^{1/2}} d^{-k}\sigma_{1/2}(d)\log 2d \sum_{q > x^{1/2}/d} q^{-k} \\ &\quad + B \sum_{d > x^{1/2}} d^{-k}\sigma_{1/2}(d)\log 2d \\ &= x^{(1-k)/2} M_1 + M_2.\end{aligned} \tag{47}$$

Hierin ist

$$M_1 = B \sum_{d \leqq x^{1/2}} d^{-1}\sigma_{1/2}(d) \log 2d = B \log x \sum_{d \leqq x^{1/2}} d^{-1}\sigma_{1/2}(d)$$

$$= B \log x \sum_{d \leqq x^{1/2}} d^{-1} \sum_{rn = d} n^{1/2}$$

$$= Bx^{1/4} \log x. \tag{48}$$

Für M_2 erhalten wir

$$M_2 = B \sum_{d > x^{1/2}} d^{-k} \log d \sum_{rn = d} n^{1/2} = B \sum_{rn > x^{1/2}} r^{-k} n^{1/2-k} \log r$$

$$+ B \sum_{rn > x^{1/2}} r^{-k} n^{1/2-k} \log n = M_3 + M_4 \tag{49}$$

mit

$$M_3 = B \sum_{r \leqq x^{1/2}} r^{-k} \log r \sum_{n > x^{1/2}/r} n^{1/2-k} + B \sum_{r > x^{1/2}} r^{-k} \log r$$

$$= Bx^{3/4-k/2} \log x \tag{50}$$

und

$$M_4 = B \sum_{n \leqq x^{1/2}} n^{1/2-k} \log n \sum_{r > x^{1/2}/n} r^{-k} + B \sum_{n > x^{1/2}} n^{1/2-k} \log n$$

$$= Bx^{3/4-k/2} \log x. \tag{51}$$

Aus (47) bis (51) folgt jetzt die Behauptung (46) unseres Hilfssatzes.

§ 3. Beweis von Satz 4

In diesem Paragraphen sollen nur Kugeln gerader Dimension $2k (k \geqq 3)$ betrachtet werden.

Ausgangspunkt bei unserem Beweis ist die Beziehung (22), die nach Hilfssatz 3 mit $l = k$, $s = x$, $h = 0$, $c = 1$ folgende Beziehung ergibt:

$$P_{2k}(x) = \frac{D_{2k}}{k} \sum_{r=1}^{j} (-1)^r \binom{k}{r} x^{k-r} \bar{B}_r(x)$$

$$+ D_{2k} \sum_{2 \leqq q \leqq x^{1/2}} \sum_{h \bmod q}{}' \left(\frac{S(h, q)}{q}\right)^{2k} \sum_{n \leqq x} n^{k-1} e\left(\frac{-nh}{q}\right) + Bx^{k/2} \log x, \tag{52}$$

wobei $j \leqq \frac{k}{2}$ ist. Ohne Beschränkung der Allgemeinheit sei $x \geqq 16$.

Für k gerade, $k \geqq 4$, ergibt sich hier für die q-Summe auf Grund von (7), (9) und (10)

$$\sum_{2 \leqq q \leqq x^{1/2}} = \sum_{3 \leqq u \leqq x^{1/2}} u^{-k} S_u(x) + (-4)^{k/2} \sum_{\substack{4 \leqq q \leqq x^{1/2} \\ q \equiv 0 (\mathrm{mod}\, 4)}} q^{-k} S_q(x), \tag{53}$$

wobei $S_q(x)$ die Funktion (24) ist. Hierbei bemerken wir, daß, wenn q alle durch 4 teilbaren Zahlen im Intervall $4 \leqq q \leqq x^{1/2}$ durchläuft, wir immer q durch $4q$ $\left(1 \leqq q \leqq \frac{x^{1/2}}{4}\right)$ ersetzen. Die Beziehung (53) wird jetzt ebenso behandelt, wie in [5], § 5.2; nur benutzt man für $S_q(x)$ statt Formel (5.1.13) die Abschätzung (25). Der Satz 4 für k gerade, $k \geqq 4$, folgt deswegen aus (52) und (53), wenn man $j = k/2 - 1$ annimmt.

Es sei k ungerade, $k \geqq 3$. Wir nehmen $j = \frac{k-1}{2}$ an.

Wegen $i^{uk} = i^k \left(\frac{-1}{u}\right)$ ist jetzt die q-Summe in (52) nach (7), (9), (10), (24) und (29) gleich

$$\sum_{2 \leqq q \leqq x^{1/2}} = A_1 + (2i)^k A_2, \tag{54}$$

wobei A_1 und A_2 die Funktionen (5.3.24) und (5.3.25) sind, d. h.

$$A_1 = \sum_{3 \leqq u \leqq x^{1/2}} \left(\frac{-1}{u}\right) u^{-k} S_u(x), \tag{55}$$

und

$$A_2 = 4^{-k} \sum_{q \leqq x^{1/2}/4} q^{-k} \sum_{n \leqq x} n^{k-1} T(-n, q). \tag{56}$$

In (55) werde $S_u(x)$ mit Hilfe von (25) ausgedrückt. Dabei ist das Restglied wegen $j = (k-1)/2$ gleich

$$Bx^{k-j-1} \sum_{3 \leqq u \leqq x^{1/2}} u^{j+1-k} = Bx^{(k-1)/2} \sum_{3 \leqq u \leqq x^{1/2}} u^{-1} = Bx^{k/2} \log x,$$

und es ergibt sich

$$A_1 = \frac{1}{k} \sum_{r \leqq (k-1)/2} (-1)^r \binom{k}{r} x^{k-r} \sum_{3 \leqq u \leqq x^{1/2}} \left(\frac{-1}{u}\right) u^{-k} \sum_{d|u} \mu\left(\frac{u}{d}\right) d^r \overline{B}_r\left(\frac{x}{d}\right)$$
$$+ Bx^{k/2} \log x. \tag{57}$$

Läßt man ferner hier u alle ungeraden Zahlen durchlaufen, so ist der Fehler nach (23) und (23_1) gleich

$$B \sum_{r \leqq (k-1)/2} x^{k-r} \sum_{u > x^{1/2}} u^{-k} \sigma_r(u) = B \sum_{r \leqq (k-1)/2} x^{k-r} \sum_{u > x^{1/2}} u^{r-k} \log 2u$$
$$= Bx^{k/2} \log x,$$

da $r \geqq 1$ gilt (vgl. Formel (5.2.5)). Daher ist nach (57)

$$A_1 = \frac{1}{k} \sum_{r \leqq (k-1)/2} (-1)^r \binom{k}{r} x^{k-r} \sum_{u=3}^{\infty} \left(\frac{-1}{u}\right) u^{-k} \sum_{d|u} \mu\left(\frac{u}{d}\right) d^r \overline{B}_r\left(\frac{x}{d}\right)$$
$$+ Bx^{k/2} \log x, \tag{58}$$

das Restglied ist hierbei besser als in Formel (5.3.26).

Es sei $q = 2^\alpha H$. Nach (30) ist die Funktion $T(-n, 2^\alpha H)$ für $n \neq 2^\alpha M$ gleich Null. Für $n = 2^\alpha M$ ist $T(-2^\alpha M, 2^\alpha H)$ durch (31) ausgedrückt. Wenn noch

$$-(2i)^k i\, 2^{1-2k} = (-1)^{(k-1)/2}\, 2^{1-k}$$

berücksichtigt wird, folgt daher aus (56)

$$(2i)^k A_2 = (-1)^{(k-1)/2}\, 2^{1-k} \sum_{2^\alpha H \leqq x^{1/2}/4} \left(\frac{-1}{H}\right) H^{-k} \sum_{M \leqq x/2^\alpha} \left(\frac{-1}{M}\right) M^{k-1}$$
$$\times \sum_{u \mid (H,M)} u\mu\left(\frac{H}{u}\right),$$

d. h.

$$(2i)^k A_2 = (-1)^{(k-1)/2}\, 2^{1-k} \sum_{2^\alpha H \leqq x^{1/2}/4} \left(\frac{-1}{H}\right) H^{-k} T_H\left(\frac{x}{2^\alpha}\right), \tag{59}$$

wobei $T_u(x)$ die Funktion (27) ist. Da aber in (59) $\dfrac{x}{2^\alpha} \geqq 4H$ ist, darf für $T_H\left(\dfrac{x}{2^\alpha}\right)$ der Hilfssatz 6 mit $j = \dfrac{k-1}{2}$ angewendet werden. Dabei ist das Restglied, wenn man noch $2^\alpha H = q$ setzt, gleich

$$Bx^{k-j-1} \sum_{2^\alpha H \leqq x^{1/2}/4} (2^\alpha H)^{-k+j+1} = Bx^{(k-1)/2} \sum_{q \leqq x^{1/2}/4} q^{-1} = Bx^{k/2} \log x,$$

und es ergibt sich

$$(2i)^k A_2 = \frac{(-1)^{(k+1)/2}}{k} 2^{1-k} \sum_{r \leqq (k-1)/2} (-1)^r \binom{k}{r} x^{k-r} \sum_{2^\alpha H \leqq x^{1/2}/4} \left(\frac{-1}{H}\right) H^{-k} 2^{\alpha(r-k)}$$
$$\times \sum_{n \mid H} \left(\frac{-1}{n}\right) \mu\left(\frac{H}{n}\right) n^r Gr\left(\frac{x}{2^\alpha n}\right). \tag{60}$$

Läßt man ferner im Hauptglied von (60) α alle nichtnegativen ganzen Zahlen und H alle positiven ungeraden Zahlen durchlaufen, so ist der Fehler nach (23) und (23_1) gleich

$$B \sum_{r \leqq (k-1)/2} x^{k-r} \sum_{2^\alpha H > x^{1/2}/4} H^{-k} 2^{\alpha(r-k)} \sigma_r(H) = B \sum_{r \leqq (k-1)/2} x^{(k-r+1)/2} \log x$$
$$= Bx^{k/2} \log x,$$

da $r \geqq 1$ ist. Daher ist nach (60)

$$(2i)^k A_2 = \frac{(-1)^{(k+1)/2}}{k}\, 2^{1-k} \sum_{r \leqq (k-1)/2} (-1)^r \binom{k}{r} x^{k-r}$$
$$\times \sum_{\alpha=0}^{\infty} \sum_{u=1}^{\infty} \left(\frac{-1}{u}\right) u^r 2^{\alpha(r-k)}\, Gr\left(\frac{x}{2^\alpha u}\right) \sum_{\substack{H=1 \\ u \mid H}}^{\infty} \left(\frac{-1}{H}\right) \mu\left(\frac{H}{u}\right) H^{-k}$$
$$+ Bx^{k/2} \log x.$$

Setzt man hier $2^\alpha u = d$ ein und berücksichtigt die Definition (4) der Funktion $\{Z(k)\}^{-1}$, so ist

$$(2i)^k A_2 = (-1)^{(k+1)/2} 2^{1-k}\{kZ(k)\}^{-1} \sum_{r \leqq (k-1)/2} (-1)^r \binom{k}{r} x^{k-r} \times \sum_{d=1}^{\infty} d^{r-k} G_r\left(\frac{x}{d}\right) + Bx^{k/2} \log x, \tag{61}$$

das Restglied ist hierbei besser als in Formel (5.3.31).

Für $j = (k-1)/2$ ergibt sich aus (52) wegen (54), (58) und (61)

$$P_{2k}(x) = \frac{D_{2k}}{k} \sum_{r \leqq (k-1)/2} (-1)^r \binom{k}{r} x^{k-r} \sum_{u=1}^{\infty} \left(\frac{-1}{u}\right) u^{-k} \sum_{d|u} \mu\left(\frac{u}{d}\right) d^r \overline{B}_r\left(\frac{x}{d}\right)$$
$$+ D_{2k}(-1)^{(k+1)/2} 2^{1-k}\{kZ(k)\}^{-1} \sum_{r \leqq (k-1)/2} (-1)^r \binom{k}{r} x^{k-r} \times \sum_{d=1}^{\infty} d^{r-k} G_r\left(\frac{x}{d}\right) + Bx^{k/2} \log x. \tag{62}$$

Hierbei ist in der u-Summe d als Teiler von u ungerade, kann also mit v bezeichnet werden. Wenn man ferner die Reihenfolge der Summationen nach u, v vertauscht, ist

$$\sum_{u=1}^{\infty} \left(\frac{-1}{u}\right) u^{-k} \sum_{d|u} \mu\left(\frac{u}{d}\right) d^r \overline{B}_r\left(\frac{x}{d}\right)$$
$$= \sum_{v=1}^{\infty} v^r \overline{B}_r\left(\frac{x}{v}\right) \sum_{\substack{u=1 \\ v|u}}^{\infty} \left(\frac{-1}{u}\right) u^{-k} \mu\left(\frac{u}{v}\right) = \{Z(k)\}^{-1} \sum_{v=1}^{\infty} \left(\frac{-1}{v}\right) v^{r-k} \overline{B}_r\left(\frac{x}{v}\right). \tag{63}$$

Der Satz 4 für k ungerade, $k \geqq 3$, folgt jetzt aus (62), (63), (15) und (16).

Der Spezialfall $r = 1$ von Satz 4 kann als besonderer Satz formuliert werden.

Satz 4a. *Es sei k ungerade, $k \geqq 3$. Dann ist*

$$P_{2k}(x) - D_{2k}\{L(k)\}^{-1}\left(-\sum_{u=1}^{\infty} \left(\frac{-1}{u}\right) u^{1-k} \psi\left(\frac{x}{u}\right) + (-1)^{(k-1)/2} 2^{1-k} \times \sum_{d=1}^{\infty} d^{1-k}\left\{\psi\left(\frac{x}{d}\right) - \psi\left(\frac{x}{2d}\right) - 2\psi\left(\frac{x-d}{4d}\right)\right\}\right) x^{k-1}$$
$$= \begin{cases} Bx^{3/2} \log x & \text{für } k = 3, \\ Bx^{k-2} & \text{für } k \geqq 5, \end{cases}$$

wobei $\psi(y)$ die Funktion (11) ist.

Beweis. Für $r = 1$ folgt aus (21), (3) und (16)

$$P_{2k}(x) + D_{2k}\{L(k)\}^{-1}\,\Phi_{k,1}(x)\,x^{k-1} = Bx^{k/2}\log x,$$

wobei $\Phi_{k,1}(x)$ die Funktion (15) ist. Wegen (11) und (14) ist ferner

$$\Phi_{k,1}(x) = \sum_{u=1}^{\infty}\left(\frac{-1}{u}\right)u^{1-k}\psi\left(\frac{x}{u}\right)$$
$$+ (-1)^{(k+1)/2}\,2^{1-k}\sum_{d=1}^{\infty} d^{1-k}\left\{\psi\left(\frac{x}{d}\right) - \psi\left(\frac{x}{2d}\right) - 2\psi\left(\frac{x-d}{4d}\right)\right\}.$$

Das Restglied ist für $k \geqq 5$ gleich

$$Bx^{k/2}\log x = Bx^{k-k/2}\log x = Bx^{k-2}.$$

Der Satz 4a ist in zweifacher Hinsicht günstiger als [5], Satz 5.3.2: Einmal wird der Wert $k = 3$ zugelassen, und das Restglied ist um $\log x$ besser.

§ 4. Beweis von Satz 5

In diesem Paragraphen sollen nur Kugeln ungerader Dimension $2k + 1$ $(k \geqq 3)$ betrachtet werden.

Wendet man Hilfssatz 3 mit $j = \frac{k}{2}$, $l = k + \frac{1}{2}$, $s = x$, $h = 0$, $c = 1$ an, so folgt aus der Beziehung (22) für $2k + 1$

$$P_{2k+1}(x) = \frac{D_{2k+1}}{k+1/2}\sum_{r\leqq k/2}(-1)^r\binom{k+1/2}{r}x^{k+1/2-r}\,\overline{B}_r(x)$$
$$+ D_{2k+1}\sum_{2\leqq q\leqq x^{1/2}}\;\sideset{}{'}\sum_{h \bmod q}\left(\frac{S(h,q)}{q}\right)^{2k+1}$$
$$\times\sum_{n\leqq x} n^{k-1/2}\,e\left(\frac{-nh}{q}\right) + Bx^{k/2+1/4}\log x. \tag{64}$$

Ohne Beschränkung der Allgemeinheit sei wie in § 3 $x \geqq 16$. Weiter berücksichtigen wir noch, daß wegen (5) und (9)

$$\sideset{}{'}\sum_{h \bmod q} S(h,q)\,e\left(\frac{-nh}{q}\right) = \sideset{}{'}\sum_{h \bmod q} e\left(\frac{-nh}{q}\right)\sum_{a \bmod q} e\left(\frac{a^2 h}{q}\right) = \sum_{a \bmod q} C(a^2 - n, q) \tag{65}$$

ist.

Es sei k gerade, $k \geqq 4$. Für ungerade h ist $i^{hk} = (-1)^{k/2}$. Deshalb ergibt sich für die q-Summe in (64) auf Grund von (7) und (65)

$$\sum_{2\leqq q\leqq x^{1/2}} = N_1 + (-4)^{k/2}\,N_2 \tag{66}$$

mit

$$N_1 = \sum_{3 \leqq u \leqq x^{1/2}} u^{-k-1} \sum_{a \bmod u} \sum_{n \leqq x} n^{k-1/2} C(a^2 - n, u) \tag{67}$$

und

$$N_2 = 4^{-k-1} \sum_{q \leqq x^{1/2}/4} q^{-k-1} \sum_{a \bmod 4q} \sum_{n \leqq x} n^{k-1/2} C(a^2 - n, 4q). \tag{68}$$

Wir werden nun das erste Glied N_1 abschätzen. Wegen (10) ist

$$\sum_{n \leqq x} n^{k-1/2} C(a^2 - n, u) = \sum_{n \leqq x} n^{k-1/2} \sum_{d \mid (a^2-n, u)} d\mu\left(\frac{u}{d}\right)$$

$$= \sum_{d \mid u} d\mu\left(\frac{u}{d}\right) \sum_{\substack{n \leqq x \\ n \equiv a^2 (\bmod d)}} n^{k-1/2}.$$

Wendet man hier den Hilfssatz 3 mit $j = \frac{k}{2}$, $l = k + \frac{1}{2}$, $s = x$, $h = a^2$, $c = d$ an, so ist

$$\sum_{n \leqq x} n^{k-1/2} C(a^2 - n, u)$$

$$= \sum_{d \mid u} d\mu\left(\frac{u}{d}\right) \left\{ \frac{1}{k + \frac{1}{2}} \frac{x^{k+1/2}}{d} + \frac{1}{k + \frac{1}{2}} \right.$$

$$\left. \times \sum_{r \leqq k/2} (-1)^r \binom{k + \frac{1}{2}}{r} x^{k+1/2-r} d^{r-1} \bar{B}r\left(\frac{x - a^2}{d}\right) + Bx^{(k-1)/2} d^{k/2} \right\}. \tag{69}$$

Da hier das erste Glied wegfällt und das Restglied nach Hilfssatz 4 gleich

$$Bx^{(k-1)/2} \sum_{d \mid u} d^{k/2+1} = Bx^{k-1/2} \sigma_{k/2+1}(u) = Bx^{k-1/2} u^{k/2+1}$$

ist, folgt aus (69)

$$\sum_{n \leqq x} n^{k-1/2} C(a^2 - n, u) = \frac{1}{k + \frac{1}{2}} \sum_{r \leqq k/2} (-1)^r \binom{k + \frac{1}{2}}{r} x^{k+1/2-r}$$

$$\times \sum_{d \mid u} d^r \mu\left(\frac{u}{d}\right) \bar{B}_r\left(\frac{x - a^2}{d}\right) + Bx^{(k-1)/2} u^{k/2+1}. \tag{70}$$

Es ist

$$Bx^{(k-1)/2} \sum_{3 \leqq u \leqq x^{1/2}} u^{-k-1} \sum_{a \bmod u} u^{k/2+1} = Bx^{(k-1)/2} \sum_{3 \leqq u \leqq x^{1/2}} u^{-1}$$

$$= Bx^{(k-1)/2} \log x, \tag{71}$$

und da $\bar{B}_r(y)$ die Periode 1 hat, ist

$$\sum_{a \bmod u} \sum_{d|u} \bar{B}_r\left(\frac{x-a^2}{d}\right) = \sum_{d|u} \frac{u}{d} \sum_{a \bmod d} \bar{B}_r\left(\frac{x-a^2}{d}\right). \tag{72}$$

Setzt man jetzt (70) in (67) ein und berücksichtigt (71) und (72), so ergibt sich

$$N_1 = \frac{1}{k+\frac{1}{2}} \sum_{r \leqq k/2} (-1)^r \binom{k+\frac{1}{2}}{r} x^{k+1/2-r} \sum_{3 \leqq u \leqq x^{1/2}} u^{-k} \sum_{d|u} d^{r-1} \mu\left(\frac{u}{d}\right)$$

$$\times \sum_{a \bmod d} \bar{B}_r\left(\frac{x-a^2}{d}\right) + Bx^{(k-1)/2} \log x.$$

Läßt man ferner hier u alle ungeraden Zahlen durchlaufen, so müssen wir den Fehler abschätzen. Für $r = 1$ benutzen wir die Formeln (43) und (46), für $r > 1$ die triviale Abschätzung (40). Dann ist der Fehler nach (23) und (23_1) gleich

$$Bx^{k/2+1/4} \log x + B \sum_{1<r \leqq k/2} x^{k+1/2-r} \sum_{u>x^{1/2}} u^{-k}\sigma_r(u) = Bx^{k/2+1/4} \log x. \tag{72$_1$}$$

Daher bekommen wir für N_1 folgende Beziehung:

$$N_1 = \frac{1}{k+\frac{1}{2}} \sum_{r \leqq k/2} (-1)^r \binom{k+\frac{1}{2}}{r} x^{k+1/2-r} \sum_{u=3}^{\infty} u^{-k} \sum_{d|u} d^{r-1} \mu\left(\frac{u}{d}\right)$$

$$\times \sum_{a \bmod d} \bar{B}_r\left(\frac{x-a^2}{d}\right) + Bx^{k/2+1/4} \log x. \tag{73}$$

Wenn man zu der u-Summe noch ein Glied mit $u = 1$ hinzufügt, folgt ähnlich wie in (63)

$$\sum_{u=1}^{\infty} u^{-k} \sum_{d|u} d^{r-1} \mu\left(\frac{u}{d}\right) \sum_{a \bmod d} \bar{B}_r\left(\frac{x-a^2}{d}\right)$$

$$= \{Z(k)\}^{-1} \sum_{u=1}^{\infty} u^{r-1-k} \sum_{a \bmod u} \bar{B}_r\left(\frac{x-a^2}{u}\right). \tag{74}$$

Daher folgt aus (73)

$$N_1 = -\frac{1}{k+\frac{1}{2}} \sum_{r \leqq k/2} (-1)^r \binom{k+\frac{1}{2}}{r} x^{k+1/2-r} \overline{B}_r(x)$$

$$+ \frac{\{Z(k)\}^{-1}}{k+\frac{1}{2}} \sum_{r \leqq k/2} (-1)^r \binom{k+\frac{1}{2}}{r} x^{k+1/2-r} \sum_{u=1}^{\infty} u^{r-1-k}$$

$$\times \sum_{a \bmod u} \overline{B}_r\left(\frac{x-a^2}{u}\right) + Bx^{k/2+1/4} \log x. \tag{75}$$

Wir werden jetzt den durch (68) definierten Wert von N_2 abschätzen. Wegen (10) ist

$$\sum_{n \leqq x} n^{k-1/2} C(a^2-n, 4q) = \sum_{d|4q} d\mu\left(\frac{4q}{d}\right) \sum_{\substack{n \leqq x \\ n \equiv a^2 (\bmod d)}} n^{k-1/2}$$

$$= 2 \sum_{d|2q} d\mu\left(\frac{2q}{d}\right) \sum_{\substack{n \leqq x \\ n \equiv a^2 (\bmod 2d)}} n^{k-1/2}, \tag{76}$$

da die Glieder $d = u$ wegen $\mu\left(\frac{4q}{d}\right) = 0$ weggelassen werden können. Ist in (76) $\frac{2q}{d} = u$, so kann d durch $2d$ ersetzt werden mit der Bedingung $\frac{q}{d} \equiv 1 (\bmod 2)$; wenn $\frac{2q}{d} = 2u$ ist, so muß $d|q$ und $\frac{q}{d} \equiv 1 (\bmod 2)$ sein. Deswegen ist

$$\sum_{d|2q} d\mu\left(\frac{2q}{d}\right) \sum_{\substack{n \leqq x \\ n \equiv a^2 (\bmod 2d)}} n^{k-1/2}$$

$$= \sum_{\substack{d|q \\ q/d \equiv 1 (\bmod 2)}} d\mu\left(\frac{q}{d}\right) \left\{ 2 \sum_{\substack{n \leqq x \\ n \equiv a^2 (\bmod 4d)}} n^{k-1/2} - \sum_{\substack{n \leqq x \\ n \equiv a^2 (\bmod 2d)}} n^{k-1/2} \right\}. \tag{77}$$

Wendet man hier den Hilfssatz 3 mit $l = k + \frac{1}{2}$, $s = x$, $h = a^2$, $c = 4d$ bzw. $2d$ an und berücksichtigt, daß das Restglied wegen Hilfssatz 4 gleich

$$Bx^{(\ -1)/2} \sum_{d|q} d^{k/2+1} = Bx^{(k-1)/2} q^{k/2+1}$$

ist, so ergibt sich aus (76) und (77)

$$\sum_{n\leq x} n^{k-1/2} C(a^2-n,4q) = \frac{2}{k+\frac{1}{2}} \sum_{r\leq k/2} (-1)^{r+1} 2^{r-1} \binom{k+\frac{1}{2}}{r} x^{k+1/2-r}$$
$$\times \sum_{\substack{d\mid q \\ q/d\equiv 1 (\bmod 2)}} \mu\left(\frac{q}{d}\right) d^r F_r\left(\frac{x-a^2}{d}\right)$$
$$+ Bx^{(k-1)/2} q^{k/2+1},$$

wobei $F_r(y)$ die Funktion (13) ist. Wir setzen diesen Wert in (68) ein. Da das Restglied gleich

$$Bx^{(k-1)/2} \sum_{q\leq x^{1/2}/4} q^{-k-1} \sum_{a \bmod 4q} q^{k/2+1} = Bx^{k/2}$$

ist und da ferner wegen der Periodizität (mit Periode 4) der Funktion $F_r(y)$

$$\sum_{a \bmod 4q} \sum_{d\mid q} F_r\left(\frac{x-a^2}{d}\right) = \sum_{d\mid q} \frac{q}{d} \sum_{a \bmod 4d} F_r\left(\frac{x-a^2}{d}\right)$$

ist, ergibt sich für N_2 folgende Beziehung:

$$N_2 = \frac{2^{-2k-1}}{k+\frac{1}{2}} \sum_{r\leq k/2} (-1)^{r+1} 2^{r-1} \binom{k+\frac{1}{2}}{r} x^{k+1/2-r} \sum_{q\leq x^{1/2}/4} q^{-k}$$
$$\times \sum_{\substack{d\mid q \\ q/d\equiv 1 (\bmod 2)}} d^{r-1} \mu\left(\frac{q}{d}\right) \sum_{a \bmod 4d} F_r\left(\frac{x-a^2}{d}\right) + Bx^{k/2}. \tag{78}$$

Wir lassen hier q alle Zahlen durchlaufen. Den dadurch entstehenden Fehler schätzen wir folgendermaßen ab: Für $r = 1$ benutzt man (44) und (46), für $r > 1$ die Abschätzung (41). Dann ist der Fehler gleich

$$Bx^{k/2+1/4} \log x \tag{79}$$

(vgl. den Beweis der Formel (72_1)). Daher folgt aus (78)

$$N_2 = \frac{2^{-2k-1}}{k+\frac{1}{2}} \sum_{r\leq k/2} (-1)^{r+1} 2^{r-1} \binom{k+\frac{1}{2}}{r} x^{k+1/2-r}$$
$$\times \sum_{q=1}^{\infty} q^{-k} \sum_{\substack{d\mid q \\ q/d\equiv 1 (\bmod 2)}} d^{r-1} \mu\left(\frac{q}{d}\right) \sum_{a \bmod 4d} F_r\left(\frac{x-a^2}{d}\right) + Bx^{k/2+1/4} \log x. \tag{80}$$

Hierin ist ähnlich wie in (63)

$$\sum_{d=1}^{\infty} q^{-k} \sum_{\substack{d \mid q \\ q/d \equiv 1 (\mathrm{mod}\, 2)}} d^{r-1} \mu\left(\frac{q}{d}\right) \sum_{a \bmod 4d} F_r\left(\frac{x-a^2}{d}\right)$$

$$= \{Z(k)\}^{-1} \sum_{d=1}^{\infty} d^{r-1-k} \sum_{a \bmod 4d} F_r\left(\frac{x-a^2}{d}\right).$$

Daher ist nach (80)

$$N_2 = \frac{2^{-2k-1}}{k+\frac{1}{2}} \{Z(k)\}^{-1} \sum_{r \leq k/2} (-1)^{r+1} 2^{r-1} \binom{k+\frac{1}{2}}{r} x^{k+1/2-r}$$

$$\times \sum_{d=1}^{\infty} d^{r-1-k} \sum_{a \bmod 4d} F_r\left(\frac{x-a^2}{d}\right) + Bx^{k/2+1/4} \log x. \tag{81}$$

Es ergibt sich jetzt aus (64) wegen (66), (75) und (81)

$$P_{2k+1}(x) = \frac{D_{2k+1}}{k+\frac{1}{2}} \{Z(k)\}^{-1} \sum_{r \leq k/2} (-1)^r \binom{k+\frac{1}{2}}{r} x^{k+1/2-r}$$

$$\times \left\{ \sum_{u=1}^{\infty} u^{r-1-k} \sum_{a \bmod u} \bar{B}_r\left(\frac{x-a^2}{u}\right) \right.$$

$$\left. + (-1)^{k/2+1} 2^{r-k-2} \sum_{d=1}^{\infty} d^{r-1-k} \sum_{a \bmod 4d} F_r\left(\frac{x-a^2}{d}\right) \right\}$$

$$+ Bx^{k/2+1/4} \log x. \tag{82}$$

Da hier

$$\frac{D_{2k+1}}{k+\frac{1}{2}} \binom{k+\frac{1}{2}}{r} = \frac{\pi^{k+1/2}}{r!\,\Gamma\left(k+\frac{3}{2}-r\right)} \tag{83}$$

ist, folgt Satz 5 für k gerade, $k \geqq 4$, aus (82), (17) und (19).

Es sei jetzt k ungerade, $k \geqq 3$. Für ungerade h gilt $i^{hk} = \left(\frac{-1}{h}\right) i^k$. Wegen (7) und (65) ist dann die q-Summe in (64) gleich

$$\sum_{2 \leq q \leq x^{1/2}} = R_1 + (2i)^k R_2 \tag{84}$$

mit

$$R_1 = \sum_{3 \leqq u \leqq x^{1/2}} \left(\frac{-1}{u}\right) u^{-k-1} \sum_{a \bmod u} \sum_{n \leqq x} n^{k-1/2} C(a^2 - n, u) \tag{85}$$

und

$$R_2 = 4^{-k-1} \sum_{q \leqq x^{1/2}/4} q^{-k-1} \sum_{n \leqq x} n^{k-1/2} \sum_{h \bmod 4q}' \left(\frac{-1}{h}\right) S(h, 4q)\, e\left(\frac{-nh}{4q}\right). \tag{86}$$

Die Funktion R_1 wird jetzt ebenso behandelt wie N_1 (siehe Formel (67)), denn sie unterscheidet sich von ihr nur durch den Multiplikator $\left(\frac{-1}{u}\right)$. Deshalb haben wir für R_1 die Abschätzung

$$R_1 = -\frac{1}{k+\frac{1}{2}} \sum_{r \leqq k/2} (-1)^r \binom{k+\frac{1}{2}}{r} x^{k+1/2-r} \bar{B}_r(x)$$

$$+ \frac{\{Z(k)\}^{-1}}{k+\frac{1}{2}} \sum_{r \leqq k/2} (-1)^r \binom{k+\frac{1}{2}}{r} x^{k+1/2-r} \sum_{u=1}^{\infty} \left(\frac{-1}{u}\right) u^{r-1-k}$$

$$\times \sum_{a \bmod u} \bar{B}_r\left(\frac{x-a^2}{u}\right) + Bx^{k/2+1/4} \log x. \tag{87}$$

Was R_2 angeht, so ist wegen (86), (5) und (29)

$$R_2 = 4^{-k-1} \sum_{q \leqq x^{1/2}/4} q^{-k-1} \sum_{a \bmod 4q} \sum_{n \leqq x} n^{k-1/2} T(a^2 - n, q). \tag{88}$$

Es sei $q = 2^\alpha H$. Die Funktion $T(a^2 - n, q)$ ist nach Hilfssatz 7 für $a^2 - n \neq 2^\alpha w$ gleich Null und für $a^2 - n = 2^\alpha w$ durch (31) ausgedrückt. Es ist deshalb in (88)

$$\sum_{n \leqq x} n^{k-1/2} T(a^2 - n, q = 2^\alpha H) = 2^{\alpha+1} i \left(\frac{-1}{H}\right) \sum_{u|H} \left(\frac{-1}{u}\right) u\mu\left(\frac{H}{u}\right)$$

$$\times \sum_{\substack{n \leqq x \\ u | a^2 - n = 2^\alpha w}} \left(\frac{-1}{uw}\right) n^{k-1/2}. \tag{89}$$

Hierin ist

$$\sum_{\substack{n \leqq x \\ u | a^2 - n = 2^\alpha w}} \left(\frac{-1}{uw}\right) n^{k-1/2} = R_3 + R_4 \tag{90}$$

mit

$$R_3 = \sum_{\substack{n \leqq x \\ u | a^2 - n = 2^\alpha w}} n^{k-1/2} = \sum_{\substack{n \leqq x \\ n \equiv a^2 (\bmod 2^\alpha u)}} n^{k-1/2} - \sum_{\substack{n \leqq x \\ n \equiv a^2 (\bmod 2^{\alpha+1} u)}} n^{k-1/2} \tag{91}$$

und

$$R_4 = \sum_{\substack{n \leqq x \\ u | a^2 - n = 2^\alpha w}} \left\{ \left(\frac{-1}{uw} \right) - 1 \right\} n^{k-1/2} = -2 \sum_{\substack{n \leqq x \\ n \equiv a^2 + 2^\alpha u (\operatorname{mod} 2^{\alpha+2} u)}} n^{k-1/2}, \tag{92}$$

da

$$\left(\frac{-1}{uw} \right) - 1 = \begin{cases} -2 & \text{für } u \equiv -w (\operatorname{mod} 4), \\ 0 & \text{für } u \equiv w (\operatorname{mod} 4) \end{cases}$$

gilt.

Es bezeichne $d = 2^\alpha u$. Für R_3 und R_4 wendet man Hilfssatz 3 mit $j = \frac{k}{2}$, $s = x$, $l = k + \frac{1}{2}$, $h = a^2$ bzw. $h = a^2 + d$, $c = d$ bzw. $c = 2d$ bzw. $c = 4d$ an. Dann folgt aus (90), (91) und (92), wenn noch die Definition (14) der Funktion $G_r(y)$ berücksichtigt wird,

$$\sum_{\substack{n \leqq x \\ u | a^2 - n = 2^\alpha w}} \left(\frac{-1}{uw} \right) n^{k-1/2} = \frac{1}{k + \frac{1}{2}} \sum_{r \leqq k/2} (-1)^r \binom{k + \frac{1}{2}}{r} x^{k+1/2-r} d^{r-1} G_r \left(\frac{x - a^2}{d} \right)$$
$$+ B d^{k/2} x^{(k-1)/2}. \tag{93}$$

Wegen $d = 2^\alpha u$ und $q = 2^\alpha H$ folgt aus $u | H$

$$d | q \quad \text{und} \quad \frac{q}{d} \equiv 1 \ (\operatorname{mod} 2).$$

Es ist dann

$$\left(\frac{-1}{H} \right) \left(\frac{-1}{u} \right) = \left(\frac{-1}{\frac{H}{u}} \right) = \left(\frac{-1}{\frac{q}{d}} \right) \tag{94}$$

und ferner wegen (23) und (23_1)

$$B x^{(k-1)/2} \sum_{u | H} (2^\alpha u)\, d^{k/2} = B x^{(k-1)/2} \sum_{d | q} d^{k/2+1} = B x^{(k-1)/2} q^{k/2+1}. \tag{95}$$

Setzt man jetzt (93) in (89) ein und berücksichtigt (94) und (95), so ergibt sich

$$\sum_{n \leqq x} n^{k-1/2}\, T(a^2 - n, q) = \frac{2i}{k + \frac{1}{2}} \sum_{r \leqq k/2} (-1)^r \binom{k + \frac{1}{2}}{r} x^{k+1/2-r}$$
$$\times \sum_{\substack{d | q \\ q/d \equiv 1 (\operatorname{mod} 2)}} \left(\frac{-1}{\frac{q}{d}} \right) d^r \mu \left(\frac{q}{d} \right) G_r \left(\frac{x - a^2}{d} \right)$$
$$+ B x^{(k-1)/2} q^{k/2+1}. \tag{96}$$

Wir sind nun imstande, R_2 abzuschätzen. Es ist

$$Bx^{(k-1)/2} \sum_{q \leqq x^{1/2}/4} q^{-k-1} \sum_{a \bmod 4q} q^{k/2+1} = Bx^{k/2}. \tag{97}$$

Da ferner wegen der Periodizität der Funktion $G_r(y)$

$$\sum_{a \bmod 4q} \sum_{d|q} G_r\left(\frac{x-a^2}{d}\right) = \sum_{d|q} \frac{q}{d} \sum_{a \bmod 4d} G_r\left(\frac{x-a^2}{d}\right) \tag{98}$$

ist, folgt aus (88), (96), (97) und (98) für R_2 folgende Abschätzung

$$R_2 = \frac{2^{-2k-1}}{k+\frac{1}{2}} i \sum_{r \leqq k/2} (-1)^r \binom{k+\frac{1}{2}}{r} x^{k+1/2-r}$$

$$\times \sum_{q \leqq x^{1/2}/4} q^{-k} \sum_{\substack{d|q \\ q/d \equiv 1 (\bmod 2)}} \left(\frac{-1}{\frac{q}{d}}\right) d^{r-1} \mu\left(\frac{q}{d}\right) \sum_{a \bmod 4d} G_r\left(\frac{x-a^2}{d}\right) + Bx^{k/2}. \tag{99}$$

Läßt man hier q alle Zahlen durchlaufen, so ist der Fehler wegen (45), (46) und (42) gleich (79) (vgl. den Beweis der Formel (72_1)). Daher liefert (99) die Abschätzung

$$R_2 = \frac{2^{-2k-1} i}{k+\frac{1}{2}} \sum_{r \leqq k/2} (-1)^r \binom{k+\frac{1}{2}}{r} x^{k+1/2-r}$$

$$\times \sum_{q=1}^{\infty} q^{-k} \sum_{\substack{d|q \\ q/d \equiv 1 (\bmod 2)}} \left(\frac{-1}{\frac{q}{d}}\right) d^{r-1} \mu\left(\frac{q}{d}\right) \sum_{a \bmod 4d} G_r\left(\frac{x-a^2}{d}\right)$$

$$+ Bx^{k/2+1/4} \log x. \tag{100}$$

Da k ungerade ist, ergibt wegen (4) eine ähnliche Überlegung wie in (63)

$$\sum_{q=1}^{\infty} q^{-k} \sum_{\substack{d|q \\ q/d \equiv 1 (\bmod 2)}} \left(\frac{-1}{\frac{q}{d}}\right) d^{r-1} \mu\left(\frac{q}{d}\right) \sum_{a \bmod 4d} G_r\left(\frac{x-a^2}{d}\right)$$

$$= \{Z(k)\}^{-1} \sum_{d=1}^{\infty} d^{r-1-k} \sum_{a \bmod 4d} G_r\left(\frac{x-a^2}{d}\right). \tag{101}$$

Aus (64), (84), (87), (100), (101) und (83) folgt

$$P_{2k+1}(x) = \pi^{k+1/2}\{Z(k)\}^{-1} \sum_{r \leqq k/2} \frac{(-1)^r}{r!} \frac{x^{k+(1/2)-r}}{\Gamma\left(k+\frac{3}{2}-r\right)}$$

$$\times \left\{\sum_{u=1}^{\infty} \left(\frac{-1}{u}\right) u^{r-1-k} \sum_{a \bmod u} \bar{B}_r\left(\frac{x-a^2}{u}\right) + (-1)^{(k+1)/2}\, 2^{-k-1}\right.$$

$$\left. \times \sum_{d=1}^{\infty} d^{r-1-k} \sum_{a \bmod 4d} G_r\left(\frac{x-a^2}{d}\right)\right\} + Bx^{k/2+1/4} \log x. \tag{102}$$

Satz 5 für ungerade k, $k \geqq 3$, folgt jetzt aus (102), (18) und (19).

Literatur

[1] E. Landau, Über Gitterpunkte in mehrdimensionalen Ellipsoiden, Math. Z. **21** (1924), 126—132.

[2] А. П. Лурсманашвили, О числе целых точек в многомерных шарах, Труды Тбилисского математ. инст. **19** (1953), 79–120.

[3] А. П. Лурсманашвили, О числе целых точек в многомерных шарах нечетной размерности, Сообщения АН ГССР **14** (1953), 513–220.

[4] H. Petersson, Über die Anzahl der Gitterpunkte in mehrdimensionalen Ellipsoiden, Abh. Math. Sem. Hamburg Univ. **5** (1926), 116—150.

[5] Arnold Walfisz, Gitterpunkte in mehrdimensionalen Kugeln, Warszawa 1957.

[6] Arnold Walfisz, Über Gitterpunkte in mehrdimensionalen Kugeln II, Acta Arithm. **6** (1960), 115—136.

[7] Arnold Walfisz, Über Gitterpunkte in mehrdimensionalen Kugeln III, Acta Arithm. **6** (1960), 193—215.

PUBLICATIONS OF EDMUND LANDAU

COMPILED BY I. J. SCHOENBERG IN MADISON (Wisc.)

1. Papers

1. Zur relativen Wertbemessung der Turnierresultate
   Deutsches Wochenschach, 11. Jahrgang (1895), 366–369.
2. Ueber das Achtdamenproblem und seine Verallgemeinerung
   Naturwissenschaftliche Wochenschrift Bd. XI, Nr. 31, 367–371.
3. Neuer Beweis der Gleichung $\sum_{1}^{\infty} \mu(k)/k = 0$
   Inaugural-Dissertation, Berlin 1899, 16 pages.
4. Contribution à la théorie de la fonction $\zeta(s)$ de Riemann
   C. R. Acad. des Sci. Paris **129** (1899), 812–815.
5. Sur la série des inverses des nombres de Fibonacci
   Bull. Soc. Math. France **27** (1899), 298–300.
6. Sur quelques problèmes rélatifs à la distribution des nombres premiers
   Bull. Soc. Math. France **28** (1900), 25–38.
7. Ueber die zahlentheoretische Function $\varphi(n)$ und ihre Beziehung zum Goldbachschen Satz
   Gött. Nachr. 1900, 177–186.
8. Sur les conditions de divisibilité d'un produit de factorielles par un autre
   Nouv. Ann. de Math. **19** (1900), 344–362.
9. Ueber einen Zahlentheoretischen Satz
   Arch. der Math. u. Phys. (3) **1** (1901), 138–142.
10. Ueber die asymptotischen Werthe einiger Zahlentheoretischer Functionen
    Math. Ann. **54** (1901), 570–591.

[1]) In the numerous references below to Academy publications we have omitted the descriptive terms like "Math.-Phys. Klasse".

11. Ueber die mittlere Anzahl der Zerlegungen aller Zahlen von 1 bis x in drei Factoren
Math. Ann. **54** (1901), 592–601.

12. Zur Theorie der Gammafunction
J. reine angew. Math. **123** (1901), 276–283.

13. Solutions de questions proposées
Nouv. Ann. de Math. (4) **1** (1901), 281–283.

14. Ein Satz über die Zerlegung homogener linearer Differentialausdrücke in irreducible Factoren
J. reine angew. Math. **124** (1901), 115–120.

15. Ueber die zu einem algebraischen Zahlkörper gehörige Zetafunction und die Ausdehnung der Tschebyschefschen Primzahltheorie auf das Problem der Vertheilung der Primideale
J. reine angew. Math. **125** (1902), 64–187.

16. Neuer Beweis des Primzahlsatzes und Beweis des Primidealsatzes
Math. Ann. **56** (1903), 645–670.

17. Über die Klassenzahl der binären quadratischen Formen von negativer Discriminante
Math. Ann. **56** (1903), 671–676.

18. Über quadrierbare Kreisbogenzweiecke
Sitzungsber. der Berliner Math. Ges., 2. Jahrgang (1903), 1–6.

19. Über die Darstellung definiter binärer Formen durch Quadrate
Math. Ann. **57** (1903), 53–64.

20. Über den Verlauf der zahlentheoretischen Funktion $\varphi(x)$
Arch. der Math. u. Phys. (3) **5** (1903), 86–91.

21. Über die Maximalordnung der Permutationen gegebenen Grades
Arch. der Math. u. Phys. (3) **5** (1903), 92–103.

22. Über die Primzahlen einer arithmetischen Progression
Wiener Sitzungsberichte **112** (1903), 493–535.

23. Über die zahlentheoretische Funktion $\mu(k)$
Wiener Sitzungsberichte **112** (1903), 537–570.

24. Solutions de questions proposées
Nouv. Ann. de Math. (4) **3** (1903), 4 pages.

25. Eine Anwendung des Eisensteinschen Satzes auf die Theorie der Gaußschen Differentialgleichung
J. reine angew. Math. **127** (1904), 92–102.

26. Über die Zerlegung definiter Funktionen in Quadrate
Arch. der Math. u. Phys. (3) **7** (1904), 271–277.

27. Bemerkungen zu der Abhandlung von Herrn Kluyver: „Reeksen afgeleid uit de reeks $\sum \mu(m)/m$"
Koninkl. Akad. v. Wetenschappen te Amsterdam **13** (1904), 71–83.

28. Über eine Verallgemeinerung des Picardschen Satzes
Berliner Sitzungsberichte **38** (1904), 1118–1133.

29. Über eine Darstellung der Anzahl der Idealklassen eines algebraischen Körpers durch eine unendliche Reihe
J. reine angew. Math. **127** (1904), 167–174.

30. Bemerkungen zu Herrn D. N. Lehmer's Abhandlung in Bd. 22 dieses Journals, S. 293–335
Amer. J. of Math. **26** (1904), 209–222.

31. Sur quelques inégalités dans la théorie de la fonction $\zeta(s)$ de Riemann
Bull. Soc. Math. France **33** (1905), 229–241.

32. Sur quelques théorèmes de M. Petrovitch rélatifs aux zéros des fonctions analytiques
Bull. Soc. Math. France **33** (1905), 251–261.

33. Über einen Satz von Herrn Phragmén
Acta Math. **30** (1906), 195–201.

34. Über einen Satz von Herrn Frobenius in der Theorie der linearen Differentialgleichungen
Arch. der Math. u. Phys. (3) **10** (1906), 45–50.

35. On a familiar theorem of the theory of functions
Bull. Amer. Math. Soc. **12** (1906), 155–156.

36. Über einen Satz von Tschebyschef
Math. Ann. **61** (1905), 527–550.

37. Über das Nichtverschwinden einer Dirichletschen Reihe
Berliner Sitzungsberichte XI (1906), 314–320.

38. Über die Darstellung definiter Funktionen durch Quadrate
Math. Ann. **62** (1906), 272–285.

39. Über die Grundlagen der Theorie der Fakultätenreihen
Sitzungsber. der Bayer. Akad. d. Wiss. **36** (1906), 151–218.

40. Sur une inégalité de M. Hadamard
Nouv. Ann. de Math. (4) **6** (1906), 135–140.

41. Über einige Ungleichheitsbeziehungen in der Theorie der analytischen Funktionen
Arch. der Math. u. Phys. (3) **11** (1906), 31–36.

42. Euler und die Funktionalgleichung der Riemannschen Zetafunktion
Bibliotheca Math. (3) **7** (1906), 69–79.

43. Über den Zusammenhang einiger neuerer Sätze der analytischen Zahlentheorie
Wiener Sitzungsberichte **115** (1906), 589–631.

44. Über den Picardschen Satz
Vierteljahrsschr. der Naturf. Ges. Zürich **51** (1906), 252–318.

45. Über die Verteilung der Primideale in den Idealklassen eines algebraischen Zahlkörpers
Math. Ann. **63** (1906), 145–204.

46. Über die Darstellung einer ganzen Zahl als Summe von Biquadraten
Rendiconti di Palermo **23** (1907), 91–96.

47. Über die Konvergenz einiger Klassen von Unendlichen Reihen am Rande des Konvergenzgebietes
Monatshefte für Math. u. Phys. **18** (1907), 8–28.

48. Der Integrallogarithmus und die Zahlentheorie
Rendiconti di Palermo **23** (1907), 126–129.

49. Über einen Konvergenzsatz
Gött. Nachr. 1907, 25–27.

50. Über einen Konvergenzsatz des Herrn Phragmén
Arkiv för Mat., Astr. o. Fys. **3** (1907), Nr. 17, 10 pages.

51. Über die größte Schwankung einer analytischen Funktion in einem Kreise (With Otto Toeplitz)
Arch. der Math. u. Phys. (3) **11** (1906), 302–307.

52. Sur quelques généralisations du théorème de M. Picard
Ann. de l'École Norm. Sup. (3) **24** (1907), 179–201.

53. Bemerkungen zu einer Arbeit des Herrn V. Furlan
Rendiconti di Palermo **23** (1907), 367–373.

54. Über die Multiplikation Dirichletscher Reihen
Rendiconti di Palermo **24** (1907), 81–160.

55. Über die Approximation einer stetigen Funktion durch eine ganze rationale Funktion
Rendiconti di Palermo **25** (1908), 337–346.

56. Beiträge zur analytischen Zahlentheorie
Rendiconti di Palermo **26** (1908), 169–302.

57. Neuer Beweis der Riemannschen Primzahlformel
Berliner Sitzungsberichte 1908, 737–745.

58. Zwei neue Herleitungen für die asymptotische Anzahl der Primzahlen unter einer gegebenen Grenze
Berliner Sitzungsberichte 1908, 746–764.

59. Neue Beiträge zur analytischen Zahlentheorie
Rendiconti di Palermo **27** (1909), 46–58.

60. Über eine Anwendung der Primzahltheorie auf das Waringsche Problem in der elementaren Zahlentheorie
Math. Ann. **66** (1908), 102–105.

61. Über die Einteilung der positiven ganzen Zahlen in vier Klassen nach der Mindestzahl der zu ihrer additiven Zusammensetzung erforderlichen Quadrate
Arch. der Math. u. Phys. (3) **13** (1908), 305–312.

62. Über einen Grenzwertsatz
Wiener Sitzungsberichte **117** (1908), 1089–1094.

63. Über die Primzahlen in einer arithmetischen Progression und die Primideale in einer Idealklasse
Wiener Sitzungsberichte **117** (1908), 1095–1107.

64. Nouvelle démonstration pour la formule de Riemann sur le nombre des nombres premiers inférieurs à une limite donnée, et démonstration d'une formule plus générale pour le cas des nombres premiers d'une progression arithmétique
Ann. de l'École Norm. Sup. (3) **25** (1908), 399–448.

65. Lösung des Lehmerschen Problems
Amer. J. of Math. **31** (1909), 86–102.

66. Über die Verteilung der Nullstellen der Riemannschen Zetafunktion und einer Klasse verwandter Funktionen
Math. Ann. **66** (1908), 419–445.

67. Bemerkung zu meinem Aufsatze: Über die Grundlagen der Theorie der Fakultätenreihen in Bd. 36 (1906) dieser Berichte, S. 151–218
Sitzungsber. der Bayer. Akad. d. Wiss. **39** (1909), 7–10.

68. Über das Konvergenzproblem der Dirichletschen Reihen
Rendiconti di Palermo **28** (1909), 113–151.

69. Über das Verhalten von $\zeta(s)$ und $\zeta_k(s)$ in der Nähe der Geraden $\sigma = 1$ (With Harald Bohr)
Gött. Nachr. 1910, 303–330.

70. Über das Nichtverschwinden der Dirichletschen Reihen, welche komplexen Charakteren entsprechen
Math. Ann. **70** (1911), 69–78.

71. Ein Satz über die ζ-Funktion
Nyt Tidsskrift for Matematik (B) **22** (1911), 1–7.

72. Über die Bedeutung einiger neuerer Grenzwertsätze der Herren Hardy und Axer
Prace Mat.-Fiz **21** (1910), 97–177.

73. Beiträge zur Konvergenz von Funktionenfolgen (With C. Carathéodory)
Berliner Sitzungsberichte **26** (1911), 587–613.

74. Sur les valeurs moyennes de certaines fonctions arithmétiques
Bull. de l'Acad. royale de Belgique 1911, 443–472.

75. Über einen zahlentheoretischen Satz und seine Anwendung auf die hypergeometrische Reihe
Sitzungsber. der Heidelberger Akad. d. Wiss. 1911, 18. Abhandlung, 1–38.

76. Über die Verteilung der Zahlen, welche aus ν Primfaktoren zusammengesetzt sind
Gött. Nachr. 1911, 361–381.

77. Über die Zerlegung positiver ganzer Zahlen in positive Kuben
Arch. d. Math. u. Phys. (3) **18** (1911), 248–252.

78. Über die Zetafunktion (With Harald Bohr)
Rendiconti di Palermo **32** (1911), 278–285.

79. Zur Theorie der Riemannschen Zetafunktion
Vierteljahrsschr. der Naturf. Ges. Zürich **56** (1911), 125–148.

80. Über die Äquivalenz zweier Hauptsätze der analytischen Zahlentheorie
Wiener Sitzungsberichte **120** (1911), 973–988.

81. Über einige Summen, die von den Nullstellen der Riemannschen Zetafunktion abhängen
Acta Math. **35** (1912), 271–294.

82. Über den Gebrauch bedingt konvergenter Integrale in der Primzahltheorie
Math. Ann. **71** (1911), 368–379.

83. Über die Zahlen mit einer gegebenen Teileranzahl
Annaes da Academia Polytechnica do Porto **6** (1911), 129–137.

84. Über eine idealtheoretische Funktion
Trans. Amer. Math. Soc. **13** (1912), 1–21.

85. Über die Nullstellen der Zetafunktion
Math. Ann. **71** (1911), 548–564.

86. Über einige neuere Grenzwertsätze
Rendiconti di Palermo **34** (1912), 121–131.

87. Über die Anzahl der Gitterpunkte in gewissen Bereichen
Gött. Nachr. 1912, 687–770.

88. Gelöste und ungelöste Probleme aus der Theorie der Primzahlverteilung und der Riemannschen Zetafunktion
Proceedings of the 5th Internat. Congress of Math., vol. I, 93–108, Cambridge 1913. Also reprinted in Jahresber. der Deutschen Math.-Ver. **21** (1912), 208–228.

89. Über die Zerlegung der Zahlen in zwei Quadrate
Annali di Mat. (3) **20** (1912), 1–28.

90. Die Bedeutung der Pfeifferschen Methode für die analytische Zahlentheorie
Wiener Sitzungsberichte **121** (1912), 2195–2332.

91. Über einen Satz des Herrn Littlewood
Rendiconti di Palermo **35** (1913), 265–276.

92. Abschätzung der Koeffizientensumme einer Potenzreihe
Arch. der Math. u. Phys. (3) **21** (1913), 42–50.

93. Über einen Satz des Herrn Sierpinski
Giornale di Mat. di Battaglini **51** (1913), 73–81.

94. Sur les séries de Lambert
C. R. Acad. des Sciences Paris **156** (1913), 1451–1454.

95. Beiträge zur Theorie der Riemannschen Zetafunktion (With Harald Bohr)
Math. Ann. **74** (1913), 3–30.

96. Ein neues Konvergenzkriterium für Integrale
Sitzungsber. der Bayer. Akad. d. Wiss. 1913, 461–467.

97. Die Identität des Cesàroschen und Hölderschen Grenzwertes für Integrale
Berichte der Sächsischen Ges. d. Wiss. Leipzig **65** (1913), 131–138.

98. Abschätzung der Koeffizientensumme einer Potenzreihe (Zweite Abhandlung)
Arch. der Math. u. Phys. (3) **21** (1913), 250–255.

99. Sur les conditions de divisibilité d'un produit de factorielles par un autre (Supplément)
Nouv. Ann. de Math. (4) **13** (1913), 353–355.

100. 46 responses parues dans l'Intermédiaire des Mathematiciens **20** (1913), 151–155, 175–181, 201, 206.

101. Über die Nullstellen Dirichletscher Reihen
Berliner Sitzungsberichte **41** (1913), 897–907.

102. Ein Satz über Dirichletsche Reihen mit Anwendung auf die ζ-Funktion und die L-Funktionen (With Harald Bohr)
Rendiconti di Palermo **37** (1914), 269–272

103. Einige Ungleichungen für zweimal differentiierbare Funktionen
Proc. London Math. Soc. (2) **13** (1913), 43–49.

104. Sur les zeros de la fonction $\zeta(s)$ de Riemann (With Harald Bohr)
C. R. Acad. des Sciences Paris **158** (1914), 158–162.

105. Sur la fonction $\zeta(s)$ dans le voisinage de la droite $\sigma = 1/2$ (With H. Bohr and J. E. Littlewood)
Bull. de l'Acad. royale de Belgique 1913, 3–35.

106. Über Preisverteilung bei Spielturnieren
Z. f. Math. u. Phys. **63** (1914), 192–202.

107. Über die Primzahlen in definiten quadratischen Formen und die Zetafunktion reiner kubischer Körper
H. A. Schwarz Festschrift, Berlin 1914, 244–273.

108. Über die Hardysche Entdeckung unendlich vieler Nullstellen der Zetafunktion mit reellem Teil 1/2
Math. Ann. **76** (1915), 212–243.

109. Über eine Aufgabe aus der Funktionentheorie
Tôhoku Math. J. **5** (1914), 97–116.

110. Über die Gitterpunkte in einem Kreise (Erste Mitteilung)
Gött. Nachr. 1915, 148–160.

111. Zur analytischen Zahlentheorie der definiten quadratischen Formen. (Über die Gitterpunkte in einem mehrdimensionalen Ellipsoid.)
Berliner Sitzungsberichte **31** (1915), 458–476.

112. Über die Gitterpunkte in einem Kreise (Zweite Mitteilung)
Gött. Nachr. 1915, 161–171.

113. Über einen Mellinschen Satz
Arch. der Math. u. Phys. (3) **24** (1915), 97–107.

114. Neuer Beweis eines analytischen Satzes des Herrn de la Vallée Poussin
Jahresber. der Deutschen Math.-Ver. **24** (1915), 250–278.

115. Anzeige von S. Wigert: Sur quelques fonctions arithmétiques
Göttinger gelehrte Anzeiger 1915, 377–414.

116. Über die Anzahl der Gitterpunkte in gewissen Bereichen (Zweite Abhandlung)
Gött. Nachr. 1915, 209–243.

117. Über Dirichlets Teilerproblem
Sitzungsber. der Bayer. Akad. d. Wiss. 1915, 317–328.

118. Über eine Aufgabe aus der Theorie der quadratischen Formen
Wiener Sitzungsberichte **124** (1915), 445–468.

119. Neue Untersuchungen über die Pfeiffersche Methode zur Abschätzung von Gitterpunktanzahlen
Wiener Sitzungsberichte **124** (1915), 469–505.

120. Abschätzung der Koeffizientensumme einer Potenzreihe (Dritte Abhandlung)
Arch. der Math. u. Phys. (3) **24** (1915), 250–260.

121. Neuer Beweis eines Hardyschen Satzes
Arch. der Math. u. Phys. (3) **25** (1917), 173–178.

122. Über die Anzahl der Gitterpunkte in gewissen Bereichen (Dritte Abhandlung)
Gött. Nachr. 1917, 96–101.

123. Über die Heckesche Funktionalgleichung
Gött. Nachr. 1917, 102–111.

124. Richard Dedekind
Gedächtnisrede, gehalten in der öffentlichen Sitzung der Königlichen Gesellschaft der Wissenschaften zu Göttingen am 12. Mai 1917, Gött. Nachr. 1917 50–70

125. Über mehrfache gliedweise Differentiation unendlicher Reihen
Arch. der Math. u. Phys. (3) **26** (1917), 69–70.

126. Über einige ältere Vermutungen und Behauptungen in der Primzahltheorie
Math. Z. **1** (1918), 1–24.

127. Neuer Beweis eines Hauptsatzes aus der Theorie der Dirichletschen Reihen
Berichte der Sächsischen Ges. d. Wiss. Leipzig **69** (1917), 336–343.

128. Ein Konvergenzkriterium für Integrale
Berichte der Sächsischen Ges. d. Wiss. Leipzig **69** (1917), 344–348.

129. Über einige ältere Vermutungen und Behauptungen in der Primzahltheorie (Zweite Abhandlung)
Math. Z. **1** (1918), 213–219.

130. Über Ideale und Primideale in Idealklassen
Math. Z. **2** (1918), 52–154.

131. Abschätzungen von Charaktersummen, Einheiten und Klassenzahlen
Gött. Nachr. 1918, 79–97.

132. Ein Satz über Riemannsche Integrale
Math. Z. **2** (1918), 350–351.

133. Über die Wigertsche asymptotische Funktionalgleichung für die Lambertsche Reihe
Arch. der Math. u. Phys. (3) **27** (1918), 144–146.

134. Über die Blaschkesche Verallgemeinerung des Vitalischen Satzes
Berichte der Sächsischen Ges. d. Wiss. Leipzig **70** (1918), 156–159.

135. Über imaginär-quadratische Zahlkörper mit gleicher Klassenzahl
Gött. Nachr. 1918, 277–284.

136. Über die Klassenzahl imaginär-quadratischer Zahlkörper
Gött. Nachr. 1918, 285–295.

137. Verallgemeinerung eines Pólyaschen Satzes auf algebraische Zahlkörper
Gött. Nachr. 1918, 478–488.

138. Auszug aus einem Briefe des Herrn Landau an den Herausgeber
Acta Math. **42** (1920), 95–98.

139. Zur Theorie der Heckeschen Zetafunktionen, welche komplexen Charakteren entsprechen
Math. Z. **4** (1919), 152–162.

140. Über die Wurzeln der Zetafunktion eines algebraischen Zahlkörpers
Math. Ann. **79** (1919), 388–401.

141. Über die Zerlegung total positiver Zahlen in Quadrate
Gött. Nachr. 1919, 392–396.

142. Über die Primfunktionen in einer arithmetischen Progression
Von Heinrich Kornblum (†), stud. math. in Göttingen. Aus dem Nachlass herausgegeben von E. Landau, Math. Z. **5** (1919), 100–111.

143. Bemerkungen zu einer Arbeit von Herrn Carleman „Über die Fourierkoeffizienten einer stetigen Funktion"
Math. Z. **5** (1919), 147–153.

144. Über die Gitterpunkte in einem Kreise
Math. Z. **5** (1919), 319–320.

145. Über die Nullstellen der Zetafunktion
Math. Z. **6** (1920), 151–154.

146. Konvergenzbeweis einer Lerchschen Reihe
Mémoires de la Soc. Royale des Sciences de Bohème, Classe des Sciences, 1919, Nr. 4, 2 pages.

147. Über Dirichlets Teilerproblem
Gött. Nachr. 1920, 13–32.

148. Über die Nichtfortsetzbarkeit einiger durch Dirichletsche Reihen definierter Funktionen (With Arnold Walfisz)
Rendiconti di Palermo **44** (1920), 82–86.

149. Note on Mr. Hardy's extension of a theorem of Mr. Pólya
Proc. of the Cambridge Phil. Soc. **20** (1920), 14–15.

150. Über die Gitterpunkte in einem Kreise (Dritte Mitteilung)
Gött. Nachr. 1920, 109–134.

151. Über Gitterpunkte in ebenen Bereichen (With J. G. van der Corput)
Gött. Nachr. 1920, 135–171.

152. On the Diophantine equation $ay^2 + by + c = dx^n$ (With Alexander Ostrowski)
Proc. London Math. Soc., Ser. 2, **19** (1920), 276–280.

153. Zur Hardy-Littlewoodschen Lösung des Waringschen Problems
Gött. Nachr. 1921, 88–92.

154. Über einen Satz des Herrn Rosenblatt
Jahresber. der Deutschen Math.-Ver. **29** (1920), 238.

155. Neuer Beweis eines Satzes von Herrn Valiron
Jahresber. der Deutschen Math.-Ver. **29** (1920), 239.

156. Über die Nullstellen Dirichletscher Reihen
Math. Z. **10** (1921), 128–129.

157. Über die Zetafunktion auf der Mittellinie des kritischen Streifens (With Harald Cramér)
Arkiv för Mat., Astr. o. Fys. **15** (1921), Nr. 28, 4 pages.

158. Neuer Beweis und Verallgemeinerung des Fabryschen Lückensatzes (With Fritz Carlson)
Gött. Nachr. 1921, 184–188.

159. Review of the first two volumes of L. E. Dickson's "History of the theory of numbers"
Gött. gelehrte Anzeiger 1921, Nr. 7–9, 2 pages.

160. Über die Nullstellen der Dirichletschen Reihen und der Riemannschen ζ-Funktion
Arkiv för Mat., Astr. o. Fys. **16** (1922), Nr. 7, 17 pages.

161. Bemerkung zu unserer Abhandlung "On the Diophantine equation $ay^2 + by + c = dx^n$" (With A. Ostrowski)
Proc. London Math. Soc., Ser. 2, **20** (1922), XXXIX.

162. Über die gleichmäßige Konvergenz Dirichletscher Reihen
Math. Z. **11** (1921), 317–318.

163. Über Dirichlets Teilerproblem (Zweite Mitteilung)
Gött. Nachr. 1921, 8–16.

164. Über die Hardy–Littlewoodschen Arbeiten zur additiven Zahlentheorie
Jahresber. der Deutschen Math.-Ver. **30** (1921), 179–185.

165. Zum Waringschen Problem
Math. Z. **12** (1922), 219–247.

166. Neuer Beweis des Schneeschen Mittelwertsatzes über Dirichletsche Reihen
Tôhoku Math. J. **20** (1922), 125–130.

167. Bemerkungen zu der Arbeit des Herrn Bieberbach „Über die Verteilung der Null- und Einsstellen analytischer Funktionen" (Math. Ann. 85)
Math. Ann. **86** (1922), 158–160.

168. Über eine Integralungleichung
Christiaan Huygens **1** (1921–22), 235–237.

169. Der Minkowskische Satz über die Körperdiskriminante
Gött. Nachr. 1922, 80–82.

170. Zum Koebeschen Verzerrungssatz
Rendiconti di Palermo **46** (1922), 347–348.

171. Über einen Bieberbachschen Satz
Rendiconti di Palermo **46** (1922), 456–462.

172. Zur additiven Primzahltheorie
Rendiconti di Palermo **46** (1922), 349–356.

173. Über Diophantische Approximationen
Scripta Universitatis Hierosolymitanarum **1** (1923), Nr. 1, 4 pages.

174. Sobre los numeros primos en progresion aritmetica
Revista Mat. Hispano-Americana **4** (1923), 56 pages.

175. Review of L. E. Dickson's History of the theory of numbers, volume 3
Gött. gelehrte Anzeiger 1923, 1 page.

176. Über das Verhalten von $1/\zeta(s)$ auf der Geraden $\sigma = 1$ (With H. Bohr)
Gött. Nachr. 1923, 71–80.

177. Über die Gitterpunkte in einem Kreise (Vierte Mitteilung)
Gött. Nachr. 1923, 58–65.

178. Bemerkungen zu einer Arbeit des Herrn Onicescu (With Karl Grandjot)
Bull de la sect. scient. de l'Acad. Roumaine **8** (1924), 166–168.

179. Über die Wurzeln der Zetafunktion
Math. Z. **20** (1924), 98–104.

180. Über die ζ-Funktion und die L-Funktionen
Math. Z. **20** (1924), 105–125.

181. The lattice points of a circle (With G. H. Hardy)
Proc. of the Royal Soc. A, **105** (1924), 244–258.

182. Über einige zahlentheoretische Funktionen
Gött. Nachr. 1924, 116–134.

183. Über die Gitterpunkte in einem Kreise (Fünfte Mitteilung)
Gött. Nachr. 1924, 135–136.

184. Über Gitterpunkte in mehrdimensionalen Ellipsoiden
Math. Z. **21** (1924), 126–132.

185. Über die Anzahl der Gitterpunkte in gewissen Bereichen (Vierte Abhandlung)
Gött. Nachr. 1924, 137–150.

186. Nachtrag zu unseren Abhandlungen aus den Jahrgängen 1910 und 1923 (With H. Bohr)
Gött. Nachr. 1924, 168–172.

187. Note added to the paper "The lattice points of a circle" by J. E. Littlewood and A. Walfisz
Proc. of the Royal Soc. A **106** (1924), 487–488.

188. Bemerkungen zu der vorstehenden Abhandlung von Herrn Franel
Gött. Nachr. 1924, 198–201.

189. Über die Möbiussche Funktion
Rendiconti di Palermo **48** (1924), 277–280.

190. Die Bedeutungslosigkeit der Pfeifferschen Methode für die analytische Zahlentheorie
Monatshefte für Math. u. Phys. **34** (1925), 1–36.

191. Bemerkungen zu der Arbeit des Herrn Walfisz: Über das Piltzsche Teilerproblem in algebraischen Zahlkörpern
Math. Z. **22** (1925), 189–205.

192. Die Ungleichungen für zweimal differentiierbare Funktionen
Kgl. Danske Videnskab. Selskab. **6** (1925), Nr. 10, 49 pages.

193. Über einen Fejérschen Satz
Gött. Nachr. 1925, 22.

194. Über Gitterpunkte in mehrdimensionalen Ellipsoiden (Zweite Abhandlung)
Math. Z. **24** (1925), 299–310.

195. Solved and unsolved problems in the elementary theory of numbers (In Hebrew)
Lecture delivered at the Inauguration of the Hebrew University in Jerusalem on April 1, 1925.

196. Einige Bemerkungen über schlichte Abbildung
Jahresber. der Deutschen Math.-Ver. **34** (1925), 239–243.

197. A note on a theorem concerning series of positive terms: Extract from a letter of Prof. E. Landau to Prof. I. Schur
J. of the London Math. Soc. **1** (1926), 38–39.

198. Die Winogradowsche Methode zum Beweise des Waring–Hilbert–Kamkeschen Satzes
Acta Math. **48** (1926), 217–253.

199. Computo asintotico dei nodi di un reticolato entro un cerchio
Semin. Mat. della Fac. di Sc. Roma (2) **3** (1926), 35–61.

200. On the representation of a number as the sum of two *k*-th power
J. of the London Math. Soc. **1** (1926), 72–74.

201. Über Dirichletsche Reihen mit komplexen Charakteren
J. reine angew. Math. **157** (1926), 26–32.

202. Über die Riemannsche Zetafunktion in der Nähe von $\sigma = 1$
Rendiconti di Palermo **50** (1926), 423–427.

203. Zum Waringschen Problem
Proc. of the London Math. Soc. (2) **25** (1926), 484–486.

204. Über das Konvergenzgebiet einer mit der Riemannschen Zetafunktion zusammenhängenden Reihe
Math. Ann. **97** (1926), 251–290.

205. Über die Zetafunktion und die Hadamardsche Theorie der ganzen Funktionen
Math. Z. **26** (1927), 170–175.

206. Der Picard-Schottkysche Satz und die Blochsche Konstante
Berliner Sitzungsberichte **32** (1926), 467–474.

207. Über die Nullstellen Dirichletscher Reihen (Zweite Abhandlung)
Berliner Sitzungsberichte 1927, 19–21.

208. Die Bedeutung der Arbeiten van der Corputs für die geometrische Zahlentheorie
De Handelingen van het XXIe Nederlandsch Natuur- en Geneeskundig Congres, April 1927, Amsterdam, 71–72.

209. Über den Picardschen Satz
De Handelingen van het XXIe Nederlandsch Natuur- en Geneeskundig Congres, April 1927, Amsterdam, 106–107.

210. Sulle funzioni intere di genere finito (With J. Hadamard)
Rendiconti dei Lincei (6) **6** (1927), 3–9.

211. Über das Vorzeichen der Gaußschen Summe
Gött. Nachr. 1928, 19–20.

212. Über eine trigonometrische Summe
Gött. Nachr. 1928, 21–24.

213. Der Picard-Schottkysche Satz und die Blochsche Konstante (Zweite Abhandlung)
Berliner Sitzungsberichte 1928, 3–8.

214. Bestimmung einer absoluten Konstante aus der Theorie der trigonometrischen Reihen (With K. Grandjot, V. Jarník and J. E. Littlewood)
Ann. di Mat. (4) **6** (1929), 1–7.

215. Über die Irreduzibilität der Kreisteilungsgleichung
Math. Z. **29** (1928), 462.

216. Über einen Egerváryschen Satz
Math. Z. **29** (1928), 461.

217. Über einen Valironschen Satz
Math. Z. **30** (1929), 205–207.

218. Über die Carathéodorysche Verschärfung des großen Picardschen Satzes
Math. Z. **30** (1929), 208–210. Also Correction on page 796, same volume.

219. Bemerkungen zu einer Arbeit von Hrn. Hoheisel über die Zetafunktion
Berliner Sitzungsberichte 1929, 271–275.

220. Über die Blochsche Konstante und zwei verwandte Weltkonstanten
Math. Z. **30** (1929), 608–634.

221. Über schlichte Funktionen
Math. Z. **30** (1929), 635–638.

222. A deduction from Schwarz's Lemma (With G. Valiron)
J. of the London Math. Soc. **4** (1929), 162–163.

223. Über einen Satz von Herrn Esclangon
Math. Ann. **102** (1929), 177–188.

224. Über einen Mordellschen Satz
Prace Mat.-Fiz. **36** (1929), 1–11.

225. Zum Waringschen Problem (Zweite Abhandlung)
Math. Z. **31** (1930), 149–150.

226. Bestimmung der genauen Konstanten in einem Koebeschen Hilfssatz
J. reine angew. Math. **161** (1929), 135–136.

227. Über die neue Winogradoffsche Behandlung des Waringschen Problems
Math. Z. **31** (1929), 319–338.

228. Über den Millouxschen Satz
Gött. Nachr. 1930, 1–9.

229. Die Goldbachsche Vermutung und der Schnirelmannsche Satz
Gött. Nachr. 1930, 255–276.

230. Zum Waringschen Problem (Dritte Abhandlung)
Math. Z. **32** (1930), 699–702.

231. Neuer Beweis eines Minkowskischen Satzes
J. reine angew. Math. **165** (1931), 1–3.

232. Über die Fareyreihe und die Riemannsche Vermutung
Gött. Nachr. 1932, 347–352.

233. Über Dirichletsche Reihen
Gött. Nachr. 1932, 525–527.

234. Über den Wienerschen neuen Weg zum Primzahlsatz
Berliner Sitzungsberichte 1932, 514–521.

235. Bemerkungen zur vorstehenden Arbeit von Herrn Bochner (With Hans Heilbronn)
Math. Z. **37** (1933), 10–16.

236. Ein Satz über Potenzreihen (With Hans Heilbronn)
Math. Z. **37** (1933), 17.

237. Anwendungen der Wienerschen Methode (With Hans Heilbronn)
Math. Z. **37** (1933), 18–21.

238. Über einen Satz von Herrn Dieudonné
Math. Z. **37** (1933), 22–27.

239. Der Paleysche Satz über Charaktere
Math. Z. **37** (1933), 28–32.

240. Über ungerade schlichte Funktionen
Math. Z. **37** (1933), 33–35.

241. Über eine trigonometrische Ungleichung
Math. Z. **37** (1933), 36.

242. Über den Wertevorrat von $\zeta(s)$ in der Halbebene $\sigma > 1$
Gött. Nachr. 1933, 81–91.

243. Eine Frage über trigonometrische Polynome
Annali di Pisa (2), **2** (1933), 209–210.

244. Bemerkungen zu der M.-P. Geppertschen Abhandlung „Approximative Darstellungen analytischer Funktionen, die durch Dirichletsche Reihen gegeben sind" im Bd. 35 dieser Zeitschrift, S. 190–211
Math. Z. **37** (1933), 314–320.

245. Bemerkungen zum Heilbronnschen Satz
Acta Arithm. **1** (1935), 1–18.

246. Verschärfung eines Romanoffschen Satzes
Acta Arithm. **1** (1935), 43–61.

247. Some inequalities satisfied by the integrals or derivatives of real or analytic functions (With G. H. Hardy and J. E. Littlewood)
Math. Z. **39** (1935), 677–695.

248. Über einige Ungleichungen von Herrn G. Grüß
Math. Z. **39** (1935), 742–744.

249. Untersuchungen über einen van der Corputschen Satz (With V. Jarník)
Math. Z. **39** (1935), 745–767.

250. Alle großen ganzen Zahlen lassen sich als Summe von höchstens 71 Primzahlen darstellen (With H. Heilbronn and P. Scherk)
Casopis Mat. a Fys. **65** (1936), 117–141.

251. Über mehrfach monotone Folgen
Prace Mat.Fiz. **44** (1936), 337–351.

252. On a Titchmarsh-Estermann sum
J. of the London Math. Soc. **11** (1936), 242–245.

253. Über das Produkt von zwei binären Linearformen (Mitgeteilt von A. Walfisz)
Travaux de l'Institut Math. de Tbilissi **5** (1938), 143–144.

254. Ausgewählte Kapitel der Funktionentheorie (Mitgeteilt von A. Walfisz)
Travaux de l'Institut Math. de Tbilissi **8** (1940), 23–68.

2. Books

I. „Handbuch der Lehre von der Verteilung der Primzahlen", Two volumes
Teubner, Leipzig 1909.

II. „Darstellung und Begründung einiger neuerer Ergebnisse der Funktionentheorie"
Springer, Berlin 1916. Second Edition 1929.

III. „Einführung in die elementare und analytische Theorie der algebraischen Zahlen und der Ideale"
Teubner, Leipzig 1918. Second Edition 1927.

IV. „Vorlesungen über Zahlentheorie", Three volumes
S. Hirzel, Leipzig 1927.

V. „Grundlagen der Analysis (Das Rechnen mit ganzen, rationalen, irrationalen, komplexen Zahlen). Ergänzung zu den Lehrbüchern der Differential- und Integralrechnung"
Akademische Verlagsgesellschaft, Leipzig 1930.

VI. „Einführung in die Differentialrechnung und Integralrechnung"
P. Noordhoff, Groningen-Batavia 1934.

VII. „Über einige neuere Fortschritte der additiven Zahlentheorie"
Cambridge Tracts in Math. and Math. Phys., No. 35, Cambridge University Press, Cambridge 1937.

3. New Editions and Translations of Landau's books[1])

I. The "Handbuch" was reprinted in 1953 by the Chelsea Publishing Co., New York, in two volumes, with a Preface and Appendix by P. T. Bateman consisting of commentary on the text and information about recent developments in the subject. Reprints of Landau's papers [179] and [234] form another Appendix. In 1961 the same work was issued with both volumes bound in one.

II. The "Darstellung" was reprinted by the Chelsea Publishing Comp. in 1946. Again reprinted as one of four monographs in the book: "Das Kontinuum, und andere Monographien", by H. Weyl, E. Landau, B. Riemann, New York 1960.

III. The "Einführung" was reprinted by the Chelsea Publishing Comp., New York 1949.

IV. The "Vorlesungen" have been reprinted, translated and revised as follows:
a) The first four parts of volume 1 were reprinted by Chelsea under the title "Elementare Zahlentheorie" in 1946.
b) The remainder of volume 1 and volumes 2 and 3 were reprinted by the same publishers in 1947.
c) The first four parts of volume 1 were translated by J. E. Goodman, with added exercises by P. T. Bateman and E. E. Kohlbecker, and appeared as "Elementary number theory", Chelsea Publishing Comp., New York 1958. Second Edition 1966.
d) A much enlarged and up-to-date version of § 4, of Kapitel 2, of Teil IX (Vorlesungen, vol. 3, 37–65) was written by the late A. Walfisz and published

---

[1]) Based on information kindly supplied by Mr. A. Galuten, President of the Chelsea Publishing Company. The information concerning each book is mentioned under the corresponding Roman numeral.

by him under Landau's name as "Diophantische Gleichungen mit endlich vielen Lösungen" (neu herausgegeben von Arnold Walfisz), VEB Deutscher Verlag der Wissenschaften, Berlin 1959.

V. The "Grundlagen" was reprinted and translated as follows:
  a) Reprinted by the Chelsea Publishing Comp. in 1946, 2nd Ed. 1948, 3rd Ed. 1960, 4th Ed. 1965.
  b) A translation by F. Steinhardt appeared as "Foundations of Analysis", Chelsea Publishing Comp., New York 1951. Second Edition 1960.

VI. A translation of Landau's book VI was prepared by M. Davis and M. Hausner and appeared as "Differential and Integral Calculus", Chelsea Publishing Comp., New York 1950, 2nd Edition 1960, 3rd Edition 1965.

4. Reprints of papers by Landau

Thirteen papers by Landau on lattice point theory, with extensive appendices freely adapted from other papers by Landau, were published by A. Walfisz under the title "Ausgewählte Abhandlungen zur Gitterpunktlehre von Edmund Landau", VEB Deutscher Verlag der Wissenschaften, Berlin 1962.

These thirteen papers can be identified in our list as bearing the following numbers: 111, 116, 122, 177, 181, 183, 184, 185, 187, 190, 191, 194, 204.